Oxford Applied Mathematics and Computing Science Series

General Editors
J. N. Buxton, R. F. Churchhouse, and A. B. Tayler

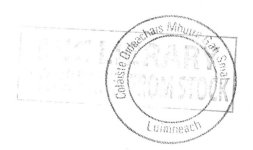

17·50

OXFORD APPLIED MATHEMATICS AND COMPUTING SCIENCE SERIES

G. D. SMITH
Brunel University

Numerical solution of partial differential equations

FINITE DIFFERENCE METHODS

THIRD EDITION

CLARENDON PRESS · OXFORD

Oxford University Press, Walton Street, Oxford OX2 6DP

Oxford New York Toronto
Delhi Bombay Calcutta Madras Karachi
Petaling Jaya Singapore Hong Kong Tokyo
Nairobi Dar es Salaam Cape Town
Melbourne Auckland

and associated companies in
Berlin Ibadan

Oxford is a trade mark of Oxford University Press

Published in the United States
by Oxford University Press, New York

© *Oxford University Press 1985*

First edition published in Oxford Mathematical Handbooks 1965
Second edition 1978, reprinted (with corrections) 1979, 1984
Third edition 1985, reprinted 1986, 1987 (with corrections), 1990

British Library Cataloguing in Publication Data

Smith, G. D.
Numerical solution of partial differential
equations: finite difference methods.—3rd ed.
—(Oxford applied mathematics and computing
science series)
1. Differential equations, Partial—Numerical
solutions
I. Title
515.3'53   QA374

ISBN 0-19-859641-3
ISBN 0-19-859650-2 (pbk)

Typeset and printed by The Universities Press (Belfast) Ltd

Preface to the third edition

As with the two previous editions this book has been written primarily as a textbook for students with no previous knowledge of numerical methods whatsoever, and because of its purpose the various methods of solution and analysis have been illustrated by worked examples. In addition, every chapter, except the first, includes exercises with solutions that often extend the theory as well as amplify points in the text.

It is intended for students taking degree courses in mathematics, physics and engineering, and is self-contained in that the basic calculus and matrix algebra needed for the development of the subject will be known to most second-year students. It is also hoped that the book will continue to be useful to those postgraduate students who need a rapid and uncomplicated introduction to this field of study.

With regard to changes, the new Chapter 2 is essentially a combination of the main ideas of Chapters 2 and 3 of the second edition; I did this because students, both undergraduate and postgraduate, said they preferred a more immediate mixture of method and analysis. The section on stability has been completely rewritten and based on the Lax–Richtmyer definition which ensures convergence when the difference equations are consistent and stable. This has necessitated an introductory section on norms.

Chapter 3 can be omitted by those who need only a quick introduction to all three types of partial differential equations because it deals mainly with an alternative derivation of difference equations approximating parabolic equations. It also includes miscellaneous methods for non-linear equations, the improvement of accuracy, the analytical solution of difference equations, and a new section on an eigenvalue–eigenvector solution that is used to give a new method of approximation particularly suited to large values of t.

Chapter 4 contains the standard work on hyperbolic equations and has been enlarged only slightly by the application of new work done in Chapter 3.

Chapter 5 on elliptic equations and iterative methods remains unchanged, except that my teaching experience has led to a simpler presentation of the theoretical work associated with the SOR method.

Finally, it gives me very great pleasure to thank all those who have been of help to me in the preparation of the three editions. Mr E. L. Albasiny and Dr D. W. Martin of the National Physical Laboratory for a stimulating set of lectures they gave on numerical methods for partial differential equations at Brunel in 1960, Professor J. Crank for criticising and improving the first edition; Dr I. Parker for computing many of the numerical examples when a research student; Professor L. Fox and Dr J. Gregory for helpful suggestions concerning the second edition, and particularly Dr N. Papamichael for his criticisms of my initial draft and his permission to use parts of his M.Sc. dissertation on consistent orderings. With regard to the third edition, I am deeply grateful to Dr E. H. Twizell and Dr A. Q. M. Khaliq, Brunel University, for their work on L_0-stable methods, and to Mr D. Drew, Brunel University, for very recent work on semi-discrete semi-analytic approximation methods.

Brunel University
December 1984

G. D. S.

Contents

3. PARABOLIC EQUATIONS: ALTERNATIVE DERIVATION OF DIFFERENCE EQUATIONS AND MISCELLANEOUS TOPICS

Contents

Notation

$i = 1(1)n$ i varies from 1 to n by intervals of 1, i.e. $i = 1, 2, 3, \ldots, n-1, n$.

$S = \{a_1, a_2, \ldots, a_n\}$ S is the set whose members are $a_1, a_2, \ldots, a_n$.

$a_r \in S$ a_r is a member of the set S.

$S = \{(x, y) : 0 \leqslant x \leqslant a, 0 \leqslant y \leqslant b\}$ S is the set of ordered pairs (x, y) such that $0 \leqslant x \leqslant a$ and $0 \leqslant y \leqslant b$.

$\mathbf{A} = [a_{ij}]$ $\mathbf{A}$ is the matrix whose element in the ith row and jth column is a_{ij}.

$\mathbf{I}_N$ The unit matrix of order N.

$\mathbf{A}^T$ The transpose of matrix $\mathbf{A}$, i.e. $\mathbf{A}^T = [a_{ji}]$.

$\mathbf{A}^{-1}$ The inverse of matrix $\mathbf{A}$, i.e., $\mathbf{A}\mathbf{A}^{-1} = \mathbf{I}$.

$\bar{\mathbf{A}}$ The matrix whose elements are the complex conjugates of the corresponding elements of matrix $\mathbf{A}$.

$\mathbf{x} = [x_1, x_2, \ldots, x_n]^T$ The column vector whose components are $x_1, x_2, \ldots, x_n$.

$|\lambda|$ The modulus of λ.

$\max_i |\lambda_i|, i = 1(1)n$.   The maximum of $|\lambda_1|, |\lambda_2|, \ldots, |\lambda_n|$.

$\rho(\mathbf{A})$ The spectral radius of matrix $\mathbf{A}$, which is the maximum of the moduli of its eigenvalues λ_i, $i = 1(1)n$.

1 Introduction and finite-difference formulae

The mathematical formulation of most problems in science involving rates of change with respect to two or more independent variables, usually representing time, length or angle, leads either to a partial differential equation or to a set of such equations. Special cases of the two dimensional second-order equation

$$a\frac{\partial^2\phi}{\partial x^2}+b\frac{\partial^2\phi}{\partial x\,\partial y}+c\frac{\partial^2\phi}{\partial y^2}+d\frac{\partial\phi}{\partial x}+e\frac{\partial\phi}{\partial y}+f\phi+g=0,$$

where a, b, c, d, e, f, and g may be functions of the independent variables x and y and of the dependent variable ϕ, occur more frequently than any other because they are often the mathematical form of one of the conservation principles of physics.

For reasons that are given in Chapter 4 this equation is said to be *elliptic* when $b^2-4ac<0$, *parabolic* when $b^2-4ac=0$, and *hyperbolic* when $b^2-4ac>0$.

Two-dimensional elliptic equations

These equations, of which the best known are Poisson's equation

$$\frac{\partial^2\phi}{\partial x^2}+\frac{\partial^2\phi}{\partial y^2}+g=0$$

and Laplace's equation

$$\frac{\partial^2\phi}{\partial x^2}+\frac{\partial^2\phi}{\partial y^2}=0,$$

are generally associated with equilibrium or steady-state problems. For example, the velocity potential for the steady flow of incompressible non-viscous fluid satisfies Laplace's equation and is the mathematical way of expressing the idea that the rate at which such fluid enters any given region is equal to the rate at which it leaves it. Similarly, the electric potential V associated with a two-dimensional electron distribution of charge density ρ

satisfies Poisson's equation $\partial^2 V/\partial x^2 + \partial^2 V/\partial y^2 + \rho/\varepsilon = 0$, where ε is the dielectric constant. This is the partial differential equation form of the well-known theorem by Gauss which states that the total electric flux through any closed surface is equal to the total charge enclosed.

The *analytical* solution of a two-dimensional elliptic equation is a function of the space co-ordinates x and y which satisfies the partial differential equation *at every point* of the area S inside a plane *closed* curve C and satisfies certain conditions *at every point* on this boundary curve C (Fig. 1.1). The function ϕ, for instance, from which we can calculate the displacements and shear stresses within a long solid elastic cylinder in a state of torsion satisfies

$$\frac{\partial^2 \phi}{\partial x^2} + \frac{\partial^2 \phi}{\partial y^2} + 2 = 0$$

at every point of a right cross-section, and has a constant value round the perimeter of the cross-section. Similarly, the steady motion of incompressible viscous fluid through a straight uniform tube can be found from a function that satisfies Laplace's equation at every point of the cross-section and equals $\frac{1}{2}(x^2 + y^2)$ at each point on the boundary.

The condition that the dependent variable must satisfy round the boundary curve C is termed the boundary condition.

To the present, only a limited number of special types of

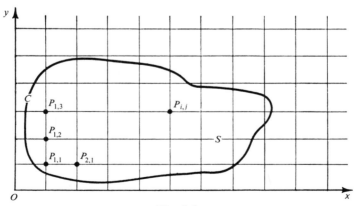

Fig. 1.1

elliptic equations have been solved analytically and the usefulness of these solutions is further restricted to problems involving shapes for which the boundary conditions can be satisfied. This not only eliminates all problems with boundary curves that are undefined in terms of equations, but also many for which the boundary conditions are too difficult to satisfy even though the equations for the boundary curves are known. In such cases approximation methods, whether analytical or numerical in character, are the only means of solution, apart from the use of analogue devices. Analytical approximation methods often provide extremely useful information concerning the character of the solution for critical values of the dependent variables but tend to be more difficult to apply than the numerical methods, and will not be discussed in this book. Of the numerical approximation methods available for solving differential equations those employing finite-differences or finite elements are more frequently used and more universally applicable than any other, although finite elements are not considered in this book. Before outlining these methods however, the reader should be aware of the manner in which the term 'approximation method' is used. Finite-difference methods are approximate in the sense that derivatives *at a point* are approximated by difference quotients *over a small interval*, i.e., $\partial\phi/\partial x$ is replaced by $\delta\phi/\delta x$ where δx is small and y is constant, but the solutions are *not* approximate in the sense of being crude estimates. The data of the problems of technology are invariably subject to errors of measurement, besides which, all arithmetical work is limited to a finite number of significant figures and contains rounding errors, so even analytical solutions provide only approximate numerical answers. Finite-difference methods generally give solutions that are either as accurate as the data warrant or as accurate as is necessary for the technical purposes for which the solutions are required. In both cases a finite-difference solution is as satisfactory as one calculated from an analytical formula. In future, all non-analytical approximation methods will be called numerical methods

They are not of course restricted to problems for which no analytical solutions can be found. The numerical evaluation of an analytical solution is often a laborious task, as can be seen by inspecting the solution of the torsion problem for a rectangular cross-section defined by $x = \pm a$, $y = \pm b$, namely

$$\phi = b^2 - y^2 - 32b^2\pi^{-3} \sum_{n=0}^{\infty} \frac{(-1)^n}{(2n+1)^3} \operatorname{sech} \frac{(2n+1)\pi a}{2b}$$

$$\times \cosh \frac{(2n+1)\pi x}{2b} \cos \frac{(2n+1)\pi y}{2b},$$

and numerical methods generally provide adequate numerical solutions more simply and efficiently. This is certainly so with finite-difference methods for solving partial differential equations.

In these methods (Fig. 1.1), the area of integration of the elliptic equation, i.e. the area S bounded by the closed curve C, is overlayed by a system of rectangular meshes formed by two sets of equally spaced lines, one set parallel to Ox and the other parallel to Oy, and an approximate solution to the differential equation is found at the points of intersection $P_{1,1}, P_{1,2}, \ldots, P_{i,j}, \ldots$ of the parallel lines, which points are called mesh points. (Other terms in common use are pivotal, nodal, grid, or lattice points.) This solution is obtained by approximating the partial differential equation over the area S by n *algebraic equations* involving the values of ϕ at the n mesh points internal to C. The approximation consists of replacing each derivative of the partial differential equation at the point $P_{i,j}$ (say) by a finite-difference approximation in terms of the values of ϕ at $P_{i,j}$ and at neighbouring mesh points and boundary points, and in writing down for each of the n internal mesh points the algebraic equation approximating the differential equation. This process clearly gives n algebraic equations for the n unknowns $\phi_{1,1}$, $\phi_{1,2}, \ldots \phi_{i,j}, \ldots$. Accuracy can usually be improved either by increasing the number of mesh points or by including 'correction terms' in the approximations for the derivatives.

Parabolic and hyperbolic equations

Problems involving time t as one independent variable lead usually to parabolic or hyperbolic equations.

The simplest parabolic equation, $\partial U/\partial t = \kappa \partial^2 U/\partial x^2$, derives from the theory of heat conduction and its solution gives, for example, the temperature U at a distance x units of length from one end of a thermally insulated bar after t seconds of heat conduction. In such a problem the temperatures at the *ends* of a bar of length l (say) are often known for all time. In other words, the *boundary conditions* are known. It is also usual for the temperature distribution along the bar to be known at some particular instant. This instant is usually taken as zero time and the temperature distribution is called the *initial condition*. The solution gives U for values of x between 0 and l and values of t from zero to infinity. Hence the area of integration S in the $x–t$ plane (Fig. 1.2), is the infinite area bounded by the x-axis and the parallel lines $x = 0, x = l$. This is described as an *open* area because the boundary curves marked C do not constitute a closed boundary in any finite region of the $x–t$ plane.

Applications of finite-difference methods of solution to parabolic equations are no different from their application to elliptic equations in so far as the integration of the differential equation over S is approximated by the solution of algebraic equations. The structure of the algebraic equations is different however in that it propagates the solution forward from one time row to the next in a step-by-step fashion.

Hyperbolic equations generally originate from vibration problems, or from problems where discontinuities can persist in time, such as with shock waves, across which there are discontinuities in speed, pressure and density. The simplest hyperbolic equation is the one-dimensional wave equation $\partial^2 U/\partial t^2 = c^2 \partial^2 U/\partial x^2$, giving, for example, the transverse displacement U at a distance x

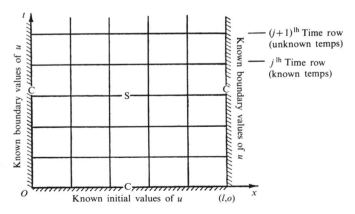

Fig. 1.2

from one end of a vibrating string of length l after a time t. As the values of U at the ends of the string are usually known for all time (the boundary conditions) and the shape and velocity of the string are prescribed at zero time (the initial conditions), it is seen (Fig. 1.2), that the solution is similar to that of a parabolic equation in that the calculation of U for a given x and t, $(0 \leqslant x \leqslant l)$, entails integration of the equation over the open area S bounded by the open curve C. Although hyperbolic equations can be solved numerically by finite-difference methods, those involving only two independent variables, x and t say, are often dealt with by the method of characteristics, especially if the initial conditions and/or boundary conditions involve discontinuities. This method finds special curves in the x–t plane, called characteristic curves, along which the solution of the *partial* differential equation is reduced to the integration of an *ordinary* differential equation. This ordinary equation is generally integrated by numerical methods.

In conclusion, it is worth noting that whereas changes to the shape of the area of integration or to the boundary and initial conditions of partial differential equations often make their analytical solutions impossible, such changes do not fundamentally affect finite-difference methods although they sometimes necessitate rather complicated modifications to the methods.

Finite-difference approximations to derivatives

When a function U and its derivatives are single-valued, finite and continuous functions of x, then by Taylor's theorem,

$$U(x+h) = U(x) + hU'(x) + \tfrac{1}{2}h^2 U''(x) + \tfrac{1}{6}h^3 U'''(x) + \ldots \tag{1.1}$$

and

$$U(x-h) = U(x) - hU'(x) + \tfrac{1}{2}h^2 u''(x) - \tfrac{1}{6}h^3 U'''(x) \ldots . \tag{1.2}$$

Addition of these expansions gives

$$U(x+h) + U(x-h) = 2U(x) + h^2 U''(x) + O(h^4), \tag{1.3}$$

where $O(h^4)$ denotes terms containing fourth and higher powers of h. Assuming these are negligible in comparison with lower powers of h it follows that,

$$U''(x) = \left(\frac{d^2U}{dx^2}\right)_{x=x} \simeq \frac{1}{h^2}\{U(x+h) - 2U(x) + U(x-h)\}, \quad (1.4)$$

with a leading error on the right-hand side of order h^2.

Subtraction of eqn (1.2) from eqn (1.1) and neglect of terms of order h^3 leads to

$$U'(x) = \left(\frac{dU}{dx}\right)_{x=x} \simeq \frac{1}{2h}\{U(x+h) - U(x-h)\}, \quad (1.5)$$

with an error of order h^2.

Equation (1.5) clearly approximates the slope of the tangent at P by the slope of the chord AB, and is called a *central-difference* approximation. We can also approximate the slope of the tangent at P by either the slope of the chord PB, giving the *forward-difference* formula,

$$U'(x) \simeq \frac{1}{h}\{U(x+h) - U(x)\}, \quad (1.6)$$

or the slope of the chord AP giving the *backward-difference* formula

$$U'(x) \simeq \frac{1}{h}\{U(x) - U(x-h)\}. \quad (1.7)$$

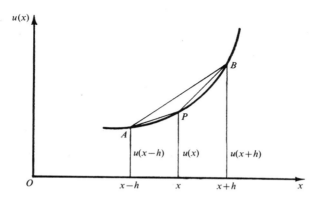

Fig. 1.3

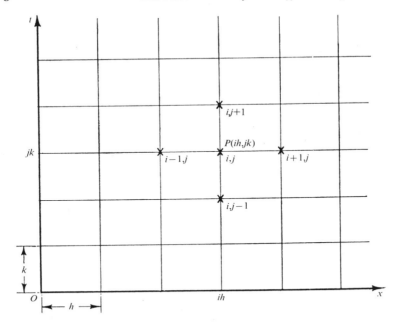

Fig. 1.4

Both (1.6) and (1.7) can be written down immediately from eqns (1.1) and (1.2) respectively, assuming second and higher powers of h are negligible. This shows that the leading errors in these forward and backward-difference formulae are both $O(h)$.

Notation for functions of several variables

Assume U is a function of the independent variables x and t. Subdivide the x–t plane into sets of equal rectangles of sides $\delta x = h$, $\delta t = k$, by equally spaced grid lines parallel to Oy, defined by $x_i = ih$, $i = 0, \pm 1, \pm 2, \ldots$, and equally spaced grid lines parallel to Ox, defined by $y_j = jk$, $j = 0, 1, 2, \ldots$, as shown in Fig. 1.4.

Denote the value of U at the representative mesh point $P(ih, jk)$ by

$$U_P = U(ih, jk) = U_{i,j}.$$

Then by eqn (1.4),

$$\left(\frac{\partial^2 U}{\partial x^2}\right)_P = \left(\frac{\partial^2 U}{\partial x^2}\right)_{i,j} \simeq \frac{U\{(i+1)h, jk\} - 2U\{ih, jk\} + U\{(i-1)h, jk\}}{h^2}.$$

i.e.

$$\frac{\partial^2 U}{\partial x^2} \simeq \frac{U_{i+1,j} - 2U_{i,j} + U_{i-1,j}}{h^2}, \qquad (1.8)$$

with a leading error of order h^2. Similarly,

$$\left(\frac{\partial^2 U}{\partial t^2}\right)_{i,j} \simeq \frac{U_{i,j+1} - 2U_{i,j} + U_{i,j-1}}{k^2}, \qquad (1.9)$$

with a leading error of order k^2.

With this notation the forward-difference approximation for $\partial U/\partial t$ at P is

$$\frac{\partial U}{\partial t} \simeq \frac{U_{i,j+1} - U_{i,j}}{k}, \qquad (1.10)$$

with a leading error of $O(k)$.

2 Parabolic equations: finite difference methods, convergence, and stability

Transformation to non-dimensional form

The computational stage of all numerical methods for solving problems of any complexity generally involves a great deal of arithmetic. It is usual therefore to arrange, whenever possible, for one solution to suffice for a variety of different problems. This can be done by expressing all equations in terms of non-dimensional variables. Then all problems with the same non-dimensional mathematical formulation can be dealt with by means of one solution. For example, the oscillation of a pendulum in a viscous medium and the discharge of electricity from a capacitance through a resistance and inductance are different problems physically, but identical mathematically when expressed in terms of non-dimensional variables. The problems need not, of course, be dimensionally different, but merely variations of the same type of problem, as we would have with the calculation of the periods of oscillation of springs of different lengths l supporting different masses m and having different stiffnesses s. A single solution of the corresponding non-dimensional equation would allow us to solve a wide variety of spring problems because a single parameter ξ, say, would replace some combination of l, m, and s.

This non-dimensionalizing process is illustrated below with the parabolic equation

$$\frac{\partial U}{\partial T} = \kappa \frac{\partial^2 U}{\partial X^2}, \quad \kappa \text{ constant}, \tag{2.1}$$

the solution of which gives the temperature U at a distance X from one end of a thin uniform rod after a time T. (This assumes the rod is heat-insulated along its length so that temperature changes occur through heat conduction along its length and heat transfer at its ends.) Let L represent the length of the rod and U_0 some particular temperature such as the maximum or minimum temperature at zero time. Put

$$x = \frac{X}{L} \quad \text{and} \quad u = \frac{U}{U_0}.$$

Then

$$\frac{\partial U}{\partial X} = \frac{\partial U}{\partial x}\frac{dx}{dX} = \frac{\partial U}{\partial x}\frac{1}{L}$$

and

$$\frac{\partial^2 U}{\partial X^2} = \frac{\partial}{\partial x}\left(\frac{\partial U}{\partial X}\right) = \frac{\partial}{\partial x}\left(\frac{1}{L}\frac{\partial U}{\partial X}\right)\frac{dx}{dX} = \frac{1}{L^2}\frac{\partial^2 U}{\partial x^2},$$

so eqn (2.1) transforms to

$$\frac{\partial(uU_0)}{\partial T} = \frac{\kappa}{L^2}\frac{\partial^2(uU_0)}{\partial x^2},$$

i.e.

$$\frac{1}{\kappa L^{-2}}\frac{\partial u}{\partial T} = \frac{\partial^2 u}{\partial x^2}.$$

Writing $t = \kappa T/L^2$ and applying the function of a function rule to the left side yields

$$\frac{\partial u}{\partial t} = \frac{\partial^2 u}{\partial x^2} \tag{2.2}$$

as the non-dimensional form of (2.1).

It should be noted that the number representing the length of the rod is 1.

An explicit method of solution

By eqns (1.10) and (1.8) one finite-difference approximation to

$$\frac{\partial U}{\partial t} = \frac{\partial^2 U}{\partial x^2} \tag{2.3}$$

is

$$\frac{u_{i,j+1} - u_{i,j}}{k} = \frac{u_{i+1,j} - 2u_{i,j} + u_{i-1,j}}{h^2},$$

where u is the exact solution of the approximating difference equations,

$$x_i = ih, \quad (i = 0, 1, 2, \ldots),$$

and

$$t_j = jk, \quad (j = 0, 1, 2, \ldots).$$

This can be written as

$$u_{i,j+1} = ru_{i-1,j} + (1 - 2r)u_{i,j} + ru_{i+1,j}, \tag{2.4}$$

where $r = \delta t/(\delta x)^2 = k/h^2$, and gives a formula for the unknown 'temperature' $u_{i,j+1}$ at the $(i, j+1)$th mesh point in terms of known 'temperatures' along the jth time-row (Fig. 2.1). Hence we can calculate the unknown pivotal values of u along the first time-row, $t = k$, in terms of known boundary and initial values along $t = 0$, then the unknown pivotal values along the second time-row in terms of the calculated pivotal values along the first, and so on. A formula such as this which expresses *one* unknown pivotal value directly in terms of known pivotal values is called an explicit formula.

Example 2.1

As a numerical example let us solve (2.4) given that the ends of the rod are kept in contact with blocks of melting ice and that the

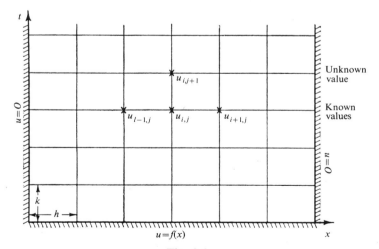

Fig. 2.1

initial temperature distribution in non-dimensional form is

$$\text{(a)} \quad U = 2x \qquad 0 \leqslant x \leqslant \tfrac{1}{2},$$
$$\text{(b)} \quad U = 2(1 - x), \quad \tfrac{1}{2} \leqslant x \leqslant 1. \tag{2.5}$$

In other words, we are seeking a numerical solution of $\partial U/\partial t = \partial^2 U/\partial x^2$ which satisfies

(i) $U = 0$ at $x = 0$ and 1 for all $t > 0$. (The boundary condition.)

(ii) $U = 2x$ for $0 \leqslant x \leqslant \tfrac{1}{2}$
and $U = 2(1 - x)$ for $\tfrac{1}{2} \leqslant x \leqslant 1.$ $\Big\} \ t = 0.$ (The initial condition.)

(This initial temperature distribution could be obtained by heating the centre of the rod for a long time and keeping the ends in contact with the ice.)

For $\delta x = h = \tfrac{1}{10}$, the initial values and boundary values are as shown in Table 2.1. The problem is symmetric with respect to $x = \tfrac{1}{2}$ so we need the solution only for $0 \leqslant x \leqslant \tfrac{1}{2}$.

Case 1

Take $\delta x = h = \tfrac{1}{10}$, $\delta t = k = \tfrac{1}{1000}$, so $r = k/h^2 = \tfrac{1}{10}$. Equation (2.4) then reads as

$$u_{i,j+1} = \tfrac{1}{10}(u_{i-1,j} + 8u_{i,j} + u_{i+1,j}). \tag{2.6}$$

For pencil and paper calculations the relationship between these four function values is represented very conveniently by the 'molecule' in Fig. 2.2. The numbers in the 'atoms' are the multipliers of the function values at the corresponding mesh points.

Application of eqn (2.6) to the data of Table 2.1 is shown in Table 2.2, and readers are recommended to check some of the

TABLE 2.1

	$x = 0$	0.1	0.2	0.3	0.4	0.5	0.6	
								x
$j = 0$	0	0.2	0.4	0.6	0.8	1.0	0.8	
$j = 1$	0							
$j = 2$	0							
$j = 3$	0							
$j = 4$	0							
	t							

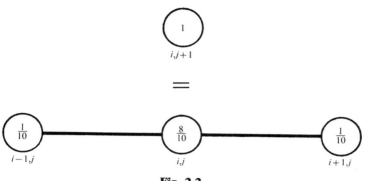

Fig. 2.2

calculations, remembering that the values of U at $x = \frac{4}{10}$ and $\frac{6}{10}$ are equal because of symmetry. (Increasing values of t, i.e. of j, are shown moving downwards for convenience of calculation.) As examples,

$$u_{5,1} = \tfrac{1}{10}\{0.8 + (8 \times 1) + 0.8\} = 0.9600.$$

$$u_{4,2} = \tfrac{1}{10}\{0.6 + (8 \times 0.8) + 0.96\} = 0.7960.$$

The analytical solution of the partial differential equation satisfying these conditions is

$$U = \frac{8}{\pi^2} \sum_{n=1}^{\infty} \frac{1}{n^2} (\sin \tfrac{1}{2}n\pi)(\sin n\pi x)\exp(-n^2\pi^2 t).$$

TABLE 2.2

	$i = 0$ $x = 0$	$i = 1$ 0.1	$i = 2$ 0.2	$i = 3$ 0.3	$i = 4$ 0.4	$i = 5$ 0.5	$i = 6$ 0.6
$(j = 0)t = 0.000$	0	0.2000	0.4000	0.6000	0.8000	1.0000	0.8000
$(j = 1)$ 0.001	0	0.2000	0.4000	0.6000	0.8000	0.9600	0.8000
$(j = 2)$ 0.002	0	0.2000	0.4000	0.6000	0.7960	0.9280	0.7960
$(j = 3)$ 0.003	0	0.2000	0.4000	0.5996	0.7896	0.9016	0.7896
$(j = 4)$ 0.004	0	0.2000	0.4000	0.5986	0.7818	0.8792	0.7818
$(j = 5)$ 0.005	0	0.2000	0.3999	0.5971	0.7732	0.8597	0.7732
$(j = 10)$ 0.01	0	0.1996	0.3968	0.5822	0.7281	0.7867	0.7281
$(j = 20)$ 0.02	0	0.1938	0.3781	0.5373	0.6486	0.6891	0.6486

TABLE 2.3

	Finite-difference solution $(x = 0.3)$	Analytical solution $(x = 0.3)$	Difference	Percentage error
$t = 0.005$	0.5971	0.5966	0.0005	0.08
$t = 0.01$	0.5822	0.5799	0.0023	0.4
$t = 0.02$	0.5373	0.5334	0.0039	0.7
$t = 0.10$	0.2472	0.2444	0.0028	1.1

Comparison of this solution with the finite-difference one at $x = 0.3$, as given above, shows that the finite-difference solution is reasonably accurate. The percentage error is the difference of the solutions expressed as a percentage of the analytical solution of the partial differential equation.

The comparison at $x = 0.5$ is not quite so good because of the discontinuity in the initial value of $\partial U / \partial x$, from $+2$ to -2, at this point (eqn 2.5). Inspection of Table 2.4 shows, however, that the effect of this discontinuity dies away as t increases.

It can be proved analytically that when the boundary values are constant the effect of discontinuities in initial values and initial derivatives upon the solution of a parabolic equation decreases as t increases.

An examination of Tables 2.19 and 2.21 given in Exercise 1 at the end of this chapter shows that the same finite-difference solution for a problem in which the initial function and all its derivatives are continuous is very close indeed to the solution of the partial differential equation.

Richtmyer, reference 25, has shown for this particular finite-difference scheme that when the initial function and its first $(p - 1)$ derivatives are continuous and the pth derivative ordinar-

TABLE 2.4

	Finite-difference solution $(x = 0.5)$	Analytical solution $(x = 0.5)$	Difference	Percentage error
$t = 0.005$	0.8597	0.8404	0.0193	2.3
$t = 0.01$	0.7867	0.7743	0.0124	1.6
$t = 0.02$	0.6891	0.6809	0.0082	1.2
$t = 0.10$	0.3056	0.3021	0.0035	1.2

TABLE 2.5

	$i = 0$ $x = 0$	1 0.1	2 0.2	3 0.3	4 0.4	5 0.5	6 0.6
$T = 0.000$	0	0.2000	0.4000	0.6000	0.8000	1.0000	0.8000
0.005	0	0.2000	0.4000	0.6000	0.8000	0.8000	0.8000
0.010	0	0.2000	0.4000	0.6000	0.7000	0.8000	0.7000
0.015	0	0.2000	0.4000	0.5500	0.7000	0.7000	0.7000
0.020	0	0.2000	0.3750	0.5500	0.6250	0.7000	0.6250
⋮							
0.100	0	0.0949	0.1717	0.2484	0.2778	0.3071	0.2778

ily discontinuous (i.e. changes by finite jumps), then the difference between the solution of the partial differential equation and a convergent solution of the difference equation is of order $(\delta t)^{(p+2)/(p+4)}$, for small δt.

In this example, $p = 1$, so the difference is of order $(\delta t)^{\frac{3}{5}}$. As $(0.001)^{\frac{3}{5}} = 0.016$, it is seen that the finite-difference solution is actually better than the estimate indicates, a feature common to most error estimates. When all the derivatives are continuous, $p \to \infty$, and the error is of order δt.

Case 2

Take $\delta x = h = \frac{1}{10}$, $\delta t = k = \frac{5}{1000}$, so $r = k/h^2 = 0.5$. Then eqn (2.4) gives

$$u_{i,j+1} = \tfrac{1}{2}(u_{i-1,j} + u_{i+1,j}), \qquad (2.7)$$

and the solution obtained by applying this finite-difference equation to the boundary and initial values is recorded in Table 2.5.

TABLE 2.6

	Finite-difference solution $(x = 0.3)$	Analytical solution $(x = 0.3)$	Difference	Percentage error
$t = 0.005$	0.6000	0.5966	0.0034	0.57
$t = 0.01$	0.6000	0.5799	0.0201	3.5
$t = 0.02$	0.5500	0.5334	0.0166	3.1
$t = 0.1$	0.2484	0.2444	0.0040	1.6

   It is seen that this finite-difference solution is not quite as good
an approximation to the solution of the partial differential equation as the previous one; nevertheless it would be adequate for
most technical purposes.

Case 3

Take $\delta x = \frac{1}{10}$, $\delta t = \frac{1}{100}$, so $r = \delta t/(\delta x)^2 = 1$. Then eqn (2.4) gives

$$u_{i,j+1} = u_{i-1,j} - u_{i,j} + u_{i+1,j}, \tag{2.9}$$

and the solution of this finite-difference scheme is as below.

TABLE 2.7

	$i = 0$	1	2	3	4	5	6
	$x = 0$	0.1	0.2	0.3	0.4	0.5	0.6
$t = 0.00$	0	0.2	0.4	0.6	0.8	1.0	0.8
0.01	0	0.2	0.4	0.6	0.8	0.6	0.8
0.02	0	0.2	0.4	0.6	0.4	1.0	0.4
0.03	0	0.2	0.4	0.2	1.2	−0.2	1.2
0.04	0	0.2	0.0	1.4	−1.2	2.6	−1.2

   Considered as a solution of the partial differential equation this is
obviously meaningless, although it is, of course, the correct
solution of eqn (2.9) with respect to the initial values and
boundary values given.
   These three cases clearly indicate that the value of r is important and it will be proved later that this explicit method is valid
only for $0 < r \leqslant \frac{1}{2}$. The conditions that must be satisfied for a valid
solution are dealt with both descriptively and analytically later in
this chapter under the headings of convergence, stability, and
consistency. Any reader who would prefer to have an introduction to these concepts at this stage could do so by reading the
descriptive treatments of these topics as they are independent of
the remainder of this chapter.
   The graphs opposite compare the analytical solution of the
partial differential equation (shown as continuous curves) with
the finite-difference solution (shown by dots) for values of r just
below and above $\frac{1}{2}$, and the same number of time-steps.

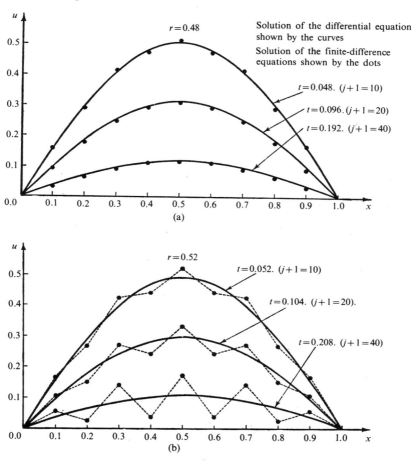

Fig. 2.3

Crank–Nicolson implicit method

Although the explicit method is computationally simple it has one serious drawback. The time step $\delta t = k$ is necessarily very small because the process is valid only for $0 < k/h^2 \leqslant \frac{1}{2}$, i.e. $k \leqslant \frac{1}{2}h^2$, and $h = \delta x$ must be kept small in order to attain reasonable accuracy. Crank and Nicolson (1947) proposed, and used, a method that reduces the total volume of calculation and is valid (i.e., con-
~t and stable) for all finite values of r. They considered the

partial differential equation as being satisfied at the midpoint $\{ih, (j+\frac{1}{2})k\}$ and replaced $\partial^2 U/\partial x^2$ by the mean of its finite-difference approximations at the jth and $(j+1)$th time-levels. In other words they approximated the equation

$$\left(\frac{\partial U}{\partial t}\right)_{i,j+\frac{1}{2}} = \left(\frac{\partial^2 U}{\partial x^2}\right)_{i,j+\frac{1}{2}}$$

by

$$\frac{u_{i,j+1}-u_{i,j}}{k} = \frac{1}{2}\left\{\frac{u_{i+1,j+1}-2u_{i,j+1}+u_{i-1,j+1}}{h^2}+\frac{u_{i+1,j}-2u_{i,j}+u_{i-1,j}}{h^2}\right\},$$

giving

$$-ru_{i-1,j+1}+(2+2r)u_{i,j+1}-ru_{i+1,j+1}=ru_{i-1,j}+(2-2r)u_{i,j}+ru_{i+1,j},$$

$$(2.10)$$

where $r = k/h^2$.

In general, the left side of eqn (2.10) contains three unknown, and the right side three known, pivotal values of u (Fig. 2.4).

If there are N internal mesh points along each time row then for $j = 0$ and $i = 1, 2, \ldots, N$, eqn (2.10) gives N simultaneous equations for the N unknown pivotal values along the first

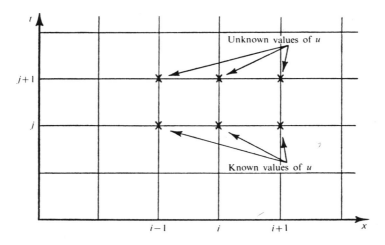

Fig. 2.4

time-row in terms of known initial and boundary values. Similarly, $j = 1$ expresses N unknown values of u along the second time-row in terms of the calculated values along the first, etc. A method such as this where the calculation of an unknown pivotal value necessitates the solution of a set of simultaneous equations is described as an *implicit* one.

For readers familiar with finite-difference notation, the Crank–Nicolson method approximates the partial differential equation at the point $\{ih, (j + \frac{1}{2})k\}$ by

$$\frac{1}{k}\, \delta_t u_{i,j+\frac{1}{2}} = \frac{1}{2h^2} \left\{ \delta_x^2 u_{i,j+1} + \delta_x^2 u_{i,j} \right\},$$

where the subscripts t and x denote differencing in the t- and x-directions respectively. Relative to the point $\{ih, (j + \frac{1}{2})k\}$, both $\partial U/\partial t$ and $\partial^2 U/\partial x^2$ have been replaced by central-difference approximations. This tends to reduce the errors introduced by the approximations.

Example 2.2

Use the Crank–Nicolson method to calculate a numerical solution of the previous worked example, namely,

$$\frac{\partial U}{\partial t} = \frac{\partial^2 U}{\partial x^2}, \quad 0 < x < 1, t > 0,$$

where (i) $U = 0$, $x = 0$ and 1, $t \geqslant 0$,
(ii) $U = 2x$, $0 \leqslant x \leqslant \frac{1}{2}$, $t = 0$,
(iii) $U = 2(1 - x)$, $\frac{1}{2} \leqslant x \leqslant 1$, $t = 0$.

Take $h = \frac{1}{10}$. Although the method is valid for all finite values of $r = k/h^2$, a large value will yield an inaccurate approximation for $\partial U/\partial t$. A suitable value is $r = 1$ and has the advantage of making the coefficient of $u_{i,j}$ zero in (2.10). Then $k = \frac{1}{100}$ and (2.10) reads as

$$-u_{i-1,j+1} + 4u_{i,j+1} - u_{i+1,j+1} = u_{i-1,j} + u_{i+1,j}. \tag{2.11}$$

The computational molecule corresponding to eqn (2.11) is shown in Fig. 2.5. Denote $u_{i,j+1}$ by u_i ($i = 1, 2, \ldots, 9$). For this problem, because of symmetry, $u_6 = u_4$, $u_7 = u_3$, etc. (Fig. 2.6).

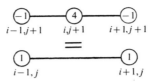

Fig. 2.5

The values of u for the first time step then satisfy

$$-0+4u_1-u_2=0 \ +0.4,$$

$$-u_1+4u_2-u_3=0.2+0.6,$$

$$-u_2 \ +4u_3-u_4=0.4 \ +0.8,$$

$$-u_3+4u_4-u_5=0.6+1.0,$$

$$-2u_4+4u_5 \ = 0.8+0.8.$$

As indicated in the next section these are easily solved by systematic eliminations to give

$$u_1=0.1989, \ u_2=0.3956, \ u_3=0.5834, \ u_4=0.7381, \ u_5=0.7691.$$

Hence the equations for the pivotal values of u along the next time row are

$$-0+4u_1-u_2=0+0.3956,$$

$$-u_1+4u_2-u_3=0.1989+0.5834,$$

$$-u_2+4u_3-u_4=0.3956+0.7381,$$

$$-u_3+4u_4-u_5=0.5834+0.7691,$$

$$-2u_4+4u_5 \ = 2\times0.7381$$

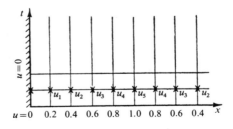

Fig. 2.6

TABLE 2.8

		$x=0$	0.1	0.2	0.3	0.4	0.5
	$t=0.00$	0	0.2	0.4	0.6	0.8	1.0
	$t=0.01$	0	0.1989	0.3956	0.5834	0.7381	0.7691
	$t=0.02$	0	0.1936	0.3789	0.5400	0.6461	0.6921
	.	0	.				
	.		.				
	.		.				
	$t=0.10$	0	0.0948	0.1803	0.2482	0.2918	0.3069
Analytical							
solution	$t=0.10$	0	0.0934	0.1776	0.2444	0.2873	0.3021

The solution of these equations is given in Table 2.8 together with figures comparing the finite-difference solution at $t = 0.1$ with the solution of the partial differential equation. The numerical solution is clearly a good one.

Table 2.9 below displays both solutions at $x = 0.5$ for various values of t. A glance at Table 2.4 shows that in this example the accuracy of this implicit method over the time-range taken is about the same as for the explicit method which uses ten times as many time-steps.

As mentioned previously the greatest difference between the two solutions occurs at $x = 0.5$ because of the ordinary discontinuity in the initial value of $\partial U/\partial x$ at this point. A glance at Table 2.25 in Exercise 3 at the end of the chapter shows that this difference is less for an initial function that is continuous together with its derivatives.

Although the Crank–Nicolson method for $\partial U/\partial t = \partial^2 U/\partial x^2$ is stable for all positive values of r in the sense that the solution and all errors eventually tend to zero as j tends to infinity, it will be shown later that large values for r, such as 40, can introduce

TABLE 2.9

	Finite-difference solution ($x = 0.5$)	Analytical solution ($x = 0.5$)	Difference	Percentage error
$t = 0.01$	0.7691	0.7743	−0.0052	−0.7
$t = 0.02$	0.6921	0.6809	+0.0112	+1.6
$t = 0.10$	0.3069	0.3021	0.0048	1.6

unwanted finite oscillations into the numerical solution. Such oscillations die away only very slowly with increasing j, and usually occur in the x-neighbourhood of points of discontinuity in the initial values or between initial values and boundary values.

Solution of the equations by Gauss's elimination method (without pivoting)

When there are $N-1$ internal mesh points along each time row the Crank–Nicolson equations (2.10) can be written very generally as

$$+b_1u_1-c_1u_2 = d_1,$$

$$-a_2u_1+b_2u_2-c_2u_3 = d_2,$$

$$\cdot \qquad \cdot \qquad \cdot$$
$$\cdot \qquad \cdot \qquad \cdot$$

$$-a_iu_{i-1}+b_iu_i-c_iu_{i+1} = d_i,$$

$$\cdot \qquad \cdot \qquad \cdot$$

$$-a_{N-1}u_{N-2}+b_{N-1}u_{N-1}=d_{N-1},$$

where the a's, b's, c's, and d's are known. The first equation can be used to eliminate u_1 from the second equation, the new second equation used to eliminate u_2 from the third equation and so on, until finally, the new last but one equation can be used to eliminate u_{N-2} from the last equation, giving one equation with only one unknown, u_{N-1}. The unknowns $u_{N-2}, u_{N-3}, \ldots u_2, u_1$ can then be found in turn by back-substitution. Noting that the coefficient c in each new equation is the same as in the corresponding old equation, assume that the following stage of the eliminations has been reached,

$$\alpha_{i-1}u_{i-1}-c_{i-1}u_i=S_{i-1},$$

$$-a_iu_{i-1}+b_iu_i-c_iu_{i+1}=d_i,$$

where $\alpha_1=b_1$, $S_1=d_1$.

Eliminating u_{i-1} leads to

$$\left(b_i-\frac{a_ic_{i-1}}{\alpha_{i-1}}\right)u_i-c_iu_{i+1}=d_i+\frac{\alpha_iS_{i-1}}{\alpha_{i-1}},$$

i.e.

$$\alpha_iu_i-c_iu_{i+1}=S_i, \tag{2.12}$$

where
$$\alpha_i = b_i - \frac{a_i c_{i-1}}{\alpha_{i-1}} \quad \text{and} \quad S_i = d_i + \frac{\alpha_i S_{i-1}}{\alpha_{i-1}} \quad (i = 2, 3, \ldots).$$

The last pair of simultaneous equations are

$$\alpha_{N-2} u_{N-2} - c_{N-2} u_{N-1} = S_{N-2}$$

and
$$-a_{N-1} u_{N-2} + b_{N-1} u_{N-1} = d_{N-1}.$$

Elimination of u_{N-2} gives

$$\left(b_{N-1} - \frac{a_{N-1} c_{N-2}}{\alpha_{N-2}} \right) u_{N-1} = d_{N-1} + \frac{a_{N-1} S_{N-2}}{\alpha_{N-2}},$$

i.e.

$$\alpha_{N-1} u_{N-1} = S_{N-1}, \tag{2.13}$$

Equations (2.12) and (2.13) show that the solution can be calculated from

$$u_{N-1} = \frac{S_{N-1}}{\alpha_{N-1}},$$

$$u_i = \frac{1}{\alpha_i} (S_i + c_i u_{i+1}) \quad (i = N-2, N-3, \ldots, 1),$$

where the α's and S's are given recursively by

$$\alpha_1 = b_1; \quad \alpha_i = b_i - \frac{a_i}{\alpha_{i-1}} c_{i-1},$$

$$S_1 = d_1; \quad S_i = d_i + \frac{a_i}{\alpha_{i-1}} S_{i-1} \quad (i = 2, 3, \ldots, N-1).$$

In many problems α_i and a_i/α_{i-1} are independent of time and need only be calculated once, irrespective of the number of time-steps.

As an illustration consider the last worked example for which the equations were

$$4u_1 - u_2 = 0.4,$$

$$-u_1 + 4u_2 - u_3 = 0.8,$$

$$-u_2 + 4u_3 - u_4 = 1.2,$$

$$-u_3 + 4u_4 - u_5 = 1.6,$$

$$-2u_4 + 4u_5 = 1.6.$$

Hence

$$a_2 = a_3 = a_4 = 1, \ a_5 = 2, \ b_1 = b_2 = b_3 = b_4 = b_5 = 4;$$
$$c_1 = c_2 = c_3 = c_4 = 1; \ d_1 = 0.4, \ d_2 = 0.8, \ d_3 = 1.2, \ d_4 = d_5 = 1.6,$$

so

$$\alpha_1 = b_1 = 4; \quad \alpha_i = b_i - \frac{a_i}{\alpha_{i-1}} c_{i-1} = 4 - \frac{a_i}{\alpha_{i-1}} \quad (i = 2, 3, 4, 5),$$

giving the following coefficients which are invariant for every time-step.

$$\alpha_1 = 4,$$

$$\frac{a_2}{\alpha_1} = \frac{1}{4} = 0.25, \qquad\qquad \alpha_2 = 4 - \frac{a_2}{\alpha_1} = 3.75,$$

$$\frac{a_3}{\alpha_2} = \frac{1}{3.75} = 0.2667, \qquad \alpha_3 = 4 - \frac{a_3}{\alpha_2} = 3.7333,$$

$$\frac{a_4}{\alpha_3} = \frac{1}{3.7333} = 0.2679, \qquad \alpha_4 = 4 - \frac{a_4}{\alpha_3} = 3.7321,$$

$$\frac{a_5}{\alpha_4} = \frac{2}{3.7321} = 0.5359, \qquad \alpha_5 = 4 - \frac{a_5}{\alpha_4} = 3.4641.$$

As

$$S_1 = d_1 = 0.4 \quad \text{and} \quad S_i = d_i + \frac{a_i}{\alpha_{i-1}} S_{i-1} \quad (i = 2, 3, 4, 5),$$

$$S_1 = 0.4,$$

$$S_2 = 0.8 + \frac{a_2}{\alpha_1} S_1 = 0.8 + (0.25)(0.4) = 0.9,$$

$$S_3 = 1.2 + \frac{a_3}{\alpha_2} S_2 = 1.4400,$$

$$S_4 = 1.6 + \frac{a_4}{\alpha_3} S_3 = 1.9858,$$

$$S_5 = 1.6 + \frac{a_5}{\alpha_4} S_4 = 2.6642,$$

and the solution for the first time-step is

$$u_5 = \frac{S_5}{\alpha_5} = 0.7691,$$

$$u_4 = \frac{1}{\alpha_4}(S_4 + c_4 u_5) = 0.7381,$$

$$u_3 = \frac{1}{\alpha_3}(S_3 + c_3 u_4) = 0.5834,$$

$$u_2 = \frac{1}{\alpha_2}(S_2 + c_2 u_3) = 0.3956,$$

$$u_1 = \frac{1}{\alpha_1}(S_1 + c_1 u_2) = 0.1989.$$

A comment on the stability of the elimination method

The non-pivoting elimination method previously described for solving the set of linear equations $\mathbf{Au} = \mathbf{d}$, with a tridiagonal matrix $\mathbf{A}$, is always stable, that is, with no growth of rounding errors, if

   (i) $a_i > 0$, $b_i > 0$ and $c_i > 0$,

   (ii) $b_i > a_{i+1} + c_{i-1}$ for $i = 1, 2, \ldots, N-1$, defining $c_0 = a_N = 0$, and

   (iii) $b_i > a_i + c_i$ for $i = 1, 2, \ldots, N-1$, defining $a_1 = c_{N-1} = 0$.

Conditions (i) and (ii), which ensure that the forward elimination is stable, state that the diagonal element must exceed the sum of the moduli of the other elements in the same column of the matrix $\mathbf{A}$ of coefficients. Conditions (i) and (iii), which ensure that the back substitution is stable, state that the diagonal element must exceed the sum of the moduli of the other elements in the same row. When these conditions are satisfied the algorithm is a very efficient one for programming on a digital computer, using a minimum of storage space.

Proof

To prove that the forward elimination procedure is stable it is necessary to show that the moduli of the multipliers $m_i = a_i/\alpha_{i-1}$

used to eliminate $u_1, u_2, \ldots$, are ≤ 1. By p. 25.

$$\alpha_i = b_i - \frac{a_i c_{i-1}}{\alpha_{i-1}} = b_i - m_i c_{i-1},$$

Therefore,

$$m_{i+1} = \frac{a_{i+1}}{\alpha_i} = \frac{a_{i+1}}{b_i - m_i c_{i-1}}.$$

Hence,

$$0 < m_2 < \frac{a_2}{b_1} < 1 \quad \text{since} \quad b_1 > a_2 > 0 = c_0.$$

Similarly,

$$0 < m_3 = \frac{a_3}{b_2 - m_2 c_1} \quad \text{since} \quad a_3 > 0, b_2 > c_1 \text{ and } 0 < m_2 < 1,$$

$$< \frac{a_3}{b_2 - c_1} \quad \text{since} \quad c_1 > 0,$$

$$< \frac{a_3}{(a_3 + c_1) - c_1} = 1 \quad \text{since} \quad b_2 > a_3 + c_1.$$

In this way, $0 < m_4, m_5, \ldots, m_{N-1} < 1$. The stability of the back substitution is proved in Exercise 4, Chapter 2.

A weighted average approximation

A more general finite-difference approximation to $\partial U/\partial t = \partial^2 U/\partial x^2$ than those considered is given by

$$\frac{u_{i,j+1} - u_{i,j}}{\delta t} = \frac{1}{(\delta x)^2} \{\theta(u_{i+1,j+1} - 2u_{i,j+1} + u_{i-1,j+1}) + (1-\theta)$$

$$\times (u_{i+1,j} - 2u_{i,j} + u_{i-1,j})\},$$

where, in practice, $0 \leq \theta \leq 1$. For readers familiar with finite-difference notation this replacement approximates the partial differential equation at the point $\{i\delta x, (j+\frac{1}{2})\delta t\}$ by the difference equation

$$\frac{1}{\delta t} \delta_t u_{i,j+\frac{1}{2}} = \frac{1}{(\delta x)^2} \{\theta \delta_x^2 u_{i,j+1} + (1-\theta)\delta_x^2 u_{i,j}\},$$

where the subscripts t and x denote differencing in the t- and x-directions respectively. $\theta = 0$ gives the explicit scheme, $\theta = \frac{1}{2}$ the Crank–Nicolson, and $\theta = 1$ a fully implicit backward time-difference method. The equations are unconditionally valid, i.e. stable and convergent for $\frac{1}{2} \leqslant \theta \leqslant 1$, but for $0 \leqslant \theta < \frac{1}{2}$ we must have

$$r = \frac{\delta t}{(\delta x)^2} \leqslant \frac{1}{2(1-2\theta)}. \quad \text{(See Exercise 20, Chapter 2.)}$$

Derivative boundary conditions

Boundary conditions expressed in terms of derivatives occur very frequently in practice. When, for example, the surface of a heat-conducting material is thermally insulated, there is no heat flow normal to the surface and the corresponding boundary condition is $\partial U/\partial n = 0$ at every point of the insulated surface, where the differentiation of the temperature U is in the direction of the normal to the surface. Similarly, the rate at which heat is transferred by radiation from an external surface at temperature U into a surrounding medium at temperature v is often assumed to be proportional to $(U-v)$. As the fundamental assumption of heat-conduction theory is that the rate of flow across any surface is equal to $-K\partial U/\partial n$ units of heat per unit time in the direction of the outward normal, the corresponding boundary condition for surface radiation is

$$-K\frac{\partial U}{\partial n} = H(U-v).$$

The constant K is the thermal conductivity of the material and the constant H its coefficient of surface heat transfer. The negative sign indicates that heat is assumed to flow in the opposite direction to that in which U increases algebraically. This equation can be written as

$$\frac{\partial U}{\partial n} = -h(U-v),$$

where h is a positive constant.

Consider a thin rod that is thermally insulated along its length and which radiates heat from the end $x = 0$. The temperature at this end at time t is now unknown and its determination requires an extra equation. This equation can be the boundary condition itself when a forward difference is used for $\partial U/\partial x$, because the boundary condition at $x = 0$, the left-hand end, namely,

$$-\frac{\partial U}{\partial x} = -h(U - v),$$

will be represented by

$$\frac{u_{1,j} - u_{0,j}}{\delta x} = h(u_{0,j} - v),$$

giving one extra equation for the temperature $u_{0,j}$. A negative sign must be associated with $\partial U/\partial x$ because the outward normal to the rod at this end is in the negative direction of the x-axis. Alternatively, in the heat-flow law, $-K\partial U/\partial n$ implies that when the positive direction of the x-axis (and of U) is to the *right*, then the quantity of heat flowing from *right to left* across unit area per unit time is $+K\partial U/\partial x$, and this is proportional to the excess temperature at $x = 0$.

If we wish to represent $\partial U/\partial x$ more accurately at $x = 0$ by a central difference formula it is necessary to introduce the 'fictitious' temperature $u_{-1,j}$ at the external mesh point $(-\delta x, j\delta t)$ (Fig. 2.7), by imagining the rod to be extended a distance δx at this end. The boundary condition can then be represented by

$$\frac{u_{1,j} - u_{-1,j}}{2\delta x} = h(u_{0,j} - v).$$

The temperature $u_{-1,j}$ is unknown and necessitates another equation. This is obtained by assuming that the heat conduction

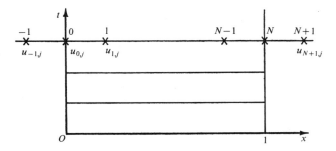

Fig. 2.7

equation is satisfied at the end $x = 0$ of the rod. The unknown $u_{-1,j}$ can then be eliminated between these equations. Similar equations can be written down for radiation from the other end of the rod.

These methods are applied below to the problem of the cooling of a homogeneous rod by radiation from its ends into air at a constant temperature, the rod being at a different constant temperature initially.

Example 2.3

Solve the equation

$$\frac{\partial U}{\partial t} = \frac{\partial^2 U}{\partial x^2}, \tag{2.14}$$

satisfying the initial condition,

$$U = 1 \text{ for } 0 \leqslant x \leqslant 1 \text{ when } t = 0,$$

and the boundary conditions,

$$\frac{\partial U}{\partial x} = U \text{ at } x = 0, \text{ for all } t,$$

$$\frac{\partial U}{\partial x} = -U \text{ at } x = 1, \text{ for all } t,$$

using an explicit method and employing central-differences for the boundary conditions.

One explicit finite-difference representation of eqn (2.14) is

$$\frac{u_{i,j+1} - u_{i,j}}{\partial t} = \frac{u_{i-1,j} - 2u_{i,j} + u_{i+1,j}}{(\delta x)^2},$$

i.e.

$$u_{i,j+1} = u_{i,j} + r(u_{i-1,j} - 2u_{i,j} + u_{i+1,j}), \tag{2.15}$$

where $r = \delta t / (\delta x)^2$.

At $x = 0$,

$$u_{0,j+1} = u_{0,j} + r(u_{-1,j} - 2u_{0,j} + u_{1,j}). \tag{2.16}$$

The boundary condition at $x = 0$, in terms of central-differences, can be written as

$$\frac{u_{1,j} - u_{-1,j}}{2\delta x} = u_{0,j}. \tag{2.17}$$

Eliminating $u_{-1,j}$ between (2.16) and (2.17) gives

$$u_{0,j+1} = u_{0,j} + 2r\{u_{1,j} - (1+\delta x)u_{0,j}\}. \tag{2.18}$$

Let $\delta x = 0.1$. Then at $x = 1$, eqn (2.15) becomes

$$u_{10,j+1} = u_{10,j} + r(u_{9,j} - 2u_{10,j} + u_{11,j}), \tag{2.19}$$

and the boundary condition is

$$\frac{u_{11,j} - u_{9,j}}{2\delta x} = -u_{10,j}. \tag{2.20}$$

Elimination of the 'fictitious' value $u_{11,j}$ between (2.19) and (2.20) yields

$$u_{10,j+1} = u_{10,j} + 2r\{u_{9,j} - (1+\delta x)u_{10,j}\}. \tag{2.21}$$

This result could have been deduced from the corresponding equation at $x = 0$ because of the symmetry with respect to $x = \frac{1}{2}$.

Later in this chapter this scheme is proved to be valid for $r \leqslant 1/(2 + h\delta x)$, i.e. $r \leqslant 1/2.1$ in this example.

Choose $r = \frac{1}{4}$. The difference equations (2.18), (2.15), then become

$$u_{0,j+1} = \tfrac{1}{2}(0.9u_{0,j} + u_{1,j}),$$

$$u_{i,j+1} = \tfrac{1}{4}(u_{i-1,j} + 2u_{i,j} + u_{i+1,j}) \quad (i = 1, 2, 3, 4),$$

and the use of symmetry rather than eqn (2.21) gives

$$u_{5,j+1} = \tfrac{1}{4}(2u_{4,j} + 2u_{5,j}).$$

As the initial temperature is $u = 1$, the values of u at the end of the first time-step when $t = r(\delta x)^2 = \frac{1}{400}$, are

$$u_{0,1} = \tfrac{1}{2}(0.9 + 1) = 0.95,$$

$$u_{1,1} = \tfrac{1}{4}(1 + 2 + 1) = 1 = u_{2,1} = u_{3,1} = u_{4,1} = u_{5,1},$$

and the values at the end of the second time-step are

$$u_{0,2} = \tfrac{1}{2}(0.9 \times 0.95 + 1) = 0.9275,$$

$$u_{1,2} = \tfrac{1}{4}(0.95 + 2 + 1) = 0.9875,$$

$$u_{2,2} = \tfrac{1}{4}(1 + 2 + 1) = 1 = u_{3,2} = u_{4,2} = u_{5,2}.$$

Similarly for subsequent time-steps. The values for several steps are recorded in Table 2.10.

The analytical solution of the partial differential equation satis-

TABLE 2.10

	$i = 0$ $x = 0$	1 0.1	2 0.2	3 0.3	4 0.4	5 0.5
$t = 0.0000$	1.0000	1.0000	1.0000	1.0000	1.0000	1.0000
0.0025	0.9500	1.0000	1.0000	1.0000	1.0000	1.0000
$t = 0.005$	0.9275	0.9875	1.0000	1.0000	1.0000	1.0000
0.0075	0.9111	0.9756	0.9969	1.0000	1.0000	1.0000
0.0100	0.8978	0.9648	0.9923	0.9992	1.0000	1.0000
0.0125	0.8864	0.9549	0.9872	0.9977	0.9998	1.0000
0.0150	0.8764	0.9459	0.9818	0.9956	0.9993	0.9999
0.0175	0.8673	0.9375	0.9762	0.9931	0.9985	0.9996
0.0200	0.8590	0.9296	0.9708	0.9902	0.9974	0.9991
. . . .						
. . . .						
0.1000	0.7175	0.7829	0.8345	0.8718	0.8942	0.9017
0.2500	0.5542	0.6048	0.6452	0.6745	0.6923	0.6983
0.5000	0.3612	0.3942	0.4205	0.4396	0.4512	0.4551
1.0000	0.1534	0.1674	0.1786	0.1867	0.1917	0.1933

fying these boundary and initial conditions is

$$U = 4 \sum_{n=1}^{\infty} \left\{ \frac{\sec \alpha_n}{(3 + 4\alpha_n^2)} e^{-4\alpha_n^2 t} \cos 2\alpha_n (x - \tfrac{1}{2}) \right\} \quad (0 < x < 1),$$

where α_n are the positive roots of

$$\alpha \tan \alpha = \tfrac{1}{2}.$$

Values of U calculated from this analytical solution are recorded in Table 2.11.

The two solutions are compared at $x = 0.2$ in Table 2.12.

The finite-difference solution is clearly very accurate for this small value of r.

Because of the symmetry with respect to $x = \tfrac{1}{2}$ the solution above is the same for a rod of length $\tfrac{1}{2}$, thermally insulated along its length and at $x = \tfrac{1}{2}$, and which cools by radiation from $x = 0$ into a medium at zero temperature.

Example 2.4

Re-solve the Worked Example 2.3 using an explicit method and employing a forward-difference for the boundary condition at $x = 0$.

TABLE 2.11

x =	0	0.1	0.2	0.3	0.4	0.5
t						
0.0025	0.9460	0.9951	0.9999	1.0000	1.0000	1.0000
0.0050	0.9250	0.9841	0.9984	0.9999	1.0000	1.0000
0.0075	0.9093	0.9730	0.9950	0.9994	1.0000	1.0000
0.0100	0.8965	0.9627	0.9905	0.9984	0.9998	1.0000
0.0125	0.8854	0.9532	0.9855	0.9967	0.9994	0.9999
0.0150	0.8755	0.9444	0.9802	0.9945	0.9988	0.9996
0.0175	0.8666	0.9362	0.9748	0.9919	0.9979	0.9992
0.0200	0.8585	0.9286	0.9695	0.9891	0.9967	0.9985
. . . .						
. . . .						
0.1000	0.7176	0.7828	0.8342	0.8713	0.8936	0.9010
0.2500	0.5546	0.6052	0.6454	0.6747	0.6924	0.6984
0.5000	0.3619	0.3949	0.4212	0.4403	0.4519	0.4558
1.0000	0.1542	0.1682	0.1794	0.1875	0.1925	0.1941

TABLE 2.12

	Finite-difference solution ($x = 0.2$)	Analytical solution ($x = 0.2$)	Percentage error
$t = 0.005$	1.0000	0.9984	0.16
0.050	0.9126	0.9120	0.07
0.100	0.8345	0.8342	0.04
0.250	0.6452	0.6454	−0.03
0.500	0.4205	0.4212	−0.16
1.000	0.1786	0.1794	−0.45

By eqn (2.15), one explicit finite-difference representation of the partial differential equation is

$$u_{i,j+1} = u_{i,j} + r(u_{i-1,j} - 2u_{i,j} + u_{i+1,j}).$$

Hence, for $i = 1$,

$$u_{1,j+1} = u_{1,j} + r(u_{0,j} - 2u_{1,j} + u_{2,j}). \tag{2.22}$$

The boundary condition at $x = 0$, namely $\partial U/\partial x = U$, in terms of a forward difference is

$$\frac{u_{1,j} - u_{0,j}}{\delta x} = u_{0,j},$$

so

$$u_{0,j} = \frac{u_{1,j}}{1+\delta x}.$$ (2.23)

Eliminating $u_{0,j}$ between (2.22) and (2.23) gives

$$u_{1,j+1} = \left(1 - 2r + \frac{r}{1+\delta x}\right)u_{1,j} + ru_{2,j}.$$ (2.24)

This scheme is valid for $0 < r \leq \frac{1}{2}$ (see Exercise 24, Chapter 2), but in order to compare its solution with the previous one, put $r = \frac{1}{4}$ and $\delta x = 0.1$. The relevant equations are then

$$u_{1,j+1} = \tfrac{8}{11}u_{1,j} + \tfrac{1}{4}u_{2,j},$$

$$u_{0,j+1} = \tfrac{10}{11}u_{1,j+1},$$

$$u_{i,j+1} = \tfrac{1}{4}(u_{i-1,j} + 2u_{i,j} + u_{i+1,j}) \quad (i = 2, 3, 4),$$

and

$$u_{5,j+1} = \tfrac{1}{4}(2u_{4,j} + 2u_{5,j}), \text{ because of symmetry.}$$

The solution of these equations for an initial value of $U = 1$ is shown in Table 2.13. A comparison with the analytical solution at $x = 0.2$ is given in Table 2.14.

TABLE 2.13

$x =$	0	0.1	0.2	0.3	0.4	0.5
$t = 0.0000$	1.0000	1.0000	1.0000	1.0000	1.0000	1.0000
0.0025	0.8884	0.9773	1.0000	1.0000	1.0000	1.0000
0.0050	0.8734	0.9607	0.9943	1.0000	1.0000	1.0000
0.0075	0.8612	0.9473	0.9873	0.9986	1.0000	1.0000
0.0100	0.8507	0.9358	0.9801	0.9961	0.9996	1.0000
0.0125	0.8415	0.9256	0.9730	0.9930	0.9989	0.9998
0.0150	0.8331	0.9164	0.9662	0.9895	0.9976	0.9993
0.0175	0.8255	0.9080	0.9596	0.9857	0.9960	0.9985
0.0200	0.8184	0.9003	0.9532	0.9817	0.9941	0.9973
. . . .						
. . . .						
0.1000	0.6869	0.7556	0.8102	0.8498	0.8738	0.8818
0.2500	0.5206	0.5727	0.6142	0.6444	0.6628	0.6689
0.5000	0.3283	0.3611	0.3873	0.4063	0.4179	0.4218
1.0000	0.1305	0.1435	0.1540	0.1615	0.1661	0.1677

TABLE 2.14

	Finite-difference solution $(x = 0.2)$	Analytical solution $(x = 0.2)$	Percentage error
$t = 0.005$	0.9943	0.9984	-0.4
0.050	0.8912	0.9120	-2.3
0.100	0.8102	0.8342	-2.9
0.250	0.6142	0.6454	-4.8
0.500	0.3873	0.4212	-8.0
1.000	0.1540	0.1794	-14.2

Although this solution is not as good as the previous one it is still sufficiently accurate for many practical purposes.

Example 2.5

Solve Example 2.3 by the Crank–Nicolson method.
    This method represents $\partial U/\partial t = \partial^2 U/\partial x^2$ by

$$\frac{u_{i,j+1} - u_{i,j}}{\delta t} = \frac{1}{2}\left\{\frac{u_{i+1,j+1} - 2u_{i,j+1} + u_{i-1,j+1}}{(\delta x)^2} + \frac{u_{i+1,j} - 2u_{i,j} + u_{i-1,j}}{(\delta x)^2}\right\},$$

which can be written as

$$-ru_{i-1,j+1} + (2+2r)u_{i,j+1} - ru_{i+1,j+1} = ru_{i-1,j} + (2-2r)u_{i,j} + ru_{i+1,j}.$$

$$(2.25)$$

    The central difference representation of the boundary condition at $x = 0$ is

$$\frac{u_{1,j} - u_{-1,j}}{2\delta x} = u_{0,j},$$

from which it follows that

$$u_{-1,j} = u_{1,j} - 2\delta x u_{0,j}$$

and

$$u_{-1,j+1} = u_{1,j+1} - 2\delta x u_{0,j+1}.$$

    The last two equations enable us to eliminate $u_{-1,j}$ and $u_{-1,j+1}$ from the equation obtained by putting $i = 0$ in (2.25).

The boundary condition at $x = 1$ can be dealt with in the same way although in this problem it is easier to make use of the symmetry with respect to $x = \frac{1}{2}$, namely, $u_{6,j} = u_{4,j}$.

This scheme is formally valid for all finite values of r but we must keep it reasonably small if we want a close approximation to the solution of the partial differential equation. Choose $r = 1$ and $\delta x = 0.1$. A small amount of algebra soon shows that the equations for the unknown pivotal values $u_{0,j+1}, u_{1,j+1}, \ldots, u_{5,j+1}$ are

$$2.1u_{0,j+1} - u_{1,j+1} = -0.1u_{0,j} + u_{1,j},$$

$$-u_{i-1,j+1} + 4u_{i,j+1} - u_{i+1,j+1} = u_{i-1,j} + u_{i+1,j} \quad (i = 1, 2, 3, 4),$$

$$-u_{4,j+1} + 2u_{5,j+1} = u_{4,j}.$$

For the first time-step these give

$$2.1u_0 - u_1 = 0.9,$$

$$-u_0 + 4u_1 - u_2 = 2.0,$$

$$-u_1 + 4u_2 - u_3 = 2.0,$$

$$-u_2 + 4u_3 - u_4 = 2.0,$$

$$-u_3 + 4u_4 - u_5 = 2.0,$$

$$-u_4 + 2u_5 = 1.0.$$

These can be solved by the direct elimination method previously described. Table 2.15 records the solution for several time-steps and Table 2.16 compares it with the analytical solution at $x = 0.2$.

TABLE 2.15

$i =$	0	1	2	3	4	5
$t = 0.00$	1.0	1.0	1.0	1.0	1.0	1.0
0.01	0.8908	0.9707	0.9922	0.9979	0.9994	0.9997
0.02	0.8624	0.9293	0.9720	0.9900	0.9964	0.9979
0.10	0.7179	0.7834	0.8349	0.8720	0.8944	0.9018
0.25	0.5547	0.6054	0.6458	0.6751	0.6929	0.6989
0.50	0.3618	0.3949	0.4212	0.4404	0.4520	0.4559
1.00	0.1540	0.1680	0.1793	0.1874	0.1923	0.1940

TABLE 2.16

	Finite-difference solution $(x = 0.2)$	Analytical solution $(x = 0.2)$	Percentage error
$t = 0.01$	0.9922	0.9905	0.17
0.05	0.9131	0.9120	0.12
0.10	0.8349	0.8342	0.08
0.25	0.6458	0.6454	0.06
0.50	0.4212	0.4212	0.00
1.00	0.1793	0.1794	−0.06

The local truncation error and consistency

The local truncation error

Let $F_{i,j}(u) = 0$ represent the difference equation approximating the partial differential equation at the (i, j)th mesh point, with exact solution u. If u is replaced by U at the mesh points of the difference equation, where U is the exact solution of the partial differential equation, the value of $F_{i,j}(U)$ is called *the local truncation error* $T_{i,j}$ at the (i, j) mesh point. $F_{i,j}(U)$ clearly measures the amount by which the exact solution values of the partial differential equation at the mesh points of the difference equation do *not* satisfy the difference equation at the point (ih, jk).

Using Taylor expansions, it is easy to express $T_{i,j}$ in terms of powers of h and k and partial derivatives of U at (ih, jk). Although U and its derivatives are generally unknown, the analysis is worthwhile because it provides a method for comparing the local accuracies of different difference schemes approximating the partial differential equation.

Example 2.6

Calculate the order of the local truncation error of the classical explicit difference approximation to

$$\frac{\partial U}{\partial t} - \frac{\partial^2 U}{\partial x^2} = 0$$

at the point (ih, jk).

$$F_{i,j}(u) = \frac{u_{i,j+1} - u_{i,j}}{k} - \frac{u_{i-1,j} - 2u_{i,j} + u_{i+1,j}}{h^2} = 0$$

Therefore,

$$T_{i,j} = F_{i,j}(U) = \frac{U_{i,j+1} - U_{i,j}}{k} - \frac{U_{i-1,j} - 2U_{i,j} + U_{i+1,j}}{h^2}.$$

By Taylor's expansion

$$U_{i+1,j} = U\{(i+1)h, jk\} = U(x_i + h, t_j)$$

$$= U_{i,j} + h\left(\frac{\partial U}{\partial x}\right)_{i,j} + \tfrac{1}{2}h^2\left(\frac{\partial^2 U}{\partial x^2}\right)_{i,j} + \tfrac{1}{6}h^3\left(\frac{\partial^3 U}{\partial x^3}\right)_{i,j} + \ldots$$

$$U_{i-1,j} = U\{(i-1)h, jk\} = U(x_i - h, t_j)$$

$$= U_{i,j} - h\left(\frac{\partial U}{\partial x}\right)_{i,j} + \tfrac{1}{2}h^2\left(\frac{\partial^2 U}{\partial x^2}\right)_{i,j} - \tfrac{1}{6}h^3\left(\frac{\partial^3 U}{\partial x^3}\right)_{i,j} + \ldots$$

$$U_{i,j+1} = U(x_i, t_j + k)$$

$$= U_{i,j} + k\left(\frac{\partial U}{\partial t}\right)_{i,j} + \tfrac{1}{2}k^2\left(\frac{\partial^2 U}{\partial t^2}\right)_{i,j} + \tfrac{1}{6}k^3\left(\frac{\partial^3 U}{\partial t^3}\right)_{i,j} + \ldots.$$

Substitution into the expression for $T_{i,j}$ then gives

$$T_{i,j} = \left(\frac{\partial U}{\partial t} - \frac{\partial^2 U}{\partial x^2}\right)_{i,j} + \tfrac{1}{2}k\left(\frac{\partial^2 U}{\partial t^2}\right)_{i,j} - \tfrac{1}{12}h^2\left(\frac{\partial^4 U}{\partial x^4}\right)_{i,j}$$

$$+ \tfrac{1}{6}k^2\left(\frac{\partial^3 U}{\partial t^3}\right)_{i,j} - \tfrac{1}{360}h^4\frac{\partial^6 U}{\partial x^6} + \ldots.$$

But U is the solution of the differential equation so

$$\left(\frac{\partial U}{\partial t} - \frac{\partial^2 U}{\partial x^2}\right)_{i,j} = 0.$$

Therefore the principal part of the local truncation error is

$$\left(\tfrac{1}{2}k\frac{\partial^2 U}{\partial t^2} - \tfrac{1}{12}h^2\frac{\partial^4 U}{\partial x^4}\right)_{i,j}.$$

Hence

$$T_{i,j} = O(k) + O(h^2).$$

When $k = rh^2$, $0 < r \leqslant \tfrac{1}{2}$, $T_{i,j}$ is $O(k)$ or $O(h^2)$, as one would expect by eqns (1.8) and (1.10).

This error may be further reduced by choosing a special value for k/h^2 because the equation for $T_{i,j}$ can be written as

$$T_{i,j} = \tfrac{1}{12}h^2\left(6\,\frac{k}{h^2}\frac{\partial^2 U}{\partial t^2} - \frac{\partial^4 U}{\partial x^4}\right)_{i,j} + O(k^2) + O(h^4).$$

By the differential equation,

$$\frac{\partial}{\partial t} = \frac{\partial^2}{\partial x^2},$$

so

$$\frac{\partial}{\partial t}\left(\frac{\partial U}{\partial t}\right) = \frac{\partial^2}{\partial x^2}\left(\frac{\partial^2 U}{\partial x^2}\right),$$

assuming that these derivatives exist. If we put $6k/h^2 = 1$, the expression in the brackets is then zero and $T_{i,j}$ is $O(k^2) + O(h^4)$. This is of little use in practice because $k = \tfrac{1}{6}h^2$ is very small for small h so the volume of arithmetic needed to advance the solution to a large time-level is substantial.

Consistency or compatibility

It is sometimes possible to approximate a parabolic or hyperbolic equation by a finite-difference scheme that is stable, (i.e. limits the amplification of all the components of the initial conditions), but which has a solution that converges to the solution of a different differential equation as the mesh lengths tend to zero. Such a difference scheme is said to be *inconsistent* or *incompatible* with the partial differential equation and an example is given in Worked Example 2.7.

The real importance of the concept of consistency lies in a theorem by Lax (reference 25), which states that if a linear finite-difference equation is consistent with a properly posed linear initial-value problem then stability guarantees convergence of u to U as the mesh lengths tend to zero. Consistency can be defined in either of two equivalent but slightly different ways.

The more general definition is as follows. Let $L(U) = 0$ represent the partial differential equation in the independent variables x and t, with exact solution U.

Let $F(u) = 0$ represent the approximating finite-difference equation with exact solution u.

Let v be a continuous function of x and t with a sufficient number of continuous derivatives to enable $L(v)$ to be evaluated at the point (ih, jk).

Then the truncation error $T_{i,j}(v)$ at the point (ih, jk) is defined by

$$T_{i,j}(v) = F_{i,j}(v) - L(v_{i,j}).$$

If $T_{i,j}(v) \to 0$ as $h \to 0$, $k \to 0$, the difference equation is said to be consistent or compatible with the partial differential equation. With this definition $T_{i,j}$ gives an indication of the error resulting from the replacement of $L(v_{i,j})$ by $F_{i,j}(v)$.

Most authors put $v = U$ because $L(U) = 0$. It then follows that

$$T_{i,j}(U) = F_{i,j}(U),$$

and the truncation error coincides with the local truncation error. The difference equation is then consistent if the limiting value of the local truncation error is zero as $h \to 0$, $k \to 0$. This is the definition that will be adopted in this book. By the Worked Example 2.6 it follows that the classical explicit approximation to $\partial U / \partial t = \partial^2 U / \partial x^2$ is consistent with the differential equation.

Example 2.7

The equation

$$\frac{\partial U}{\partial t} - \frac{\partial^2 U}{\partial x^2} = 0$$

is approximated at the point (ih, jk) by the difference equation

$$\frac{u_{i,j+1} - u_{i,j-1}}{2k} - \frac{u_{i+1,j} - 2\{\theta u_{i,j+1} + (1-\theta)u_{i,j-1}\} + u_{i-1,j}}{h^2} = 0.$$

Show that the local truncation error at this point is

$$\frac{k^2}{6} \frac{\partial^3 U}{\partial t^3} - \frac{h^2}{12} \frac{\partial^4 U}{\partial x^4} + (2\theta - 1)\frac{2k}{h^2} \frac{\partial U}{\partial t} + \frac{k^2}{h^2} \frac{\partial^2 U}{\partial t^2} + O\left(\frac{k^3}{h^2}, h^4, k^4\right).$$

Discuss the consistency of this scheme with the partial differential equation when:

$$\text{(i)} \quad k = rh \quad \text{and} \quad \text{(ii)} \quad k = rh^2,$$

where r is a positive constant and θ a variable parameter.

Expansion of the terms $U_{i,j+1}$, $U_{i,j-1}$, $U_{i+1,j}$, and $U_{i-1,j}$ about the point (ih, jk) by Taylor's series, as in Example 2.6, and substitution into

$$T_{i,j} = \frac{U_{i,j+1} - U_{i,j-1}}{2k} - \frac{U_{i+1,j} - 2\{\theta U_{i,j+1} + (1-\theta)U_{i,j-1}\} + U_{i-1,j}}{h^2}$$

leads to

$$T_{i,j} = \left(\frac{\partial U}{\partial t} - \frac{\partial^2 U}{\partial x^2}\right)_{i,j} + \left\{\frac{k^2}{6}\frac{\partial^3 U}{\partial t^3} - \frac{h^2}{12}\frac{\partial^4 U}{\partial x^4}\right.$$
$$\left. + (2\theta - 1)\frac{2k}{h^2}\frac{\partial U}{\partial t} + \frac{k^2}{h^2}\frac{\partial^2 U}{\partial t^2}\right\} + O\left(\frac{k^3}{h^2}, h^4, k^4\right).$$

Hence the result since $\partial U/\partial t - \partial^2 U/\partial x^2 = 0$.

Case (i) $k = rh$

As $h \to 0$,

$$T_{i,j} = F_{i,j}(U) \to \left\{\frac{\partial U}{\partial t} - \frac{\partial^2 U}{\partial x^2} + (2\theta - 1)\frac{2r}{h}\frac{\partial U}{\partial t} + r^2\frac{\partial^2 U}{\partial t^2}\right\}_{i,j}.$$

When $\theta \neq \frac{1}{2}$ the third term tends to infinity. When $\theta = \frac{1}{2}$ the limiting value of $T_{i,j}$ is

$$\frac{\partial U}{\partial t} - \frac{\partial^2 U}{\partial x^2} + r^2\frac{\partial^2 U}{\partial t^2}.$$

In this case the finite-difference equation is consistent with the *hyperbolic* equation

$$\frac{\partial U}{\partial t} - \frac{\partial^2 U}{\partial x^2} + r^2\frac{\partial^2 U}{\partial t^2} = 0.$$

Hence the difference equation is always inconsistent with $\partial U/\partial t - \partial^2 U/\partial x^2 = 0$ when $k = rh$.

Case (ii) $k = rh^2$

As $h \to 0$,

$$T_{i,j} \to \frac{\partial U}{\partial t} - \frac{\partial^2 U}{\partial x^2} + 2(2\theta - 1)r\frac{\partial U}{\partial t}.$$

When $\theta \neq \frac{1}{2}$ the difference scheme is consistent with the parabolic equation

$$\{1+2(2\theta-1)r\}\frac{\partial U}{\partial t}-\frac{\partial^2 U}{\partial x^2}=0.$$

It is only when $\theta=\frac{1}{2}$ that the difference scheme is consistent with the given differential equation. This is then the well-known Du Fort and Frankel three-level explicit scheme which is also stable for all $r>0$. (See Worked Example 3.2). It was devised to overcome the unconditional instability of the early Richardson explicit scheme

$$\frac{u_{i,j+1}-u_{i,j-1}}{2k}-\frac{u_{i-1,j}-2u_{i,j}+u_{i+1,j}}{h^2}=0,$$

but to retain the advantage of the central-difference approximation to the time-derivative which gives a local truncation error of $O(k^2)+O(h^2)$ as opposed to $O(k)+O(h^2)$ for the classical explicit approximation to $\partial U/\partial t-\partial^2 U/\partial x^2=0$.

Convergence and stability

The following sections are concerned with the conditions that must be satisfied if the solution of the finite-difference equations is to be a reasonably accurate approximation to the solution of the corresponding parabolic or hyperbolic partial differential equation.

These conditions are associated with two different but interrelated problems. The first concerns the convergence of the exact solution of the approximating difference equations to the solution of the differential equation; the second concerns the unbounded growth, or controlled decay or boundedness of the exact solution of the finite-difference equations, and therefore of all rounding errors introduced during the computation because the errors and exact solution are processed by the same arithmetic operations. (The stability problem.)

Descriptive treatment of convergence

Let U represent the *exact* solution of a partial differential equation with independent variables x and t, and u the *exact* solution of the difference equations used to approximate the partial differential equation. Then the finite-difference equation is said to be convergent when u tends to U at a fixed point or along a fixed t-level as δx and δt both tend to zero.

Although the conditions under which u converges to U have been established for linear elliptic, parabolic and hyperbolic second-order partial differential equations with solutions satisfying fairly general boundary and initial conditions, they are not yet known for non-linear equations except in a few particular cases. (The equation,

$$a\frac{\partial^2 U}{\partial x^2} + b\frac{\partial^2 U}{\partial x \partial t} + c\frac{\partial^2 U}{\partial t^2} + d\frac{\partial U}{\partial x} + e\frac{\partial U}{\partial t} + fU + g = 0,$$

is linear when the coefficients $a, b, \ldots, g$, are constants or functions of x and t only. Otherwise it is non-linear. If the coefficients of the second-order derivatives are functions of x, t, U, $\partial U/\partial x$ and $\partial U/\partial t$ but not of second-order derivatives the equation is described as quasi-linear even though it is non-linear. The important feature of linear equations is that the sum of separate solutions is also a solution.)

The difference $(U - u)$ is called the *discretization error*. Some texts call it the truncation error but in this book the latter term will be reserved for the difference between the differential equation and its approximating difference equation. The magnitude of the discretization error at any mesh point depends on the finite-sizes of the mesh lengths, δx and δt, i.e. on the distances between consecutive, discrete grid-points, and on the number of terms in the truncated series of differences used to approximate the derivatives. Readers familiar with the calculus of finite-differences will have recognized the approximation used earlier for $\partial U/\partial t$ as the first term in either the series

$$(\delta t)\left(\frac{\partial U}{\partial t}\right)_{i,j} = (\Delta_t - \tfrac{1}{2}\Delta_t^2 + \tfrac{1}{3}\Delta_t^3 - \ldots)U_{i,j}$$

or the series

$$(\delta t)\left(\frac{\partial U}{\partial t}\right)_{i,j+\frac{1}{2}} = (\delta_t - \tfrac{1}{24}\delta_t^3 + \tfrac{3}{640}\delta_t^5 + \ldots)U_{i,j+\frac{1}{2}},$$

and the approximation for $\partial^2 U/\partial x^2$ as the first term in the series

$$(\delta x)^2 \frac{\partial^2 U_{i,j}}{\partial x^2} = (\delta_x^2 - \tfrac{1}{12}\delta_x^4 + \tfrac{1}{90}\delta_x^6 - \ldots)U_{i,j},$$

where the subscripts t and x denote the directions in which the differences are calculated. The symbols Δ and δ are the forward and central difference operators defined by $\Delta_t u_{i,j} = u_{i,j+1} - u_{i,j}$ and $\delta_t u_{i,j+\frac{1}{2}} = u_{i,j+1} - u_{i,j}$, so that $\delta_x^2 u_{i,j} = \delta_x(\delta_x u_{i,j}) = \delta_x(u_{i+\frac{1}{2},j} - u_{i-\frac{1}{2},j}) = u_{i+1,j} - 2u_{i,j} + u_{i-1,j}$. Better approximations can be obtained by truncating the series after two or more terms but have the disadvantage of involving more pivotal values of u. It will be shown later that the discretization error can be analysed in terms of preceding local truncation errors. (See p. 73.)

The discretization error can usually be diminished by decreasing δx and δt, subject invariably to some relationship between them, but as this leads to an increase in the number of equations to be solved, this method of improvement is limited by such factors as cost of computation and computer storage requirements, etc.

In general, the problem of convergence is a difficult one to investigate usefully because the final expression for the discretization error is usually in terms of unknown derivatives for which no bounds can be estimated. Fortunately, however, the convergence of difference equations approximating *linear* parabolic and hyperbolic differential equations can be investigated in terms of stability and consistency, which are easier to deal with. (Reference Lax's equivalence theorem, p. 72).

Analytical treatment of convergence *(A direct method)*

The convergence of the solution of an approximating set of *linear*-difference equations to the solution of a *linear* partial differential equation is dealt with most easily via Lax's equivalence theorem. Explicit difference schemes however, can be investigated directly by deriving a difference equation for the discretization error e. Denote the exact solution of the partial differential equation by U and the exact solution of the finite-difference equation by u. Then $e = U - u$.

Consider the equation

$$\frac{\partial U}{\partial t} = \frac{\partial^2 U}{\partial x^2}, \quad 0 < x < 1, t > 0, \tag{2.26}$$

where U is known for $0 \le x \le 1$ when $t = 0$, and at $x = 0$ and 1 when $t > 0$.

The simplest explicit finite-difference approximation to (2.26) is

$$\frac{u_{i,j+1} - u_{i,j}}{k} = \frac{u_{i-1,j} - 2u_{i,j} + u_{i+1,j}}{h^2}. \tag{2.27}$$

At the mesh points,

$$u_{i,j} = U_{i,j} - e_{i,j}, \quad u_{i,j+1} = U_{i,j+1} - e_{i,j+1}, \text{ etc.}$$

Substitution into (2.27) leads to

$$e_{i,j+1} = re_{i-1,j} + (1 - 2r)e_{i,j} + re_{i+1,j} + U_{i,j+1} - U_{i,j}$$
$$+ r(2U_{i,j} - U_{i-1,j} - U_{i+1,j}). \tag{2.28}$$

By Taylor's theorem,

$$U_{i+1,j} = U(x_i + h, t_j) = U_{i,j} + h\left(\frac{\partial U}{\partial x}\right)_{i,j} + \frac{h^2}{2!}\frac{\partial^2 U}{\partial x^2}(x_i + \theta_1 h, t_j),$$

$$U_{i-1,j} = U(x_i - h, t_j) = U_{i,j} - h\left(\frac{\partial U}{\partial x}\right)_{i,j} + \frac{h^2}{2!}\frac{\partial^2 U}{\partial x^2}(x_i - \theta_2 h, t_j),$$

$$U_{i,j+1} = U(x_i, t_j + k) = U_{i,j} + k\frac{\partial U(x_i, t_j + \theta_3 k)}{\partial t}$$

where $0 < \theta_1 < 1$, $0 < \theta_2 < 1$ and $0 < \theta_3 < 1$. Substitution into eqn (2.28) gives

$$e_{i,j+1} = re_{i-1,j} + (1 - 2r)e_{i,j} + re_{i+1,j}$$
$$+ k\left\{\frac{\partial U(x_i, t_j + \theta_3 k)}{\partial t} - \frac{\partial^2 U(x_i + \theta_4 h, t_j)}{\partial x^2}\right\}, \tag{2.29}$$

where $-1 < \theta_4 < 1$.

This is a difference equation for $e_{i,j}$ which fortunately we need not solve.

Let E_j denote the maximum value of $|e_{i,j}|$ along the jth time-row and M the maximum modulus of the expression in the braces for all i and j. When $r \leqslant \frac{1}{2}$, all the coefficients of e in eqn (2.29) are positive or zero, so

$$|e_{i,j+1}| \leqslant r|e_{i-1,j}| + (1 - 2r)|e_{i,j}| + r|e_{i+1,j}| + kM$$
$$\leqslant rE_j + (1 - 2r)E_j + rE_j + kM$$
$$= E_j + kM.$$

As this is true for all values of i it is true for $\max|e_{i,j+1}|$. Hence

$$E_{j+1} \leqslant E_j + kM \leqslant (E_{j-1} + kM) + kM = E_{j-1} + 2kM,$$

etc., from which it follows that

$$E_j \leqslant E_0 + jkM = tM,$$

because the initial values for u and U are the same, i.e. $E_0 = 0$. When h tends to zero, $k = rh^2$ also tends to zero and M tends to

$$\left(\frac{\partial U}{\partial t} - \frac{\partial^2 U}{\partial x^2} \right)_{i,j}.$$

Since U is a solution of eqn (2.26) the limiting value of M and therefore of E_j is zero. As $|U_{i,j} - u_{i,j}| \leqslant E_j$, this proves that u converges to U as h tends to zero when $r \leqslant \frac{1}{2}$ and t is finite.

When $r > \frac{1}{2}$ it can be shown that the complementary function of the difference equation (2.29) tends to infinity as h tends to zero. There is no need, however, to do this when our main purpose is to find the conditions necessary for a useful numerical solution because we shall prove later that this finite-difference scheme is stable for $r \leqslant \frac{1}{2}$ but unstable for $r > \frac{1}{2}$.

The proof above implies that $\partial U / \partial t$ and $\partial^2 U / \partial x^2$ are uniformly continuous and bounded throughout the solution domain. This was so in the Worked Example 2.1 in which $\partial^2 U / \partial x^2$ was initially zero in spite of the discontinuity in $\partial U / \partial x$. If it is assumed that U possesses continuous bounded derivatives up to order three in t and order six in x, Exercise 13 shows that the discretization error is of order h^2, except when $r = \frac{1}{6}$ in which case it is of order h^4.

Descriptive treatment of stability

The equations that are actually solved are, of course, the finite-difference equations, and their application and solution to successive time-rows advances the finite-difference solution from the initial line, on which initial values are known, to time-levels k, $2k, \ldots Jk = T$, say, where T is finite. If no rounding errors were introduced into this numerical process then the exact solution $u_{i,j}$ of the finite-difference equations would be obtained at each mesh point (i, j), $i = 0(1)N$, $0 < j \leqslant J$.

The essential idea defining stability is that this numerical process, applied exactly, should limit the amplification of all components of the initial conditions.

For linear initial-value boundary-value problems, Lax and

Richtmyer have related stability to convergence via Lax's Equivalence Theorem (p. 72) by defining stability, in effect, in terms of the boundedness of the solution of the finite-difference equations at a fixed time-level T as $k \to 0$, i.e. as $J \to \infty$, it being assumed that $\delta x = h$ is related to k in such a way that $h \to 0$ as $k \to 0$.

Assume that the vector of solution values $\boldsymbol{u}_{j+1} = [u_{1,j+1}, u_{2,j+1}, \ldots, u_{N-1,j+1}]^T$ of the finite-difference equations at the $(j+1)$th time-level is related to the vector of solution values at the jth time-level by the equation

$$\boldsymbol{u}_{j+1} = \mathbf{A}\boldsymbol{u}_j + \mathbf{b}_j,$$

where $\mathbf{b}_j$ is a column vector of known boundary-values and zeros, and matrix $\mathbf{A}$ an $(N-1) \times (N-1)$ matrix of known elements. Then it will be shown that the practical consequence of this definition of stability is that a norm of matrix $\mathbf{A}$ compatible with a norm of $\boldsymbol{u}$ must satisfy

$$\|\mathbf{A}\| \leq 1$$

when the solution of the partial differential equation does not increase as t increases, or

$$\|\mathbf{A}\| \leq 1 + O(k)$$

when the solution of the partial differential increases as t increases.

These conditions also ensure the boundedness of all rounding errors because they are subject to the same arithmetic operations as the finite-difference solution.

In an actual computation, however, k and h are normally kept constant as the solution is propagated forward time-level by time-level from $t = 0$ to $t_j = jk$, and in many textbooks and papers stability is defined in terms of the boundedness of this numerical solution as $j \to \infty$, k fixed. In this process the order $(N-1)$ of matrix $\mathbf{A}$ remains constant, unlike the matrix $\mathbf{A}$ associated with Lax and Richtmyer's definition. The matrix method of analysis then shows that the equations are stable if the largest of the moduli of the eigenvalues of matrix $\mathbf{A}$, i.e. the spectral radius $\rho(\mathbf{A})$ of $\mathbf{A}$, satisfies

$$\rho(\mathbf{A}) \leq 1,$$

when the solution of the differential equation does not increase with increasing t.

Although this condition ensures the boundedness of the computed solution it does not guarantee convergence unless the eigenvalues of $\mathbf{A}$ are restricted to satisfy $\rho(\mathbf{A}) \leq \|\mathbf{A}\| \leq 1$, as $N \to \infty$. In practice, assuming that the difference equations are consistent, it is usually only in the immediate neighbourhood of $\rho(\mathbf{A}) = 1$ that non-convergence might occur. An illuminating discussion of these points is given in reference 19. (If the solution of the partial differential equation increases as $t \to \infty$, the condition for stability with fixed h and k is then $\rho(\mathbf{A}) \leq 1 + O(k)$. See p. 66.)

Vector and matrix norms

This section is needed for the Lax–Richtmyer definition of stability.

Vector norms

The norm of vector $\mathbf{x}$ is a real positive number giving a measure of the 'size' of the vector and is denoted by $\|\mathbf{x}\|$. It must satisfy the following axioms.

    (i) $\|\mathbf{x}\| > 0$ if $\mathbf{x} \neq \mathbf{0}$ and $\|\mathbf{x}\| = 0$ if $\mathbf{x} = \mathbf{0}$.

    (ii) $\|c\mathbf{x}\| = |c| \|\mathbf{x}\|$ for a real or complex scalar c.

    (iii) $\|\mathbf{x} + \mathbf{y}\| \leq \|\mathbf{x}\| + \|\mathbf{y}\|$.

If the $n \times 1$ vector $\mathbf{x}$ has components $x_1, x_2, \ldots, x_n$, then the three most commonly used norms are defined as follows.

The 1-norm of $\mathbf{x}$ is the sum of the moduli of the components of $\mathbf{x}$, i.e.

$$\|\mathbf{x}\|_1 = |x_1| + |x_2| + \ldots + |x_n| = \sum_{i=1}^{n} |x_i|.$$

The infinity norm of $\mathbf{x}$ is the maximum of the moduli of the components of $\mathbf{x}$, i.e.

$$\|\mathbf{x}\|_\infty = \max_i |x_i|.$$

The 2-norm of $\mathbf{x}$ is the square root of the sum of the squares of the moduli of the components of $\mathbf{x}$, i.e.

$$\|\mathbf{x}\|_2 = (|x_1|^2 + |x_2|^2 + \cdots |x_n|^2)^{\frac{1}{2}} = \left[\sum_{i=1}^{n} |x_i|^2 \right]^{\frac{1}{2}}.$$

The 2-norm gives the 'length' of the vector.

Matrix norms

The norm of matrix $\mathbf{A}$ is a real positive number giving a measure of the 'size' of the matrix and must satisfy the following axioms.
  (i) $\|\mathbf{A}\| > 0$ if $\mathbf{A} \neq \mathbf{0}$ and $\|\mathbf{A}\| = 0$ if $\mathbf{A} = \mathbf{0}$.
  (ii) $\|c\mathbf{A}\| = |c| \|\mathbf{A}\|$ for a real or complex scalar c.
  (iii) $\|\mathbf{A} + \mathbf{B}\| \leq \|\mathbf{A}\| + \|\mathbf{B}\|$.
  (iv) $\|\mathbf{AB}\| \leq \|\mathbf{A}\| \|\mathbf{B}\|$.

Compatible or consistent norms

Vectors and matrices occur together so it is essential that they satisfy a condition equivalent to (iv). As a consequence, matrix and vector norms are said to be compatible or consistent if

$$\|\mathbf{Ax}\| \leq \|\mathbf{A}\| \|\mathbf{x}\|, \quad \mathbf{x} \neq \mathbf{0}.$$

Subordinate matrix norms

Let $\mathbf{A}$ be an $n \times n$ matrix and $\mathbf{x}$ a member of the set S of $n \times 1$ vectors whose norms are unity, i.e. $\mathbf{x} \in S$ if $\|\mathbf{x}\| = 1$. In general, the norm of the vector $\mathbf{Ax}$ will vary as $\mathbf{x}$ varies, $\mathbf{x} \in S$. Let $\mathbf{x}_0$ be a member of S that makes $\|\mathbf{Ax}\|$ attain its maximum value. Then the norm of matrix $\mathbf{A}$ is defined by

$$\|\mathbf{A}\| = \|\mathbf{Ax}_0\| = \max_{\|\mathbf{x}\|=1} \|\mathbf{Ax}\|.$$

This matrix norm is said to be subordinate to the vector norm and automatically satisfies the compatibility condition, because, if $\mathbf{x} = \mathbf{x}_1$ is any other member of S, $\|\mathbf{Ax}_1\| \leq \|\mathbf{Ax}_0\| = \|\mathbf{A}\| = \|\mathbf{A}\| \|\mathbf{x}_1\|$, since $\|\mathbf{x}_1\| = 1$. It also follows that for all subordinate matrix norms,

$$\|\mathbf{I}\| = \max_{\|\mathbf{x}\|=1} \|\mathbf{Ix}\| = \max_{\|\mathbf{x}\|=1} \|\mathbf{x}\| = 1$$

where $\mathbf{I}$ is the unit matrix. The definitions of the 1, 2, and ∞ norms with $\|\mathbf{x}\| = 1$ lead to the following results which are proved in most linear algebra books. (See Exercise 15.)

The 1-norm of matrix $\mathbf{A}$ is the maximum column sum of the moduli of the elements of $\mathbf{A}$.

The infinity norm of matrix $\mathbf{A}$ is the maximum row sum of the moduli of the elements of $\mathbf{A}$.

The 2-norm of matrix $\mathbf{A}$ is the square root of the spectral radius of $\mathbf{A}^H\mathbf{A}$, where $\mathbf{A}^H = (\bar{\mathbf{A}})^T$, the transpose of the conjugate complex of $\mathbf{A}$. For example, if

$$\mathbf{A} = \begin{bmatrix} -1 & 1 \\ 3 & -2 \end{bmatrix}, \text{ then } \mathbf{A}^H\mathbf{A} = \mathbf{A}^T\mathbf{A} = \begin{bmatrix} 10 & -7 \\ -7 & 5 \end{bmatrix}$$

with eigenvalues 14.93 and 0.067. Hence $\|\mathbf{A}\|_1 = 1+3 = 4$, $\|\mathbf{A}\|_\infty = 3+2 = 5$ and $\|\mathbf{A}\|_2 = \sqrt{14.93} = 3.86$.

When matrix $\mathbf{A}$ is real and symmetric, $\mathbf{A}^H = \mathbf{A}$, and

$$\|\mathbf{A}\|_2 = [\rho(\mathbf{A}^2)]^{\frac{1}{2}} = [\rho^2(\mathbf{A})]^{\frac{1}{2}} = \rho(\mathbf{A}) = \max_i |\lambda_i|.$$

A bound for the spectral radius

Let λ_i be an eigenvalue of the $n \times n$ matrix $\mathbf{A}$ and $\mathbf{x}_i$ the corresponding eigenvector. Hence

$$\mathbf{A}\mathbf{x}_i = \lambda_i \mathbf{x}_i$$

and

$$\|\mathbf{A}\mathbf{x}_i\| = \|\lambda_i \mathbf{x}_i\| = |\lambda_i| \, \|\mathbf{x}_i\|.$$

For all compatible matrix and vector norms it follows that

$$|\lambda_i| \, \|\mathbf{x}_i\| = \|\mathbf{A}\mathbf{x}_i\| \leqslant \|\mathbf{A}\| \, \|\mathbf{x}_i\|.$$

Therefore,

$$|\lambda_i| \leqslant \|\mathbf{A}\|, \quad i = 1(1)n.$$

Hence,

$$\rho(\mathbf{A}) \leqslant \|\mathbf{A}\|.$$

A necessary and sufficient condition for stability. (Constants coefficients)

Let the solution domain of the partial differential equation be the finite rectangle $0 \leqslant x \leqslant 1$, $0 \leqslant t \leqslant T$, and subdivide it into uniform rectangular meshes by the lines $x_i = ih$, $i = 0(1)N$, where $Nh = 1$, and the lines $t_j = jk$, $j = 0(1)J$ where $Jk = T$. It will be assumed that h is related to k by some relationship such as $k = rh$ or $k = rh^2$, $r > 0$ and finite, so that $h \to 0$ as $k \to 0$.

Assume that the finite-difference equation relating the mesh-point values along the $(j+1)$th and jth time-rows is

$$b_{i-1}u_{i-1,j+1} + b_i u_{i,j+1} + b_{i+1}u_{i+1,j+1} = c_{i-1}u_{i-1,j} + c_i u_{i,j} + c_{i+1}u_{i+1,j}$$

where the coefficients are constants.

If the boundary values at $i = 0$ and N, $j > 0$, are known, these $(N-1)$ equations for $i = 1(1)N-1$ can be written in matrix form as

$$
\begin{bmatrix}
b_1 & b_2 & & & \\
b_1 & b_2 & b_3 & & \\
& \cdot & \cdot & \cdot & \\
& & \cdot & \cdot & \cdot \\
& & b_{N-3} & b_{N-2} & b_{N-1} \\
& & & b_{N-2} & b_{N-1}
\end{bmatrix}
\begin{bmatrix}
u_{1,j+1} \\
u_{2,j+1} \\
\vdots \\
\\
u_{N-2,j+1} \\
u_{N-1,j+1}
\end{bmatrix}
$$

$$
=
\begin{bmatrix}
c_1 & c_2 & & & \\
c_1 & c_2 & c_3 & & \\
& \cdot & \cdot & \cdot & \\
& & \cdot & \cdot & \cdot \\
& & c_{N-3} & c_{N-2} & c_{N-1} \\
& & & c_{N-2} & c_{N-1}
\end{bmatrix}
\begin{bmatrix}
u_{1,j} \\
u_{2,j} \\
\vdots \\
\\
u_{N-2,j} \\
u_{N-1,j}
\end{bmatrix}
+
\begin{bmatrix}
c_0 u_{0,j} - b_0 u_{0,j+1} \\
0 \\
\vdots \\
\\
0 \\
c_N u_{N,j} - b_N u_{N,j+1}
\end{bmatrix}
$$

i.e. as $\mathbf{Bu}_{j+1} = \mathbf{Cu}_j + \mathbf{d}_j$, where the matrices $\mathbf{B}$ and $\mathbf{C}$ of order $(N-1)$ are as shown, $\mathbf{u}_{j+1}$ denotes the column vector with components $u_{1,j+1}, u_{2,j+1}, \ldots, u_{N-1,j+1}$ and $\mathbf{d}_j$ denotes the column vector of known boundary values and zeros.
Hence,

$$\mathbf{u}_{j+1} = \mathbf{B}^{-1}\mathbf{Cu}_j + \mathbf{B}^{-1}\mathbf{d}_j,$$

which may be expressed more conveniently as

$$\mathbf{u}_{j+1} = \mathbf{Au}_j + \mathbf{f}_j,$$

where $\mathbf{A} = \mathbf{B}^{-1}\mathbf{C}$ and $\mathbf{f}_j = \mathbf{B}^{-1}\mathbf{d}_j$. Applied recursively, this leads to

$$
\begin{aligned}
\mathbf{u}_j &= \mathbf{Au}_{j-1} + \mathbf{f}_{j-1} = \mathbf{A}(\mathbf{Au}_{j-2} + \mathbf{f}_{j-2}) + \mathbf{f}_{j-1} \\
&= \mathbf{A}^2\mathbf{u}_{j-2} + \mathbf{Af}_{j-2} + \mathbf{f}_{j-1} \\
&= \ldots \\
&= \mathbf{A}^j\mathbf{u}_0 + \mathbf{A}^{j-1}\mathbf{f}_0 + \mathbf{A}^{j-2}\mathbf{f}_1 + \ldots + \mathbf{f}_{j-1}, \tag{2.30}
\end{aligned}
$$

where $\mathbf{u}_0$ is the vector of initial values and $\mathbf{f}_0, \mathbf{f}_1, \ldots, \mathbf{f}_{j-1}$ are vectors of known boundary-values. When we are concerned more with a property of the equations, such as stability, than with a numerical solution, the constant vectors can be eliminated by investigating the propagation of a perturbation.

Perturb the vector of initial values $\mathbf{u}_0$ to $\mathbf{u}_0^*$. The exact solution at the jth time-row will then be

$$\mathbf{u}_j^* = \mathbf{A}^j \mathbf{u}_0^* + \mathbf{A}^{j-1}\mathbf{f}_0 + \mathbf{A}^{j-2}\mathbf{f}_1 + \ldots + \mathbf{f}_{j-1}. \tag{2.31}$$

If the perturbation or 'error' vector $\mathbf{e}$ is defined by

$$\mathbf{e} = \mathbf{u}^* - \mathbf{u},$$

it follows by eqns (2.30) and (2.31) that

$$\mathbf{e}_j = \mathbf{u}_j^* - \mathbf{u}_j = \mathbf{A}^j(\mathbf{u}_0^* - \mathbf{u}_0) = \mathbf{A}^j \mathbf{e}_0, \quad j = 1(1)J. \tag{2.32}$$

In other words, a perturbation $\mathbf{e}_0$ of the initial values will be propagated according to the equation

$$\mathbf{e}_j = \mathbf{A}\mathbf{e}_{j-1} = \mathbf{A}^2 \mathbf{e}_{j-2} = \ldots = \mathbf{A}^j \mathbf{e}_0, \quad j = 1(1)J.$$

Hence, for compatible matrix and vector norms,

$$\|\mathbf{e}_j\| \leqslant \|\mathbf{A}^j\| \, \|\mathbf{e}_0\|.$$

Lax and Richtmyer define the difference scheme to be stable when there exists a positive number M, independent of j, h, and k, such that

$$\|\mathbf{A}^j\| \leqslant M, \quad j = 1(1)J.$$

This clearly limits the amplification of any initial perturbation, and therefore of any arbitrary initial rounding errors, because

$$\|\mathbf{e}_j\| \leqslant M \, \|\mathbf{e}_0\|.$$

In reference 25, this definition of stability is related to convergence via Lax's equivalence theorem (p. 72).

Since

$$\|\mathbf{A}^j\| = \|\mathbf{A}\mathbf{A}^{j-1}\| \leqslant \|\mathbf{A}\| \, \|\mathbf{A}^{j-1}\| \leqslant \ldots \leqslant \|\mathbf{A}\|^j,$$

it follows that the Lax–Richtmyer definition of stability is satisfied by

$$\|\mathbf{A}\| \leqslant 1.$$

This is the necessary and sufficient condition for the difference equations to be stable when the solution of the partial differential equation does not increase as t increases.

When this condition is satisfied it follows automatically that the spectral radius $\rho(\mathbf{A}) \leq 1$ since $\rho(\mathbf{A}) \leq \|\mathbf{A}\|$.

If, however, $\rho(\mathbf{A}) \leq 1$, it does not follow that $\|\mathbf{A}\| \leq 1$. This is demonstrated by the simple example

$$\mathbf{A} = \begin{bmatrix} -0.8 & 0 \\ 0.4 & 0.7 \end{bmatrix},$$

for which $\lambda_1 = -0.8$, $\lambda_2 = 0.7$, $\rho(\mathbf{A}) = 0.8$, $\|\mathbf{A}\|_1 = 1.2$ and $\|\mathbf{A}\|_\infty = 1.1$. If, however, $\mathbf{A}$ is real and symmetric, then

$$\|\mathbf{A}\|_2 = \sqrt{\rho(\mathbf{A}^T\mathbf{A})} = \sqrt{\rho(\mathbf{A}^2)} = \sqrt{\rho^2(\mathbf{A})} = \rho(\mathbf{A}).$$

Example 2.8

Consider the stability of the classical explicit equations

$$u_{i,j+1} = ru_{i-1,j} + (1-2r)u_{i,j} + ru_{i+1,j}, \quad i = 1(1)N-1,$$

for which the $(N-1) \times (N-1)$ matrix $\mathbf{A}$ is

$$\begin{bmatrix} (1-2r) & r & & \\ r & (1-2r)r & & \\ & r & (1-2r) & r \\ & & r & (1-2r), \end{bmatrix}$$

where $r = k/h^2 > 0$, and it is assumed that the boundary values $u_{0,j}$ and $u_{N,j}$ are known for $j = 1, 2, \ldots$.

When $1 - 2r \geq 0$, then $0 < r \leq \frac{1}{2}$

and

$$\|\mathbf{A}\|_\infty = r + (1-2r) + r = 1.$$

When $1 - 2r < 0$, $r > \frac{1}{2}$, $|1-2r| = 2r - 1$

and

$$\|\mathbf{A}\|_\infty = r + 2r - 1 + r = 4r - 1 > 1.$$

Therefore the scheme is stable for $0 < r \leq \frac{1}{2}$.

Alternatively, since matrix $\mathbf{A}$ is real and symmetric,

$$\|\mathbf{A}\|_2 = \rho(\mathbf{A}) = \max_s |\mu_s|,$$

where μ_s is the sth eigenvalue of $\mathbf{A}$. Now $\mathbf{A}$ can be written as

$$\begin{bmatrix} 1 & & & \\ & 1 & & \\ & & \ddots & \\ & & & 1 \\ & & & & 1 \end{bmatrix} + r \begin{bmatrix} -2 & 1 & & \\ 1 & -2 & 1 & \\ & & \ddots & \\ & & 1 & -2 & 1 \\ & & & 1 & -2 \end{bmatrix} = \mathbf{I}_{N-1} + r\mathbf{T}_{N-1},$$

where $\mathbf{I}_{N-1}$ is the unit matrix of order $(N-1)$ and $\mathbf{T}_{N-1}$ an $(N-1)\times(N-1)$ matrix whose eigenvalues λ_s are given by

$$\lambda_s = -4\sin^2 s\pi/2N, \quad s = 1(1)N-1. \text{ (See p. 59.)}$$

Hence the eigenvalues of $\mathbf{A}$, as shown later in 'A note on eigenvalues and eigenvectors' are $\mu_s = 1 - 4r\sin^2 s\pi/2N$.

Therefore the equations will be stable when

$$\|\mathbf{A}\|_2 = \max_s |1 - 4r\sin^2 s\pi/2N| \leqslant 1,$$

i.e.,

$$-1 \leqslant 1 - 4r\sin^2 s\pi/2N \leqslant 1, \quad s = 1(1)N-1.$$

The left-hand inequality gives that

$$r \leqslant 1/2\sin^2 (N-1)\pi/2N.$$

As $h \to 0$, $N \to \infty$ and $\sin^2(N-1)\pi/2N \to 1$.
Hence $r \leqslant \frac{1}{2}$.

It has been shown that these equations are also consistent. Hence by Lax's equivalence theorem they are also convergent for $0 < r \leqslant \frac{1}{2}$.

Example 2.9

The Crank–Nicolson equations (2.10) are

$$-ru_{i-1,j+1} + (2+2r)u_{i,j+1} - ru_{i+1,j+1}$$
$$= ru_{i-1,j} + (2-2r)u_{i,j} + ru_{i+1,j}, \, i = 1(1)N-1.$$

In matrix form, for known boundary values, these give

$$
\begin{bmatrix}
(2+2r) & -r \\
-r & (2+2r) & -r \\
& & \ddots \\
& & & -r & (2+2r) & -r \\
& & & & -r & (2+2r)
\end{bmatrix}
\begin{bmatrix}
u_{1,j+1} \\
u_{2,j+1} \\
\vdots \\
\vdots \\
u_{N-2,j+1} \\
u_{N-1,j+1}
\end{bmatrix}
$$

$$
=
\begin{bmatrix}
(2-2r) & r \\
r & (2-2r) & r \\
& & \ddots \\
& & \ddots \\
& & r & (2-2r) & r \\
& & & r & (2-2r)
\end{bmatrix}
\begin{bmatrix}
u_{1,j} \\
u_{2,j} \\
\vdots \\
\vdots \\
u_{N-2,j} \\
u_{N-1,j}
\end{bmatrix}
+ \mathbf{b}_j,
$$

where $\mathbf{b}_j$ is a vector of known boundary values and zeros. This can be written as

$$(2\mathbf{I}_{N-1} - r\mathbf{T}_{N-1})u_{j+1} = (2\mathbf{I}_{N-1} + r\mathbf{T}_{N-1})\mathbf{u}_j + \mathbf{b}_j$$

from which it follows that matrix $\mathbf{A}$ of eqn (2.30) is

$$\mathbf{A} = (2\mathbf{I}_{N-1} - r\mathbf{T}_{N-1})^{-1}(2\mathbf{I}_{N-1} + r\mathbf{T}_{N-1}).$$

In Exercise 14 it is proved that if the $n \times n$ symmetric matrices $\mathbf{B}$ and $\mathbf{C}$ commute then $\mathbf{B}^{-1}\mathbf{C}$, $\mathbf{B}\mathbf{C}^{-1}$ and $\mathbf{B}^{-1}\mathbf{C}^{-1}$ are symmetric. Matrix $\mathbf{T}_{N-1}$ is symmetric so $2\mathbf{I}_{N-1} - r\mathbf{T}_{N-1}$ and $2\mathbf{I}_{N-1} + r\mathbf{T}_{N-1}$ are symmetric. They also commute as their multiplication immediately shows. Hence matrix $\mathbf{A}$ is symmetric. Since the eigenvalues of $\mathbf{T}_{N-1}$ are $\lambda_s = -4\sin^2 s\pi/2N$, $s = 1(1)N-1$, it follows that the eigenvalues of $\mathbf{A}$ are $(2+4r\sin^2 s\pi/2N)^{-1}(2-4r\sin^2 s\pi/2N)$. (See p. 59.)
Therefore

$$\|\mathbf{A}\|_2 = \rho(\mathbf{A}) = \max_s \left| \frac{1 - 2r\sin^2 s\pi/2N}{1 + 2r\sin^2 s\pi/2N} \right| < 1 \text{ for all } r > 0,$$

proving that the Crank–Nicolson equations are unconditionally stable. They are also consistent, reference Exercise 11, so they are also convergent, although it will be shown under A_0-stability that r must be restricted in order to avoid the possibility of finite oscillations near points of discontinuity.

Matrix method of analysis, fixed mesh lengths

The following method of analysis will establish the conditions necessary for the boundedness of the analytical solution of the finite-difference equations as $t_j = jk$ tends to infinity for fixed mesh lengths h and k. These conditions may not, however, be sufficient to ensure convergence when the equations are also consistent and large finite errors can occur near the end points of the range of values of some parameter, such as $r = k/h^2$, for which the equations are bounded.

Consider the classical explicit finite-difference equations incorporating known boundary values, namely,

$$
\begin{bmatrix} u_{1,j+1} \\ u_{2,j+1} \\ \cdot \\ \cdot \\ \cdot \\ u_{N-2,j+1} \\ u_{N-1,j+1} \end{bmatrix} = \begin{bmatrix} (1-2r)\ r & & & \\ r(1-2r)r & & & \\ & \cdot\ \ \cdot\ \ \cdot & & \\ & \cdot\ \ \cdot\ \ \cdot & & \\ & & \cdot\ \ \cdot\ \ \cdot & \\ & & r & (1-2r)r \\ & & & r(1-2r) \end{bmatrix} \begin{bmatrix} u_{1,j} \\ u_{2,j} \\ \cdot \\ \cdot \\ \cdot \\ u_{N-2,j} \\ u_{N-1,j} \end{bmatrix} + \begin{bmatrix} ru_{0,j} \\ 0 \\ \cdot \\ \cdot \\ \cdot \\ 0 \\ ru_{N,j} \end{bmatrix},
$$

i.e.

$$\mathbf{u}_{j+1} = \mathbf{A}\mathbf{u}_j + \mathbf{b}_j.$$

As shown by eqn (2.32), if the vector of initial values $\mathbf{u}_0$ is perturbed to $\mathbf{u}_0^*$ and no further perturbations or errors are introduced into the subsequent calculations, the perturbation vector $\mathbf{e} = \mathbf{u}^* - \mathbf{u}$ will be propagated forward in time, according to the equation

$$\mathbf{e}_j = \mathbf{A}^j \mathbf{e}_0,$$

a procedure that eliminates the boundary values.

For fixed mesh lengths h and k the difference equations will be stable if $\mathbf{e}_j$ remains bounded as j increases indefinitely. This can always be investigated by expressing the initial perturbation vector in terms of the eigenvectors of $\mathbf{A}$, which remain fixed as j increases.

Assume that matrix $\mathbf{A}$ is non-deficient, i.e. has $(N-1)$ linearly independent eigenvectors $\mathbf{v}_s$, which will be so if the eigenvalues λ_s of $\mathbf{A}$ are all distinct or $\mathbf{A}$ is real and symmetric. Then these eigenvectors can be used as a basis for our $(N-1)$ dimensional vector space and the perturbation vector $\mathbf{e}_0$, with its $(N-1)$ components, can be expressed uniquely as a linear combination of them, namely,

$$\mathbf{e}_0 = \sum_{s=1}^{N} c_s \mathbf{v}_s,$$

where the c_s, $s = 1(1)N - 1$, are known scalars.

The perturbations along time-level $t = k$, resulting from the initial perturbation $\mathbf{e}_0$ will be

$$\mathbf{e}_1 = \mathbf{A}\mathbf{e}_0 = \mathbf{A} \sum_{s=1}^{N-1} c_s \mathbf{v}_s = \sum c_s \mathbf{A}\mathbf{v}_s.$$

But $\mathbf{A}\mathbf{v}_s = \lambda_s \mathbf{v}_s$ by the definition of an eigenvalue. Therefore,

$$\mathbf{e}_1 = \sum c_s \lambda_s \mathbf{v}_s.$$

Similarly,

$$\mathbf{e}_j = \sum_{s=1}^{N-1} c_s \lambda_s^j \mathbf{v}_s.$$

This shows that the perturbations will not increase exponentially with j provided

$$\max_s |\lambda_s| \leqslant 1, \quad s = 1(1)N - 1.$$

By p. 59, $\lambda_s = 1 - 4r \sin^2 s\pi/2N$.

Therefore any perturbation, rounding errors and $u_{i,j}$ will be bounded as j increases if

$$-1 \leqslant 1 - 4r \sin^2 s\pi/2N \leqslant 1, \quad \text{where } r > 0.$$

This is satisfied by $r \leqslant \frac{1}{2}$.

A note on eigenvalues and eigenvectors

Let $\mathbf{x}$ be an eigenvector of the matrix $\mathbf{A}$ corresponding to the eigenvalue λ. Then $\mathbf{A}\mathbf{x} = \lambda\mathbf{x}$. Hence $\mathbf{A}(\mathbf{A}\mathbf{x}) = \mathbf{A}^2\mathbf{x} = \lambda\mathbf{A}\mathbf{x} = \lambda^2\mathbf{x}$, showing that the matrix $\mathbf{A}^2$ has an eigenvalue λ^2 corresponding to the eigenvector $\mathbf{x}$. Similarly $\mathbf{A}^p\mathbf{x} = \lambda^p\mathbf{x}$, $p = 3, 4, \ldots$.

(i) If $f(\mathbf{A}) = a_p \mathbf{A}^p + a_{p-1} \mathbf{A}^{p-1} + \ldots + a_0 \mathbf{I}$ is a polynomial in $\mathbf{A}$ with scalar coefficients $a_p, \ldots, a_0$, then $f(\mathbf{A})\mathbf{x} = (a_p \lambda^p + \ldots + a_0)\mathbf{x} = f(\lambda)\mathbf{x}$, showing that $f(\mathbf{A})$ has an eigenvalue $f(\lambda)$ corresponding to the eigenvector $\mathbf{x}$.

(ii) The eigenvalue of $[f_1(\mathbf{A})]^{-1}f_2(\mathbf{A})$ corresponding to the eigenvector $\mathbf{x}$ is $f_2(\lambda)/f_1(\lambda)$, where $f_1(\mathbf{A})$ and $f_2(\mathbf{A})$ are polynomials in $\mathbf{A}$. The proof is as follows. By (i),

$$f_1(\mathbf{A})\mathbf{x} = f_1(\lambda)\mathbf{x} \quad \text{and} \quad f_2(\mathbf{A})\mathbf{x} = f_2(\lambda)\mathbf{x}.$$

Premultiply both equations by $[f_1(\mathbf{A})]^{-1}$ and write as

$$[f_1(\mathbf{A})]^{-1}\mathbf{x} = \mathbf{x}/f_1(\lambda) \quad \text{and} \quad [f_1(\mathbf{A})]^{-1}f_2(\mathbf{A})\mathbf{x} = f_2(\lambda)[f_1(\mathbf{A})]^{-1}\mathbf{x}.$$

Then the elimination of $[f_1(\mathbf{A})]^{-1}\mathbf{x}$ between these two equations shows that

$$[f_1(\mathbf{A})]^{-1}f_2(\mathbf{A})\mathbf{x} = \{f_2(\lambda)/f_1(\lambda)\}\mathbf{x},$$

which states, by the definition of an eigenvalue, that $f_2(\lambda)/f_1(\lambda)$ is an eigenvalue of $[f_1(\mathbf{A})]^{-1}f_2(\mathbf{A})$ corresponding to the eigenvector $\mathbf{x}$. In a similar manner the eigenvalue of $f_2(\mathbf{A})[f_1(\mathbf{A})]^{-1}$ corresponding to the eigenvector $\mathbf{x}$ is $f_2(\lambda)/f_1(\lambda)$.

In particular, the eigenvalue of $[f_1(\mathbf{A})]^{-1}$ corresponding to the eigenvector $\mathbf{x}$ is $1/f_1(\lambda)$.

The eigenvalues of a common tridiagonal matrix

The eigenvalue of the $N \times N$ matrix

$$\begin{bmatrix} a & b & & & & \\ c & a & b & & & \\ & c & a & b & & \\ & & \cdot & \cdot & \cdot & \\ & & & c & a & b \\ & & & & c & a \end{bmatrix}$$

are

$$\lambda_s = a + 2\{\sqrt{(bc)}\}\cos\frac{s\pi}{N+1}, \quad s = 1(1)N,$$

where a, b, and c may be real or complex. A proof is given on p. 154.

Another useful result is the following. If a real tridiagonal matrix has either all its off-diagonal elements positive or all its off-diagonal elements negative, then all its eigenvalues are real. A proof is given in Exercise 16.

Useful theorems on bounds for eigenvalues

Gerschgorin's first theorem

The largest of the moduli of the eigenvalues of the square matrix
A cannot exceed the largest sum of the moduli of the elements
along any row or any column. In other words

$$\rho(A) \leqslant \|\mathbf{A}\|_1 \quad \text{or} \quad \|\mathbf{A}\|_\infty.$$

Proof

Let λ_i be an eigenvalue of the $N \times N$ matrix **A**, and $\mathbf{x}_i$ the
corresponding eigenvector with components $v_1, v_2 \ldots v_n$. Then
the equation
$$\mathbf{A}\mathbf{x}_i = \lambda_i \mathbf{x}_i$$

in detail, is

$$a_{1,1}v_1 + a_{1,2}v_2 + \ldots + a_{1,n}v_n = \lambda_i v_1,$$

$$a_{2,1}v_1 + a_{2,2}v_2 + \ldots + a_{2,n}v_n = \lambda_i v_2,$$

$$\cdot \qquad \cdot \qquad \qquad \cdot \qquad \cdot$$
$$\cdot \qquad \cdot \qquad \qquad \cdot \qquad \cdot$$

$$a_{s,1}v_1 + a_{s,2}v_2 + \ldots + a_{s,n} v_n = \lambda_i v_s,$$

$$\cdot \qquad \cdot \qquad \qquad \cdot \qquad \cdot$$
$$\cdot \qquad \cdot \qquad \qquad \cdot \qquad \cdot$$

Let v_s be the largest in modulus of $v_1, v_2, \ldots, v_n$. Select the sth
equation and divide by v_s, giving

$$\lambda_i = a_{s,1}\left(\frac{v_1}{v_s}\right) + a_{s,2}\left(\frac{v_2}{v_s}\right) + \ldots + a_{s,n}\left(\frac{v_n}{v_s}\right).$$

Therefore

$$|\lambda_i| \leqslant |a_{s,1}| + |a_{s,2}| + \ldots + |a_{s,n}|,$$

because

$$\left|\frac{v_i}{v_s}\right| \leqslant 1, \quad i = 1, 2, \ldots, n.$$

If this is not the largest row sum then $|\lambda_i| <$ the largest row sum.
In particular this holds for $|\lambda_i| = \max|\lambda_s|$, $s = 1(1)N$.
Since the eigenvalues of the transpose of **A** are the same as those
of **A** the theorem is also true for columns.

Gerschgorin's circle theorem or Brauer's theorem

Let P_s be the sum of the moduli of the elements along the sth row excluding the diagonal element $a_{s,s}$. Then each eigenvalue of **A** lies inside or on the boundary of at least one of the circles $|\lambda - a_{s,s}| = P_s$.

Proof

By the previous proof,

$$\lambda_i = a_{s,1}\left(\frac{v_1}{v_s}\right) + a_{s,2}\left(\frac{v_2}{v_s}\right) + \ldots + a_{s,s} + \ldots + a_{s,n}\left(\frac{v_n}{v_s}\right).$$

Hence

$$|\lambda_i - a_{s,s}| = \left|a_{s,1}\left(\frac{v_1}{v_s}\right) + \ldots + 0 + \ldots + a_{s,n}\left(\frac{v_s}{v_s}\right)\right|$$

$$\leqslant |a_{s,1}| + |a_{s,2}| + \ldots + 0 + \ldots + |a_{s,n}|$$

$$= P_s,$$

which completes the proof.

As an illustrative example consider the Crank–Nicolson equations with known boundary-values, namely

$$(2\mathbf{I}_{N-1} - r\mathbf{T}_{N-1})\mathbf{u}_{j+1}$$
$$= (2\mathbf{I}_{N-1} + r\mathbf{T}_{N-1})\mathbf{u}_j + \mathbf{b}_j = \{4\mathbf{I}_{N-1} - (2\mathbf{I}_{N-1} - r\mathbf{T}_{N-1})\}\mathbf{u}_j + \mathbf{b}_j,$$

which can be written as

$$\mathbf{B}\mathbf{u}_{j+1} = (4\mathbf{I}_{N-1} - \mathbf{B})\mathbf{u}_j + \mathbf{b}_j,$$

giving

$$\mathbf{u}_{j+1} = (4\mathbf{B}^{-1} - \mathbf{I}_{N-1})\mathbf{u}_j + \mathbf{B}^{-1}\mathbf{b}_j,$$

where

$$\mathbf{B} = \begin{bmatrix} (2+2r) & -r & & & \\ -r & (2+2r) & -r & & \\ & \cdot & \cdot & \cdot & \\ & & \cdot & -r(2+2r)-r \\ & & & -r & (2+2r) \end{bmatrix}$$

The equations will be stable in the Lax–Richtmyer sense when $\|4\mathbf{B}^{-1}-\mathbf{I}_{N-1}\|\leqslant 1$. Since $\mathbf{B}$ is real and symmetric it follows by Exercise 14 that $4\mathbf{B}^{-1}-\mathbf{I}_{N-1}$ is real and symmetric, so $\|4\mathbf{B}^{-1}-\mathbf{I}_{N-1}\|=\rho(4\mathbf{B}^{-1}-\mathbf{I}_{N-1})$. The stability condition will therefore be satisfied when the modulus of every eigenvalue of $4\mathbf{B}^{-1}-\mathbf{I}_{N-1}$ does not exceed one; that is, when

$$\left|\frac{4}{\lambda}-1\right|\leqslant 1, \quad \text{implying} \quad -1\leqslant\frac{4}{\lambda}-1\leqslant 1,$$

where λ is an eigenvalue of $\mathbf{B}$. This states that $\lambda\geqslant 2$.

For the matrix $\mathbf{B}$, $a_{s,s}=2+2r$, max $P_s=2r$, so Gerschgorin's circle theorem leads to

$$|\lambda-2-2r|\leqslant 2r,$$

from which it follows that

$$-2r\leqslant\lambda-2-2r\leqslant 2r,$$

or

$$2\leqslant\lambda\leqslant 2+4r,$$

proving that the equations are unconditionally stable, since $\lambda\geqslant 2$ for all $r>0$.

Gerschgorin's circle theorem and the norm of matrix A

It should be noted that when the eigenvalues λ_i of matrix $\mathbf{A}$ are estimated by the circle theorem, the condition $|\lambda_i|\leqslant 1$ is equivalent to $\|\mathbf{A}\|_\infty\leqslant 1$ or $\|\mathbf{A}\|_1\leqslant 1$. The theorem states that

$$|\lambda-a_{s,s}|\leqslant P_s.$$

Hence,

$$-P_s\leqslant\lambda-a_{ss}\leqslant P_s,$$

so that

$$-P_s+a_{ss}\leqslant\lambda\leqslant P_s+a_{ss}.$$

The eigenvalue λ will therefore satisfy $-1\leqslant\lambda\leqslant 1$ if

$$-1\leqslant-P_s+a_{ss}\leqslant P_s+a_{ss}\leqslant 1, \quad s=1(1)N-1.$$

Remembering that P_s is the sum of the moduli of the elements of $\mathbf{A}$ along the sth row and that a_{ss} may be positive or negative, this inequality is equivalent to

$$\sum_{j=1}^{N-1} |a_{sj}| \leqslant 1, \quad s = 1(1)N - 1,$$

i.e. to $\|\mathbf{A}\|_\infty \leqslant 1$ for rows or $\|\mathbf{A}\|_1 \leqslant 1$ for columns.

When both inequalities hold, it follows automatically that $\|\mathbf{A}\|_2 \leqslant 1$ because it is shown in Exercise 15 that $\|\mathbf{A}\|_2^2 \leqslant \|\mathbf{A}\|_1 \|\mathbf{A}\|_\infty$. Therefore, Gerschgorin's circle theorem can be used to establish conditions satisfying the Lax–Richtmyer definition of stability, a situation that frequently arises with derivative boundary-conditions.

Gerschgorin's third theorem

If p of the circles of Gerschgorin's circle theorem form a connected domain that is isolated from the other circles, then there are precisely p eigenvalues of matrix $\mathbf{A}$ within this connected domain.

In particular, an isolated Gerschgorin circle contains one eigenvalue.

A proof is given in reference 30.

Stability criteria for derivative boundary conditions

Consider the equation

$$\frac{\partial U}{\partial t} = \frac{\partial^2 U}{\partial x^2}, \quad 0 < x < 1,$$

and the conditions,

$$\frac{\partial U}{\partial x} = h_1(U - v_1) \quad \text{at } x = 0, \quad t \geqslant 0,$$

$$\frac{\partial U}{\partial x} = -h_2(U - v_2) \quad \text{at } x = 1, t \geqslant 0,$$

where h_1, h_2, v_1, v_2 are constants, $h_1 \geqslant 0$, $h_2 \geqslant 0$.

When the boundary conditions are approximated by the central-difference equations

$$(u_{1,j} - u_{-1,j})/2\delta x = h_1(u_{0,j} - v_1),$$

$$(u_{N+1,j} - u_{N-1,j})/2\delta x = -h_2(u_{N,j} - v_2), \quad (N\delta x = 1),$$

and the differential equation by the explicit scheme

$$u_{i,j+1} = ru_{i-1,j} + (1-2r)u_{i,j} + ru_{i+1,j},$$

elimination of $u_{-1,j}$, $u_{N+1,j}$, leads to the equations

$$\begin{bmatrix} u_{0,j+1} \\ u_{1,j+1} \\ \cdot \\ u_{N-1,j+1} \\ u_{N,j+1} \end{bmatrix}$$

$$= \begin{bmatrix} \{1-2r(1+h_1\delta x)\} & 2r & & & \\ r & (1-2r) & r & & \\ & & \cdot & \cdot & \cdot \\ & & r & (1-2r) & r \\ & & & 2r & \{1-2r(1+h_2\delta x)\} \end{bmatrix}$$

$$\times \begin{bmatrix} u_{0,j} \\ u_{1,j} \\ \vdots \\ u_{N-1,j} \\ u_{N,j} \end{bmatrix} + \begin{bmatrix} 2rh_1v_1\delta x \\ 0 \\ \vdots \\ 0 \\ 2rh_2v_2\delta x \end{bmatrix}$$

As each component of the last column vector is a constant the matrix determining the propagation of the error is

$$\begin{bmatrix} \{1-2r(1+h_1\delta x)\} & 2r & & & \\ r & (1-2r) & r & & \\ & & \cdot & & \\ & & r & (1-2r) & r \\ & & & 2r & \{1-2r(1+h_2\delta x)\} \end{bmatrix}$$

Since the off-diagonal elements of this real matrix are one-signed, all its eigenvalues are real. (See Exercise 16.) Application of the circle theorem to this matrix, with

$$a_{ss} = 1-2r(1+h_1\delta x) \quad \text{and} \quad P_s = 2r,$$

shows that some of its eigenvalues λ may lie on or within the circle

$$|\lambda - \{1-2r(1+h_1\delta x)\}| \leq 2r.$$

Using Fig. 2.8,

$$\lambda_1 = 1-2r(2+h_1\delta x), \quad \lambda_2 = 1-2rh_1\delta x,$$

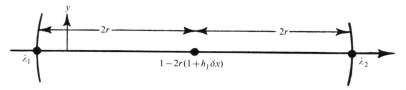

Fig. 2.8

and for stability,

$$|\lambda_1| \le 1, \quad |\lambda_2| \le 1.$$

Hence

$$-1 \le 1 - 2r(2 + h_1\delta x) \le 1, \quad \text{giving } r \le 1/(2 + h_1\delta x),$$

and

$$-1 \le 1 - 2rh_1\delta x, \quad \text{giving } r \le 1/h_1\delta x.$$

The least of these is $r \le 1/(2 + h_1\delta x)$.
Similarly, we require $r \le 1/(2 + h_2\delta x)$.
For rows $2(1)N - 1$, $a_{s,s} = 1 - 2r$, $P_s = 2r$, giving $r \le \frac{1}{2}$.
For overall stability, $r \le \min\{1/(2 + h_1\delta x); 1/(2 + h_2\delta x)\}$.

Direct application of $\|\mathbf{A}\|_\infty \le 1$

As mentioned earlier, the circle theorem and $|\lambda| \le 1$ is equivalent
to $\|\mathbf{A}\|_\infty \le 1$. This is easily verified for this example.
Row 1. If $1 - 2r(1 + h_1\delta x) \ge 0$, the sum of the moduli of the
terms along row $1 = 1 - 2rh_1\delta x$.
  Since $\|\mathbf{A}\|_\infty \le 1$ for Lax–Richtmyer stability,

$$0 < 1 - 2rh_1\delta x \le 1 \quad \text{implies} \quad r < 1/2h_1\delta x.$$

If $1 - 2r(1 + h_1\delta x) \le 0$, the sum of the moduli of the elements
along row $1 = 2r(1 + h_1\delta x) - 1 + 2r = 2r(2 + h_1\delta x) - 1$.
  As before, $0 < 2r(2 + h_1\delta x) - 1 \le 1$ implies that $r \le 1/(2 + h_1\delta x)$.
Similarly for the other rows.

Crank–Nicolson equations

It is easily shown that the Crank–Nicolson equations for the
problem just considered propagate any perturbation or rounding
error by the recursion formula

$$\mathbf{e}_{j+1} = (4\mathbf{B}^{-1} - \mathbf{I})\mathbf{e}_j,$$

where

$$\mathbf{B} = \begin{bmatrix} \{2+2r(1+h_1\delta x)\} & -2r & & & \\ -r & (2+2r) & -r & & \\ & \cdot & \cdot & & \cdot \\ & -r & (2+2r) & & -r \\ & & -2r & \{2+2r(1+h_2\delta x)\} \end{bmatrix}$$

As shown previously, the spectral radius of $4\mathbf{B}^{-1}-\mathbf{I} \leqslant 1$ if the eigenvalues of $\mathbf{B} \geqslant 2$. By the circle theorem,

$$|\lambda - \{2+2r(1+h_1\delta x)\}| \leqslant 2r,$$

giving

$$2+2rh_1\delta x \leqslant \lambda.$$

Similarly, the remaining rows give that $2 \leqslant \lambda$ and $2+2rh_2\delta x \leqslant \lambda$. Hence the equations are unconditionally stable.

Stability condition allowing exponential growth (Constant coefficients)

If the solution of the partial differential equation increases exponentially with increasing t then the exact solution of the difference equations must do so also. In order to allow this we need to return to the original definition of stability for the equations

$$\mathbf{e}_j = \mathbf{A}^j\mathbf{e}_0, \quad j = 1(1)J, \quad Jk = T,$$

which was that

$$\|\mathbf{A}^j\| \leqslant M,$$

where M is a positive number independent of j, h, and k. Since $\|\mathbf{A}^j\| = \|\mathbf{A} \cdot \mathbf{A}^{j-1}\| \leqslant \|\mathbf{A}\| \|\mathbf{A}^{j-1}\| \leqslant \ldots \leqslant \|\mathbf{A}\|^j$, the definition will be satisfied by

$$\|\mathbf{A}\|^j \leqslant M, \quad 0 < j \leqslant J.$$

Assuming $M > 1$,

$$\|\mathbf{A}\| \leqslant M^{1/j} = \exp(\ln M^{1/j}) = \exp\left(\frac{1}{j}\ln M\right)$$

$$= 1 + \frac{1}{j}\ln M + \ldots$$

In particular, this must satisfy

$$\|\mathbf{A}\| \leqslant 1 + \frac{1}{J}\ln M + \ldots$$

$$\geqslant 1 + \frac{1}{J}\ln M$$

$$= 1 + \frac{k}{T}\ln M.$$

In other words, a necessary and sufficient condition for stability is that

$$\|\mathbf{A}\| \leqslant 1 + Kk = 1 + O(k),$$

where the positive number K is independent of h and k. (See Worked Example 2.11.) A much more rigorous treatment of this condition based on infinite sets of uniformly bounded linear operators is given in reference 25.

Stability by the Fourier series method (von Neumann's method)

Assume we are concerned with the stability of a linear two time-level difference equation in $u(x, t)$ in the time interval $0 \leqslant t \leqslant T = Jk$, T finite, as $\delta x = h \to 0$ and $\delta t = k \to 0$, i.e. as $J \to \infty$. The Fourier series or von Neumann method, first discussed in detail in reference 21, expresses the initial values at the mesh points along $t = 0$ in terms of a finite Fourier series, then considers the growth of a function that reduces to this series for $t = 0$ by a 'variables separable' method identical to that commonly used for solving partial differential equations.

The Fourier series can be formulated in terms of sines and cosines but the algebra is easier if the complex exponential form is used, i.e. with $\sum a_n \cos(n\pi x/l)$ or $\sum b_n \sin(n\pi x/l)$ replaced by the equivalent $\sum A_n e^{in\pi x/l}$, where $i = \sqrt{-1}$ and l is the x-interval throughout which the function is defined. Clearly we need to change our usual notation $u_{i,j}$ to $u(ph, qk) = u_{p,q}$. In terms of this notation,

$$A_n e^{in\pi x/l} = A_n e^{in\pi ph/Nh} = A_n e^{i\beta_n ph}$$

where $\beta_n = n\pi/Nh$ and $Nh = l$.

Denote the initial values at the pivotal points along $t = 0$ by

$u(ph, 0) = u_{p,0}$, $p = 0(1)N$. Then the $N + 1$ equations

$$u_{p,0} = \sum_{n=0}^{N} A_n e^{i\beta_n ph}, \quad p = 0, 1, \ldots, N, \tag{2.33}$$

are sufficient to determine the $(N + 1)$ unknowns $A_0, A_1, \ldots, A_N$ uniquely, showing that the initial mesh values can be expressed in this complex exponential form. As we are considering only linear-difference equations we need investigate the propagation of only one initial value, such as $e^{i\beta ph}$, because separate solutions are additive. The coefficient A_n is a constant and can be neglected.

To investigate the propagation of this term as t increases, put

$$u_{p,q} = e^{i\beta x} e^{\alpha t} = e^{i\beta ph} e^{\alpha qk} = e^{i\beta ph} \xi^q, \tag{2.34}$$

where $\xi = e^{\alpha k}$ and α, in general, is a complex constant. ξ is often called the amplification factor.

The finite-difference equations will be stable by the Lax–Richtmyer definition if $|u_{p,q}|$ remains bounded for all $q \leqslant J$ as $h \to 0$ and $k \to 0$, and for all values of β needed to satisfy the initial conditions.

If the exact solution of the difference equations does not increase exponentially with time, then a necessary and sufficient condition for stability is that

$$|\xi| \leqslant 1,$$

i.e.

$$-1 \leqslant \xi \leqslant 1. \tag{2.35}$$

If, however, $u_{p,q}$ does increase with t, (see Worked Example 2.11), then the necessary and sufficient condition for stability is, as shown in the preceding section,

$$|\xi| \leqslant 1 + Kk = 1 + O(k), \tag{2.36}$$

where the positive number K is independent of h, k and β.

It should be noted that this method applies only to linear equations with constant coefficients, and strictly speaking only to initial value problems with periodic initial data, of period l. For difference equations involving three or more time-levels or two or more dependent variables, the von Neumann conditions (2.35) and (2.36) are always necessary but may not be sufficient, reference 25. In practice, the method often gives useful results even when its application is not fully justified.

Example 2.10

Investigate the stability of the fully implicit difference equation

$$\frac{1}{k}(u_{p,q+1} - u_{p,q}) = \frac{1}{h^2}(u_{p-1,q+1} - 2u_{p,q+1} + u_{p+1,q+1})$$

approximating $\partial U/\partial t = \partial^2 U/\partial x^2$ at (ph, qk).

Substitution of $u_{p,q} = e^{i\beta ph}\xi^q$ into the difference equation shows that

$$e^{i\beta ph}\xi^{q+1} - e^{i\beta ph}\xi^q = r\{e^{i\beta(p-1)h}\xi^{q+1} - 2e^{i\beta ph}\xi^{q+1} + e^{i\beta(p+1)h}\xi^{q+1}\},$$

where $r = k/h^2$. Division by $e^{i\beta ph}\xi^q$ leads to

$$\xi - 1 = r\xi(e^{-i\beta h} - 2 + e^{i\beta ph})$$
$$= r\xi(2\cos\beta h - 2)$$
$$= -4r\xi\sin^2(\beta h/2).$$

Hence,

$$\xi = \frac{1}{1 + 4r\sin^2\beta h/2}.$$

Clearly, $0 < \xi \leqslant 1$ for all $r > 0$ and all β. Therefore the equations are unconditionally stable.

Example 2.11

Investigate the stability of the linear difference equation

$$\frac{1}{k}(u_{p,q+1} - u_{p,q}) = \frac{a}{h^2}(u_{p-1,q} - 2u_{p,q} + u_{p,+1,q}) + bu_{p,q}$$

approximating the parabolic equation

$$\frac{\partial U}{\partial t} = a\frac{\partial^2 U}{\partial x^2} + bU$$

at the point (ph, qk), where a and b are positive constants.

The analytical solution of this equation increases with t because it is eventually dominated by the term

$$e^{bt}\int_0^1 f(x)\,dx, \quad \text{where} \quad U = f(x) \text{ when } t = 0,\ 0 \leqslant x \leqslant 1.$$

Substitution of $u_{p,q} = e^{i\beta ph}\xi^q$ into the difference equation leads to

$$\xi - 1 = ra(e^{-i\beta h} - 2 + e^{i\beta h}) + kb, \quad r = k/h^2.$$

This leads to

$$\xi = 1 - 4ra\,\sin^2(\beta h/2) + kb, \quad b > 0.$$

Hence,

$$|\xi| \leq |1 - 4ra\,\sin^2(\beta h/2)| + kb$$

satisfies the von Neumann condition (2.36) if

$$|1 - 4ra\,\sin^2(\beta h/2)| \leq 1.$$

This leads to $r \leq 1/2a$, which is the same restriction on r as for $b = 0$. It is easily shown that the difference equation is consistent so it follows that it is also convergent.

Example 2.12

The hyperbolic equation $\partial^2 U/\partial t^2 = \partial^2 U/\partial x^2$ is approximated at (ph, qk) by the explicit difference scheme

$$(u_{p,q+1} - 2u_{p,q} + u_{p,q-1})/k^2 = (u_{p+1,q} - 2u_{p,q} + u_{p-1,q})/h^2.$$

Investigate its stability.

It is easily shown by the procedure of Example 2.10 that the equation for ξ is

$$\xi^2 - 2A\xi + 1 = 0,$$

where

$$A = 1 - 2r^2\sin^2(\beta h/2) \quad r = k/h. \tag{2.37}$$

Hence the values of ξ are

$$\xi_1 = A + (A^2 - 1)^{\frac{1}{2}} \quad \text{and} \quad \xi_2 = A - (A^2 - 1)^{\frac{1}{2}}.$$

As U does not increase exponentially with t and because the difference equation is a three time-level approximation, a necessary condition for stability is that

$$|\xi| \leq 1.$$

As r, k and β are real, $A \leq 1$ by (2.37).
  When $A < -1$, $|\xi_2| > 1$, giving instability.
  When

$$-1 \leq A \leq 1, \ A^2 \leq 1, \ \xi_1 = A + i(1 - A^2)^{\frac{1}{2}}, \ \xi_2 = A - i(1 - A^2)^{\frac{1}{2}},$$

hence

$$|\xi_1| = |\xi_2| = \{A^2 + (1 - A^2)\}^{\frac{1}{2}} = 1,$$

showing that a necessary condition for stability is $-1 \leq A \leq 1$. By

(2.37),

$$-1 \leqslant 1 - 2r^2 \sin^2(\beta h/2) \leqslant 1.$$

The only useful inequality is

$$-1 \leqslant 1 - 2r^2 \sin^2(\beta h/2),$$

giving

$$r = k/h \leqslant 1.$$

In reference 25, this condition is also shown to be sufficient.

The global rounding error

For simplicity, assume that all boundary values are zero so that the finite-difference equations approximating the initial-value differential equation in the solution domain $0 < x < 1$, $t > 0$, can be written as

$$\mathbf{u}_j = \mathbf{A}\mathbf{u}_{j-1},$$

where $\mathbf{u}_0 = \mathbf{U}_0$ is the vector of known initial values and $\mathbf{A}$ is a square matrix of known elements of order $(N-1)$.

In general, the computer will not store the initial value $u_{i,0}$ exactly, but a numerical approximation $N_{i,0}$, so that

$$N_{i,0} = u_{i,0} - r_{i,0}, \quad \text{i.e.} \quad \mathbf{N}_0 = \mathbf{u}_0 - \mathbf{r}_0,$$

where $\mathbf{r}_0$ is the vector of initial rounding errors. As rounding errors will be introduced at every stage of the calculations the numerical solution values calculated by the computer at the first time-level will be

$$\mathbf{N}_1 = \mathbf{A}\mathbf{N}_0 - \mathbf{r}_1 = \mathbf{A}\mathbf{u}_0 - \mathbf{A}\mathbf{r}_0 - \mathbf{r}_1.$$

Finally, at the jth time-level, the computed solution will be

$$\mathbf{N}_j = \mathbf{A}^j\mathbf{u}_0 - \mathbf{A}^j\mathbf{r}_0 - \mathbf{A}^{j-1}\mathbf{r}_1 - \ldots - \mathbf{r}_j.$$

If there were no rounding errors the exact solution of the difference equations would be

$$\mathbf{u}_j = \mathbf{A}^j\mathbf{u}_0.$$

Hence the difference between the exact solution and the computed solution, i.e., the global rounding error $\mathbf{R}_j$, at the jth time-level is

$$\mathbf{u}_j - \mathbf{N}_j = \mathbf{A}^j\mathbf{r}_0 + \mathbf{A}^{j-1}\mathbf{r}_1 + \ldots + \mathbf{r}_j.$$

This shows that the local rounding error vector at each time-level propagates forward in the same way as the exact solution vector at that time-level. As proved earlier, the effect of each local rounding error will diminish with increasing j if $\max_i |\lambda_i| < 1$, where λ_i, $i = 1(1)(N-1)$, are the eigenvalues of $\mathbf{A}$, but the global rounding error cannot possibly tend to zero because of the terms $\mathbf{r}_j$, $\mathbf{Ar}_{j-1}, \ldots$.

Lax's equivalence theorem

Given a properly posed linear initial-value problem and a linear finite-difference approximation to it that satisfies the consistency condition, stability is the necessary and sufficient condition for convergence.

The proof of this theorem is beyond the scope of this book and interested readers should consult reference 25.

A simple example demonstrating the relationship between convergence, stability, and consistency

The classical explicit approximation to the heat conduction equation provides a comparatively simple illustration of Lax's equivalence theorem.

In general, a problem is properly posed if:

(i) The solution is unique when it exists.

(ii) The solution depends continuously on the initial data.

(iii) A solution always exists for initial data that is arbitrarily close to initial data for which no solution exists. (In heat flow problems, for example, discontinuous temperature distributions can be approximated by a sum of N continuous functions whose limiting value, as N tends to infinity, equals the discontinuous distribution except at the points of discontinuity.)

Let U satisfy the equation

$$\frac{\partial U}{\partial t} = \frac{\partial^2 U}{\partial x^2}, \quad 0 < x < 1, \quad t > 0, \tag{2.38}$$

have known continuous initial values when $t = 0$, $0 \le x \le 1$, and known continuous boundary values at $x = 0$ and 1, $t > 0$.

The classical explicit approximation to (2.38) is

$$\frac{u_{i,j+1} - u_{i,j}}{k} = \frac{u_{i-1,j} - 2u_{i,j} + u_{i+1,j}}{h^2}, \quad i = 1(1)(N-1), \quad (2.39)$$

where $x = ih$, $t = jk$, and $Nh = 1$. The local truncation error of this difference scheme is defined by

$$T_{i,j} = \frac{U_{i,j+1} - U_{i,j}}{k} - \frac{U_{i-1,j} - 2U_{i,j} + U_{i+1,j}}{h^2},$$

which may be written as

$$U_{i,j+1} = kT_{i,j} + rU_{i-1,j} + (1-2r)U_{i,j} + rU_{i+1,j}, \quad (2.40)$$

where $r = k/h^2$. For $i = 1(1)(N-1)$, eqns (2.40) can be written in matrix form as

$$
\begin{bmatrix} U_{1,j+1} \\ U_{2,j+1} \\ \vdots \\ U_{N-1,j+1} \end{bmatrix} = k \begin{bmatrix} T_{1,j} \\ T_{2,j} \\ \vdots \\ T_{N-1,j} \end{bmatrix} + \begin{bmatrix} (1-2r) & r & & \\ r & (1-2r) & r & \\ & & \ddots & \\ & & r & (1-2r) \end{bmatrix}
$$

$$
\times \begin{bmatrix} U_{1,j} \\ U_{2,j} \\ \vdots \\ U_{N-1,j} \end{bmatrix} + \begin{bmatrix} rU_{0,j} \\ 0 \\ \vdots \\ 0 \\ rU_{N,j} \end{bmatrix}
$$

i.e. as

$$\mathbf{U}_{j+1} = k\mathbf{T}_j + \mathbf{A}\mathbf{U}_j + \mathbf{c}_j, \quad (2.41)$$

where $\mathbf{c}_j$ is a vector of known boundary values. Applying this recursively,

$$
\begin{aligned}
\mathbf{U}_{j+1} &= k\mathbf{T}_j + \mathbf{A}(k\mathbf{T}_{j-1} + \mathbf{A}\mathbf{U}_{j-1} + \mathbf{c}_{j-1}) + \mathbf{c}_j \\
&= k(\mathbf{T}_j + \mathbf{A}\mathbf{T}_{j-1}) + \mathbf{A}^2\mathbf{U}_{j-1} + (\mathbf{c}_j + \mathbf{A}\mathbf{c}_{j-1}) \\
&= \ldots \\
&= k(\mathbf{T}_j + \mathbf{A}\mathbf{T}_{j-1} + \ldots + \mathbf{A}^j\mathbf{T}_0) + \mathbf{A}^{j+1}\mathbf{U}_0 \\
&\quad + (\mathbf{c}_j + \mathbf{A}\mathbf{c}_{j-1} + \ldots + \mathbf{A}^j\mathbf{c}_0).
\end{aligned}
\quad (2.42)
$$

As the boundary and initial values for u are the same as for U, it follows from eqn (2.39), which can be written as

$$u_{i,j+1} = ru_{i-1,j} + (1-2r)u_{i,j} + ru_{i+1,j}, \quad i = 1(1)(N-1),$$

that

$$\mathbf{u}_{j+1} = \mathbf{A}\mathbf{u}_j + \mathbf{c}_j.$$

This leads as before to

$$\mathbf{u}_{j+1} = \mathbf{A}^{j+1}\mathbf{u}_0 + (\mathbf{c}_j + \mathbf{A}\mathbf{c}_{j-1} + \ldots + \mathbf{A}^j\mathbf{c}_0). \tag{2.43}$$

Subtraction of (2.43) from (2.42) shows that

$$\mathbf{U}_{j+1} - \mathbf{u}_{j+1} = k(\mathbf{T}_j + \mathbf{A}\mathbf{T}_{j-1} + \ldots + \mathbf{A}^j\mathbf{T}_0) + \mathbf{A}^{j+1}(\mathbf{U}_0 - \mathbf{u}_0).$$

But $\mathbf{U}_0 = \mathbf{u}_0 = $ vector of initial values. Hence

$$\mathbf{U}_{j+1} - \mathbf{u}_{j+1} = k(\mathbf{T}_j + \mathbf{A}\mathbf{T}_{j-1} + \ldots + \mathbf{A}^j\mathbf{T}_0). \tag{2.44}$$

This equation shows that the difference between the *exact* solution of the partial differential equation and the *exact* solution of the approximating difference equation at, say, the $(i, j+1)$th mesh point depends on the local truncation errors at certain mesh points on every preceding time-level and on the difference scheme used. Whether or not the accumulative effect at the $(i, j+1)$th mesh point of these preceding errors is a catastrophic build-up or a hoped-for decay as j increases depends clearly on the matrix $\mathbf{A}$ and the nature of the $T_{i,j}$, $i = 1(1)N$, $j = 0(1)J$, if the field of integration of the differential equation is the rectangle $0 \leq x \leq 1$, $0 \leq t = jk \leq T = Jk$, T finite.

Hence eqn (2.44) gives

$$\|\mathbf{U}_{j+1} - \mathbf{u}_{j+1}\| \leq k\{\|\mathbf{T}_j\| + \|\mathbf{A}\|\|\mathbf{T}_{j-1}\| + \ldots + \|\mathbf{A}^j\|\|\mathbf{T}_0\|\}, \quad 0 \leq j \leq J-1.$$

By the Lax–Richtmyer definition of stability,

$$\|\mathbf{A}^j\| \leq M, \quad j = 1(1)J,$$

where M is a positive number independent of j, h and k. Therefore,

$$\|\mathbf{U}_{j+1} - \mathbf{u}_{j+1}\| \leq k\|\mathbf{T}_j\| + kM(\|\mathbf{T}_{j-1}\| + \ldots + \|\mathbf{T}_0\|).$$

If the maximum of all of the moduli of the components of $\mathbf{T}_0$, $\mathbf{T}_1, \ldots, \mathbf{T}_j$ is C, then the infinity norm of each $\mathbf{T}_s$, $s = 1(1)j$, will be $\leq C$. Using this norm it follows that

$$\|\mathbf{U}_{j+1} - \mathbf{u}_{j+1}\| \leq kC + kMjC = kC + MtC.$$

But $jk = t \leqslant T = Jk$ is finite and $k \to 0$ as $J \to \infty$, so the first term of the right-hand side tends to zero irrespective of the magnitudes of M and C. The second term, however, tends to zero if and only if $C = \max_{i,j} |T_{i,j}|$ tends to zero as k and $h = (k/r)^{\frac{1}{2}}$ tend to zero, which, by definition, is the condition for consistency. This proves, in this particular case, that the difference scheme is convergent when it is stable and consistent.

Finite difference approximations to $\partial U/\partial t = \nabla^2 U$ in cylindrical and spherical polar co-ordinates

The non-dimensional form of the equation for heat-conduction in three dimensions is $\partial U/\partial t = \nabla^2 U$, which, in cylindrical polar coordinates (r, θ, z) is

$$\frac{\partial U}{\partial t} = \frac{\partial^2 U}{\partial r^2} + \frac{1}{r}\frac{\partial U}{\partial r} + \frac{1}{r^2}\frac{\partial^2 U}{\partial \theta^2} + \frac{\partial^2 U}{\partial z^2}.$$

Assuming, for simplicity, that U is independent of z, this reduces to the two-dimensional equation

$$\frac{\partial U}{\partial t} = \frac{\partial^2 U}{\partial r^2} + \frac{1}{r}\frac{\partial U}{\partial r} + \frac{1}{r^2}\frac{\partial^2 U}{\partial \theta^2}. \tag{2.45}$$

For non-zero values of r there is no difficulty in expressing each derivative in terms of standard finite-difference approximations, as shown in Chapter 5, but at $r = 0$ the right side appears to contain singularities. This complication can be dealt with by replacing the polar co-ordinate form of $\nabla^2 U$ by its Cartesian equivalent which transforms eqn (2.45) to the equation

$$\frac{\partial U}{\partial t} = \frac{\partial^2 U}{\partial x^2} + \frac{\partial^2 U}{\partial y^2}. \tag{2.46}$$

Now construct a circle of radius δr, centre the origin, and denote the four points in which Ox, Oy meet this circle by 1, 2, 3, 4. Denote the corresponding function values by u_1, u_2, u_3, and u_4 and the value at the origin by u_0. Then

$$\nabla^2 U = \frac{(u_1 + u_2 + u_3 + u_4 - 4u_0)}{(\delta r)^2} + O\{(\delta r)^2\}.$$

Rotation of the axes through a small angle clearly leads to a similar equation. Repetition of this rotation and the addition of all such equations then gives

$$\nabla^2 U = \frac{4(u_M - u_0)}{(\delta r)^2} + O\{(\delta r)^2\},$$

where u_M is a mean value of U round the circle. The best mean value available is given, of course, by adding all values and dividing by their number.

When a two-dimensional problem in cylindrical co-ordinates possesses circular symmetry, then $\partial^2 U/\partial \theta^2 = 0$, and eqn (2.45) simplifies to

$$\frac{\partial U}{\partial t} = \frac{\partial^2 U}{\partial r^2} + \frac{1}{r}\frac{\partial U}{\partial r}. \tag{2.47}$$

Assuming $\partial U/\partial r = 0$ at $r = 0$, which it will be if the problem is symmetrical with respect to the origin, it is seen that $(1/r)\partial U/\partial r$ assumes the indeterminate form $0/0$ at this point.

By Maclaurin's expansion,

$$U'(r) = U'(0) + rU''(0) + \tfrac{1}{2}r^2 U'''(0) + \ldots,$$

but $U'(0) = 0$, so the limiting value of $(1/r)\partial U/\partial r$ as r tends to zero is the value of $\partial^2 U/\partial r^2$ at $r = 0$. Hence eqn (2.47) at $r = 0$ can be replaced by

$$\frac{\partial U}{\partial t} = 2\frac{\partial^2 U}{\partial r^2}. \tag{2.48}$$

This result can also be deduced from eqn (2.46) because $\partial^2 U/\partial x^2 = \partial^2 U/\partial y^2$ from the circular symmetry, and we can make the x-axis coincide with the direction of r. The finite-difference representation of (2.48) is further simplified by the condition $\partial U/\partial r = 0$ at $r = 0$ because this gives $u_{-1,j} = u_{1,j}$. For example, the explicit approximation

$$\frac{(u_{0,j+1} - u_{0,j})}{\delta t} = \frac{2(u_{1,j} - 2u_{0,j} + u_{-1,j})}{(\delta r)^2}$$

to eqn (2.48) simplifies to

$$\frac{(u_{0,j+1} - u_{0,j})}{\delta t} = \frac{4(u_{1,j} - u_{0,j})}{(\delta r)^2}. \text{ (See Example 2.13)}$$

A complication identical to the one above also arises at $r = 0$ with the spherical polar form of $\nabla^2 U$, namely

$$\frac{\partial^2 U}{\partial r^2} + \frac{2}{r} \frac{\partial U}{\partial r} + \frac{\cot \theta}{r} \frac{\partial U}{\partial \theta} + \frac{1}{r^2} \frac{\partial^2 U}{\partial \theta^2} + \frac{1}{r^2 \sin^2 \theta} \frac{\partial^2 U}{\partial \phi^2}.$$

By the same argument as in the two-dimensional case, this can be replaced at $r = 0$ by $\partial^2 U/\partial x^2 + \partial^2 U/\partial y^2 + \partial^2 U/\partial z^2$ and approximated by $6(u_M - u_0)/(\delta r)^2$, where u_M is the mean value of U over the sphere of radius δr, centre the origin. The factor 6 occurs because Ox, Oy, Oz meet the sphere in six points. If, however, the problem is symmetrical with respect to the origin, i.e. independent of θ and ϕ, $\nabla^2 U$ reduces to $\partial^2 U/\partial r^2 + (2/r)\partial U/\partial r$, with $\partial U/\partial r$ zero at $r = 0$. By either of the previous arguments it follows that the heat conduction equation at $r = 0$ becomes

$$\frac{\partial U}{\partial t} = 3 \frac{\partial^2 U}{\partial r^2}.$$

It is of interest to note that symmetrical heat flow problems for hollow cylinders and spheres that *exclude* $r = 0$ can be solved by simpler equations than those considered because the change of independent variable defined by $R = \log r$ transforms the cylindrical equation

$$\frac{\partial U}{\partial t} = \frac{\partial^2 U}{\partial r^2} + \frac{1}{r} \frac{\partial U}{\partial r} \quad \text{to} \quad e^{2R} \frac{\partial U}{\partial t} = \frac{\partial^2 U}{\partial R^2},$$

and the change of dependent variable given by $U = w/r$ transforms the spherical equation

$$\frac{\partial U}{\partial t} = \frac{\partial^2 U}{\partial r^2} + \frac{2}{r} \frac{\partial U}{\partial r} \quad \text{to} \quad \frac{\partial w}{\partial t} = \frac{\partial^2 w}{\partial r^2}.$$

Example 2.13

The function U is a solution of the equation

$$\frac{\partial U}{\partial t} = \frac{\partial^2 U}{\partial x^2} + \frac{2}{x} \frac{\partial U}{\partial x}, \quad 0 < x < 1, t > 0,$$

and satisfies the initial conditions

$$U = 1 - x^2 \text{ when } t = 0, \quad 0 \leqslant x \leqslant 1,$$

and the boundary conditions

$$\frac{\partial U}{\partial x} = 0 \text{ at } x = 0, \quad t > 0; \quad U = 0 \text{ at } x = 1, \quad t > 0.$$

Using a rectangular grid defined by $\delta x = 0.1$ and $\delta t = 0.001$, calculate a finite-difference solution to $4D$ by an explicit method at the points $(0, 0.001)$, $(0.1, 0.001)$ and $(0.9, 0.001)$ in the x–t plane. (See Chapter 2, Exercise 23, for the stability of the difference scheme.) At $x = 0$, $(2/x)(\partial U/\partial x)$ is indeterminate. As

$$\lim_{x \to 0} \frac{2}{x} \frac{\partial U}{\partial x} = \lim_{x \to 0} 2 \frac{\partial^2 U}{\partial x^2},$$

the equation can be replaced at $x = 0$ by

$$\frac{\partial U}{\partial t} = 3 \frac{\partial^2 U}{\partial x^2}.$$

This may be approximated by the difference equation

$$\frac{u_{0,j+1} - u_{0,j}}{\delta t} = \frac{3(u_{-1,j} - 2u_{0,j} + u_{1,j})}{(\delta x)^2}. \tag{2.49}$$

If $(\partial U/\partial x)_{i,j}$ is approximated by $(u_{i+1,j} - u_{i-1,j})/2(\delta x)$, it follows that $u_{-1,j} = u_{1,j}$ since $(\partial U/\partial x)_{0,j} = 0$. Hence eqn (2.49) reduces to

$$u_{0,j+1} = u_{0,j} + 3r(2u_{1,j} - 2u_{0,j}) = (1 - 6r)u_{o,j} + 6ru_{1,j},$$

where $r = \delta t/(\delta x)^2 = 0.1$ in this example. Therefore

$$u_{0,j+1} = \tfrac{1}{5}(2u_{0,j} + 3u_{1,j}). \tag{2.50}$$

At $x \neq 0$ the differential equation can be approximated by

$$\frac{1}{\delta t}(u_{i,j+1} - u_{i,j}) = \frac{1}{(\delta x)^2}(u_{i-1,j} - 2u_{i,j} + u_{i+1,j}) + \frac{2}{2i(\delta x)^2}(u_{i+1,j} - u_{i-1,j})$$

giving

$$u_{i,j+1} = r\left(1 - \frac{1}{i}\right)u_{i-1,j} + (1 - 2r)u_{i,j} + r\left(1 + \frac{1}{i}\right)u_{i+1,j}.$$

Therefore

$$u_{1,j+1} = \tfrac{1}{5}(4u_{1,j} + u_{2,j}) \tag{2.51}$$

and

$$u_{9,j+1} = \tfrac{4}{5}(\tfrac{1}{9}u_{8,j} + u_{9,j}) \text{ since } u_{10,j} = 0. \tag{2.52}$$

By eqns (2.50), (2.51), and (2.52) the solution values at the points in question are as shown below

$x =$	0	0.1	0.2	...	0.8	0.9
$t = 0$	1.0000	0.9900	0.9600	...	0.3600	0.1900
$t = 0.001$	0.9940	0.9840				0.1840

Exercises and solutions

F.D.S. *means finite difference solution.*
A.S. *means analytical solution of the partial differential equation.*

1. Calculate a finite-difference solution of the equation

$$\frac{\partial U}{\partial t} = \frac{\partial^2 U}{\partial x^2} \quad (0 < x < 1, t > 0)$$

satisfying the initial condition

$$U = \sin \pi x \text{ when } t = 0 \text{ for } 0 \leqslant x \leqslant 1,$$

and the boundary condition

$$U = 0 \text{ at } x = 0 \text{ and } 1 \text{ for } t > 0,$$

using an explicit method with $\delta x = 0.1$ and $r = 0.1$.
    Show by the method of separation of the variables, or merely verify, that the analytical solution is $U = e^{-\pi^2 t} \sin \pi x$. Hence check the accuracy of the numerical solution for $t = 0.005$.

Solutions (Tables 2.18–2.21)

Solutions for $r = 0.1$ and 0.5 and comparisons with the analytical solution at $x = 0.5$ are given overpage. They show clearly that when the initial function and all its derivatives are continuous, and the boundary values at $(0, 0)$ and $(1, 0)$ remain equal to the initial values at these points, then the finite-difference solution can be very accurate indeed. Only one-half of the solution is shown because the problem is symmetric with respect to $x = 0.5$.

2. Calculate a numerical solution of the equation $\partial U/\partial t = \partial^2 U/\partial x^2$, $0 < x < 1$, satisfying the initial condition $U = 1$ when $t = 0$, $0 < x < 1$, and the boundary condition $U = 0$ at $x = 0$ and 1, $t \geqslant 0$.

TABLE 2.18

	t	$x=0$	0.1	0.2	0.3	0.4	0.5
(F.D.).	0.005	0	0.2942	0.5596	0.7702	0.9054	0.9520
(A.S.)	0.005	0	0.2941	0.5595	0.7701	0.9053	0.9519
	. . .	.					
(F.D.S.)	0.01	0	0.2801	0.5327	0.7332	0.8602	0.9063
(A.S.)	0.01	0	0.2800	0.5325	0.7330	0.8617	0.9060
	. . .	.					
F.D.S.	0.02	0	0.2538	0.4828	0.6645	0.7812	0.8214
(A.S.)	0.02	0	0.2537	0.4825	0.6641	0.7807	0.8209
	. . .	.					
(F.D.S.)	0.10	0	0.1156	0.2198	0.3025	0.3556	0.3739
(A.S.)	0.10	0	0.1152	0.2191	0.3015	0.3545	0.3727

TABLE 2.19

	Finite-difference solution ($x=0.5$)	Analytical solution ($x=0.5$)	Percentage error
$t=0.005$	0.9520	0.9519	negligible
$t=0.01$	0.9063	0.9060	negligible
$t=0.02$	0.8214	0.8209	negligible
$t=0.10$	0.3739	0.3727	0.3
r = 0.5	$u_{i,j+1}=\frac{1}{2}(u_{i-1,j}+u_{i+1,j}).$		

TABLE 2.20

	t	$x=0$	0.1	0.2	0.3	0.4	0.5
(F.D.S.)	0.005	0	0.2939	0.5590	0.7694	0.9045	0.9511
(F.D.S.)	0.010	0	0.2795	0.5317	0.7318	0.8602	0.9045
	.	0					
(F.D.S.)	0.02	0	0.2528	0.4809	0.6619	0.7781	0.8181
	.						
	.						
(F.D.S.)	0.10	0	0.1133	0.2154	0.2965	0.3486	0.3665
(A.S.)	0.10	0	0.1152	0.2191	0.3015	0.3545	0.3727

TABLE 2.21

	Finite-difference solution ($x = 0.5$)	Analytical solution ($x = 0.5$)	Percentage error
$t = 0.005$	0.9511	0.9519	−0.08
$t = 0.01$	0.9045	0.9060	−0.17
$t = 0.02$	0.8181	0.8209	−0.34
$t = 0.10$	0.3665	0.3727	−1.66

2 (continued)

Show by the method of separation of the variables that the analytical solution is

$$U = \frac{4}{\pi} \sum_{n=0}^{\infty} \frac{1}{(2n+1)} \, e^{-(2n+1)^2 \pi^2 t} \sin(2n+1)\pi x.$$

Compare these solutions at $x = 0.1$ for small values of t. (See the comments below the tables in the solution on p. 82.)

This problem concerns the temperature changes in a uniform heat-insulated rod that is initially at a constant temperature and which is cooled by having its ends reduced to zero temperature at zero time and subsequently kept at zero.

Solution

The finite-difference solution given by eqn (2.6) with $r = 0.1$ is shown in Table 2.22, and the analytical solution in Table 2.23. Only one-half the solution is given because the problem is symmetrical with respect to $x = \frac{1}{2}$.

TABLE 2.22. Finite-difference solution

	$x = 0$	0.1	0.2	0.3	0.4	0.5
$t = 0.000$	0	1.0000	1.0000	1.0000	1.0000	1.0000
0.001	0	0.9000	1.0000	1.0000	1.0000	1.0000
0.002	0	0.8200	0.9900	1.0000	1.0000	1.0000
0.005	0	0.6566	0.9335	0.9927	0.9996	1.0000
0.010	0	0.5113	0.8283	0.9566	0.9919	0.9979
0.050	0	0.2429	0.4589	0.6263	0.7313	0.7669
0.100	0	0.1460	0.2776	0.3820	0.4490	0.4721
0.200	0	0.0546	0.1038	0.1428	0.1679	0.1766

TABLE 2.23. Analytical solution

$x = 0$	0.1	0.2	0.3	0.4	0.5
$t = 0.000$ 0	1.0000	1.0000	1.0000	1.0000	1.0000
0.001 0	0.9747	1.0000	1.0000	1.0000	1.0000
0.002 0	0.8862	0.9984	1.0000	1.0000	1.0000
0.005 0	0.6827	0.9545	0.9973	0.9999	1.0000
0.010 0	0.5205	0.8427	0.9661	0.9953	0.9992
0.050 0	0.2442	0.4616	0.6304	0.7363	0.7723
0.100 0	0.1467	0.2790	0.3839	0.4513	0.4745
0.200 0	0.0547	0.1040	0.1431	0.1682	0.1769

Comments

An obvious difficulty arises in the solution domain at $(0, 0)$
because the limiting value of the initial temperature is unity as x
tends to zero, whereas the limiting value of the boundary temperatures is zero as t tends to zero. In other words the temperature
is discontinuous at $(0, 0)$ and its value could equally well have
been chosen as 1 or $\frac{1}{2}$ instead of zero. Because of this discontinuity the finite-difference solution is a poor one near $x = 0$ for
small values of t. It will be noticed, however, on comparing
Tables 2.22 and 2.23, that the accuracy of the finite-difference
solution near $x = 0$ improves as t increases. This is characteristic
of parabolic equations and indicates that an implicit method
would give a better solution in the neighbourhood of $(0, 0)$ than
an explicit one because it would not draw its information exclusively from the first row and column. In general, however, we
cannot calculate an accurate solution near a point of discontinuity
by finite-difference methods unless we remove the discontinuity
by a suitable transformation such as below. An alternative approach is to calculate an analytical solution that is continuous
near the discontinuity.

For the problem above it is possible to remove the discontinuity at $(0, 0)$ by changing the independent variables from (x, t)
to (X, T) by means of the equations

$$X = xt^{-\frac{1}{2}}; \quad T = t^{\frac{1}{2}}.$$

This expands the origin $x = t = 0$ into the positive half of the
X-axis and collapses the positive half of the x-axis into the point

at infinity along the x-axis. Consequently the discontinuity in U at $x = t = 0$ is transformed into a smooth change along the positive half of the X-axis. Further details are given in reference 8.

3. Derive the Crank–Nicolson equations for the problem in Exercise 1 and solve them directly for at least two time-steps.

Evaluate the corresponding analytical solution and calculate the percentage error in the numerical solution.

Solution

The solution for $x = 0(0.1)0.5$ and $r = 1$ is given in Table 2.24.

TABLE 2.24

		$x = 0$	0.1	0.2	0.3	0.4	0.5
	t						
(F.D.S.)	0.01	0	0.2802	0.5329	0.7335	0.8623	0.9067
(A.S.)	0.01	0	0.2800	0.5325	0.7330	0.8617	0.9060
(F.D.S.)	0.02	0	0.2540	0.4832	0.6651	0.7818	0.8221
(A.S.)	0.02	0	0.2537	0.4825	0.6641	0.7807	0.8209
(F.D.S.)	0.10	0	0.1160	0.2207	0.3037	0.3571	0.3754
(A.S.)	0.10	0	0.1152	0.2191	0.3015	0.3545	0.3727

TABLE 2.25

	Finite-difference solution ($x = 0.5$)	Analytical solution ($x = 0.5$)	Percentage error
$t = 0.01$	0.9067	0.9060	0.08
0.02	0.8221	0.8209	0.15
0.10	0.3754	0.3727	0.72

4. Prove that the back-substitution procedure of the non-pivoting elimination algorithm given on p. 24 for solving a tridiagonal system of linear equations is stable when $a_i > 0$, $b_i > 0$, $c_i > 0$, and $b_i > a_i + c_i$, $i = 1, 2, \ldots, (N-1)$, where $a_1 = c_{N-1} = 0$.

Solution

$$u_i = \frac{1}{\alpha_i}(S_i + c_i u_{i+1}) = p_{i+1} u_{i+1} + \frac{S_i}{\alpha_i}, \quad \text{(say)},$$

where

$$\alpha_1 = b_1, \quad \alpha_i = b_i - \frac{a_i c_{i-1}}{\alpha_{i-1}}, \quad S_1 = d_1, \quad S_i = d_i + \frac{a_i S_{i-1}}{\alpha_{i-1}},$$

$$a_1 = 0 = c_{N-1}, \quad i = 1, 2, \ldots, N-1.$$

There will be no build-up of errors in the back-substitution process if $|p_{i+1}| < 1$, where

$$p_{i+1} = \frac{c_i}{\alpha_i} = \frac{c_i}{b_i - a_i p_i}, \quad i = 1(1)N - 1).$$

Now $p_2 = (c_1/b_1)$, since $a_1 = 0$. Also $b_1 > c_1$ by hypothesis. Therefore, $0 < p_2 < 1$.

$$p_3 = \frac{c_2}{b_2 - a_2 p_2}.$$

As $c_2 > 0$, $0 < p_2 < 1$, and $b_2 > a_2 > 0$, it follows that

$$0 < p_3 < \frac{c_2}{b_2 - a_2}.$$

By hypothesis, $b_2 > a_2 + c_2$. Hence

$$0 < p_3 < \frac{c_2}{(a_2 + c_2) - a_2} = 1.$$

Similarly, $0 < p_4, p_5, \ldots, p_{N-1} < 1$.

5. Use the Gaussian elimination algorithm on p. 24 to calculate a finite-difference approximation to Worked Example 2.1 for one time-step taking $\delta x = 0.1$ and $r = 1$ using either
   (i) the fully implicit backward time-difference equation

$$u_{i,j+1} - u_{i,j} = r(u_{i-1,j+1} - 2u_{i,j+1} + u_{i+1,j+1}),$$

or (ii) the Douglas equations (3.45) of Chapter 3.

Solution

(i) $-r u_{i-1,j+1} + (1+2r) u_{i,j+1} - r u_{i+1,j+1} = u_{i,j}.$

The equations for the first time-step are $3u_1 - u_2 = 0.2,$

$-u_1 + 3u_2 - u_3 = 0.4$, $\quad -u_2 + 3u_3 - u_4 = 0.6$, $\quad -u_3 + 3u_4 - u_5 = 0.8$,
and $-2u_4 + 3u_5 = 1.0$.

$\alpha_1 = 3$, $\quad \alpha_2 = 2.66667$, $\quad \alpha_3 = 2.62500$, $\quad \alpha_4 = 2.61905$, $\quad \alpha_5 = 2.23636$. $S_1 = 0.2$, $S_2 = 0.46667$, $S_3 = 0.77500$, $S_4 = 1.09524$, $S_5 = 1.83636$. $u_5 = S_5 / \alpha_5$ and $u_i = (S_i + c_i u_{i+1}) / \alpha_i$ giving $u_5 = 0.8211$, $u_4 = 0.7317$, $u_3 = 0.5740$, $u_2 = 0.3902$, $u_1 = 0.1967$.

(ii) $\quad (1 - 6r)u_{i-1,j+1} + (10 + 12r)u_{i,j+1} + (1 - 6r)u_{i+1,j+1}$
$$= (1 + 6r)u_{i-1,j} + (10 - 12r)u_{i,j} + (1 - 6r)u_{i+1,j}.$$

The equations for the first time-step are $22u_1 - 5u_2 = 2.4$, $-5u_1 + 22u_2 - 5u_3 = 4.8$, $-5u_2 + 22u_3 - 5u_4 = 7.2$, $-5u_3 + 22u_4 - 5u_5 = 9.6$, and $-10u_4 + 22u_5 = 9.2$.

$\alpha_1 = 22$, $\quad \alpha_2 = 20.86364$, $\quad \alpha_3 = 20.80174$, $\quad \alpha_4 = 20.79818$, $\quad \alpha_5 = 19.59594$.

$S_1 = 2.4$, $\quad S_2 = 5.34545$, $\quad S_3 = 8.48104$, $\quad S_4 = 11.63855$, $\quad S_5 = 14.79594$.

$u_5 = 0.7551$, $\quad u_4 = 0.7411$, $\quad u_3 = 0.5858$, $\quad u_2 = 0.3966$, and $u_1 = 0.1992$.

6. The function U satisfies the equation

$$\frac{\partial U}{\partial t} = \frac{\partial^2 U}{\partial x^2} \quad (0 < x < 1, \, t > 0)$$

and the boundary conditions

$$\frac{\partial U}{\partial x} = h_1(U - v_1) \text{ at } x = 0, \quad \frac{\partial U}{\partial x} = -h_2(U - v_2) \text{ at } x = 1,$$

where h_1, h_2, v_1, v_2 are positive constants.

(a) When the boundary conditions are approximated by central differences (see Worked Example 2.3) show that one explicit difference scheme is

$$u_{0,j+1} = \{1 - 2r(1 + h_1\delta x)\}u_{0,j} + 2ru_{1,j} + 2rh_1v_1\delta x,$$

$$u_{i,j+1} = ru_{i-1,j} + (1 - 2r)u_{i,j} + ru_{i+1,j}, \quad (i = 1, 2, \ldots, N-1),$$

$$u_{N,j+1} = 2ru_{N-1,j} + \{1 - 2r(1 + h_2\delta x)\}u_{N,j} + 2rh_2v_2\delta x,$$

where $N\delta x = 1$ and $r = \delta t / (\delta x)^2$.

(b) When the boundary conditions are approximated by forward-differences at $x = 0$ and backward-differences at $x = 1$ (see Worked Example 2.4), show that another explicit difference scheme is

$$u_{1,j+1} = \{1 - 2r + r/(1 + h_1\delta x)\}u_{1,j} + ru_{2,j} + rh_1v_1\delta x/(1 + h_1\delta x),$$

$$u_{0,j+1} = (u_{1,j+1} + h_1v_1\delta x)/(1 + h_1\delta x),$$

$$u_{i,j+1} = ru_{i-1,j} + (1 - 2r)u_{i,j} + ru_{i+1,j}, \quad (i = 2, 3, \ldots, N-2),$$

$$u_{N-1,j+1} = \{1 - 2r + r/(1 + h_2\delta x)\}u_{N-1},$$
$$+ ru_{N-2,j} + rh_2v_2\delta x/(1 + h_2\delta x),$$

$$u_{N,j+1} = (u_{N-1,j+1} + h_2v_2\delta x)/(1 + h_2\delta x).$$

7. A uniform solid rod of one-half a unit of length is thermally insulated along its length and its initial temperature at zero time is 0 °C. One end is thermally insulated and the other supplied with heat at a steady rate. Show that the subsequent temperatures at points within the rod are given, in non-dimensional form, by the solution of the equation

$$\frac{\partial U}{\partial t} = \frac{\partial^2 U}{\partial x^2} \quad (0 < x < \tfrac{1}{2}, t > 0)$$

satisfying the initial condition

$$U = 0 \text{ when } t = 0 \quad (0 \leq x \leq \tfrac{1}{2}),$$

and the boundary conditions

$$\frac{\partial U}{\partial x} = 0 \text{ at } x = 0, \quad t > 0, \quad \frac{\partial U}{\partial x} = f \text{ at } x = \tfrac{1}{2}, \quad t > 0$$

where f is a constant.

Solve this problem numerically for $f = 1$ using
(a) an explicit method with $\delta x = 0.1$ and $r = \tfrac{1}{4}$,
(b) an implicit method with $\delta x = 0.1$ and $r = 1$.

Solution

The solution given by the explicit method of Worked Example 2.3, for which the equations are

$$u_{0,j+1} = \tfrac{1}{2}(u_{0,j} + u_{1,j}),$$

$$u_{i,j+1} = \tfrac{1}{4}(u_{i-1,j} + 2u_{i,j} + u_{i+1,j}) \quad (i = 1, 2, 3, 4),$$

$$u_{5,j+1} = \tfrac{1}{2}(u_{4,j} + u_{5,j} + 0.1),$$

TABLE 2.26. Explicit method

x =	0	0.1	0.2	0.3	0.4	0.5
t = 0.005	0.0000	0.0000	0.0000	0.0000	0.0125	0.0750
0.0075	0.0000	0.0000	0.0000	0.0031	0.0250	0.0938
0.01	0.0000	0.0000	0.0008	0.0078	0.0367	0.1094
0.02	0.0009	0.0027	0.0103	0.0313	0.0767	0.1571
0.03	0.0062	0.0104	0.0248	0.0554	0.1095	0.1934
0.05	0.0291	0.0364	0.0594	0.1007	0.1636	0.2509
0.10	0.1169	0.1265	0.1556	0.2044	0.2735	0.3631
0.20	0.3150	0.3250	0.3550	0.4050	0.4750	0.5650
0.50	0.9150	0.9250	0.9550	1.0050	1.0750	1.1650
1.00	1.9150	1.9250	1.9550	2.0050	2.0750	2.1650

is recorded in Table 2.26. The Crank–Nicolson solution, as in the Worked Example 2.5, is shown in Table 2.27. The analytical solution of the differential equation is

$$U = 2t + \frac{1}{2}\left\{\frac{12x^2 - 1}{6} - \frac{2}{\pi^2}\sum_{n=1}^{\infty}\frac{(-1)^n}{n^2}e^{-4\pi^2 n^2 t}\cos 2n\pi x\right\}$$

$$= 2\sqrt{t}\sum_{n=0}^{\infty}\left\{i\,\mathrm{erfc}\,\frac{(2n+1-2x)}{4\sqrt{t}} + i\,\mathrm{erfc}\,\frac{(2n+1+2x)}{4\sqrt{t}}\right\},$$

and is evaluated in Table 2.28. Comparisons are made in Table 2.29.

The finite-difference solutions are clearly very accurate except for small values of t.

A simple calculation shows that the effect of the exponential

TABLE 2.27. Crank–Nicolson method

x =	0	0.1	0.2	0.3	0.4	0.5
t = 0.01	0.0003	0.0006	0.0022	0.0083	0.0309	0.1155
0.02	0.0023	0.0039	0.0108	0.0302	0.0770	0.1540
0.03	0.0077	0.0115	0.0252	0.0552	0.1080	0.1925
0.05	0.0301	0.0373	0.0597	0.1004	0.1627	0.2499
0.10	0.1172	0.1268	0.1557	0.2043	0.2732	0.3628
0.20	0.3150	0.3250	0.3550	0.4050	0.4750	0.5650
0.50	0.9150	0.9250	0.9550	1.0050	1.0750	1.1650
1.00	1.9150	1.9250	1.9550	2.0050	2.0750	2.1650

TABLE 2.28. Analytical solution

$x =$	0	0.1	0.2	0.3	0.4	0.5
$t = 0.0025$	0.0000	0.0000	0.0000	0.0001	0.0050	0.0564
0.0050	0.0000	0.0000	0.0001	0.0017	0.0167	0.0798
0.0075	0.0000	0.0000	0.0006	0.0053	0.0286	0.0977
0.01	0.0000	0.0002	0.0017	0.0101	0.0399	0.1128
0.02	0.0016	0.0035	0.0117	0.0333	0.0791	0.1596
0.03	0.0074	0.0117	0.0264	0.0573	0.1115	0.1954
0.05	0.0307	0.0381	0.0610	0.1023	0.1653	0.2526
0.10	0.1186	0.1282	0.1573	0.2061	0.2751	0.3647
0.20	0.3167	0.3267	0.3567	0.4067	0.4766	0.5666
0.50	0.9167	0.9267	0.9567	1.0067	1.0767	1.1667
1.00	1.9167	1.9267	1.9567	2.0067	2.0767	2.1667

component of the analytical solution is negligible for values of t in excess of 0.1.

A point of interest in this example is that the difference between the analytical solution and both finite-difference solutions for values of t in excess of 0.1 is $0.0017 = 0.01/6 = (\delta x)^2/6$. It can be proved that the transient component of the solution of any explicit or implicit finite-difference scheme for a parabolic equation satisfying the boundary conditions above does not tend to zero as t increases, as does the transient, i.e. exponential component of the solution of the differential equation, but tends to a value of $k(\delta x)^2$, k constant. In this example $k = \frac{1}{6}$.

TABLE 2.29

t	Analytical solution $x = 0.3$	Explicit solution $x = 0.3$	Percentage error	Crank–Nicolson solution $x = 0.3$	Percentage error
0.0075	0.0053	0.0031	−41.5		
0.01	0.0101	0.0078	−22.8	0.0083	−17.8
0.05	0.1023	0.1007	−1.56	0.1004	−1.85
0.10	0.2061	0.2044	−0.82	0.2043	−0.87
0.50	1.0067	1.0050	−0.17	1.0050	−0.17
1.00	2.0067	2.0050	−0.08	2.0050	−0.08

8. The function U satisfies the equation

$$\frac{\partial U}{\partial t} = x \frac{\partial^2 U}{\partial x^2}, \quad 0 < x < \tfrac{1}{2}, \quad t > 0,$$

the boundary conditions

$$U = 0 \text{ at } x = 0, \quad t > 0, \quad \frac{\partial U}{\partial x} = -\tfrac{1}{2}U \text{ at } x = \tfrac{1}{2}, \quad t > 0,$$

and the initial condition $U = x(1-x)$ when $t = 0$, $0 \leqslant x \leqslant \tfrac{1}{2}$.

When all the derivatives with respect to x are expressed in terms of central-difference formulae show that the simplest explicit difference equations approximating this problem at the point (ih, jk) in the $x - t$ plane can be written as

$$u_{i,j+1} = irhu_{i-1,j} + (1 - 2irh)u_{i,j} + irhu_{i+1,j}, \quad i = 1(1)(N-1),$$

and

$$u_{N,j+1} = 2Nrhr_{N-1,j} + (1 - 2Nrh - Nrh^2)u_{N,j},$$

where $r = k/h^2$ and $Nh = \tfrac{1}{2}$.

Take $h = 0.1$ and $r = 0.5$ and calculate a numerical solution to 4D at the points along the first time-level corresponding to $i = 3$ and 5. Will the solution at points near the point $(\tfrac{1}{2}, 0)$ be very accurate? Give a reason for your answer. (The stability of these equations is considered in Chapter 2, Exercise 25.)

Solution

The simplest explicit approximation is

$$\frac{u_{i,j+1} - u_{i,j}}{k} = \frac{ih(u_{i-1,j} - 2u_{i,j} + u_{i+1,j})}{h^2},$$

giving

$$u_{i,j+1} = irhu_{i-1,j} + (1 - 2irh)u_{i,j} + irhu_{i+1,j}, \quad i = 1(1)(N-1).$$

Mentally extend the interval of integration $0 \leqslant x \leqslant \tfrac{1}{2}$ to $0 \leqslant x \leqslant \tfrac{1}{2} + h$. Then the previous equation holds for $i = N$. Eliminate $u_{N+1,j}$ by means of the approximating boundary condition equation

$$(u_{N+1,j} - u_{N-1,j})/2h = -\tfrac{1}{2}u_{N,j}$$

to give

$$u_{N,j+1} = 2Nrhu_{N-1,j} + (1 - 2Nrh - Nrh^2)u_{N,j}.$$

For $h = 0.1$ and $r = \frac{1}{2}$,

$$u_{i,j+1} = 0.05iu_{i-1,j} + (1 - 0.1i)u_{i,j} + 0.05iu_{i+1,j}, \quad i = 1(1)4,$$

and

$$u_{5,j+1} = 0 \cdot 5u_{4,j} + 0 \cdot 475u_{5,j}.$$

$i =$	0	1	2	3	4	5
$t = 0$	0	0.0900	0.1600	0.2100	0.2400	0.2500
$t = 0.005$	0			0.2070		0.2388

As $(x, 0) \to (\frac{1}{2}, 0)$, $\partial U/\partial x = 1 - 2x \to 0$. At $(\frac{1}{2}, 0)$, $U = \frac{1}{2}(1 - \frac{1}{2}) = \frac{1}{4}$. As $(\frac{1}{2}, t) \to (\frac{1}{2}, 0)$, $\partial U/\partial x = -\frac{1}{2}U \to -\frac{1}{2} \cdot \frac{1}{4} = -\frac{1}{8}$. Therefore $\partial U/\partial x$ is discontinuous at $(\frac{1}{2}, 0)$ so the finite-difference solution will not be very accurate at mesh points near $(\frac{1}{2}, 0)$. The solution of the differential equation is continuous at $(\frac{1}{2}, 0)$ and the effect of this discontinuous derivative on the difference solution is very small a few mesh lengths from $(\frac{1}{2}, 0)$.

9. The function V satisfies the non-linear equation

$$\frac{\partial V}{\partial t} = \frac{\partial^2 V}{\partial x^2} + \left(\frac{\partial V}{\partial x}\right)^2, \quad 0 < x < 1, \quad t > 0,$$

the initial condition $V = 0$ when $t = 0$, $0 \le x \le 1$, and the boundary conditions

$$\frac{\partial V}{\partial x} = 1 \text{ at } x = 0, t > 0; \quad V = 0 \text{ at } x = 1, \quad t > 0.$$

Show that the change of dependent variable defined by $V = \log U$, $U \ne 0$, transforms the problem to the solution of the equation

$$\frac{\partial U}{\partial t} = \frac{\partial^2 U}{\partial x^2}, \quad 0 < x < 1, \quad t > 0,$$

where U satisfies the conditions

$$U = 1 \text{ when } t = 0, \quad 0 \le x \le 1,$$

$$\frac{\partial U}{\partial x} = U \text{ at } x = 0, t > 0; \quad U = 1 \text{ at } x = 1, \quad t > 0.$$

Using a rectangular mesh defined by $\delta x = 0.1$ and $\delta t = 0.0025$, approximate the equation for U by the classical explicit scheme, the derivative boundary condition being approximated by a central-difference formula.

Hence calculate a numerical solution for V at the points $(0, 0.005)$ and $(0, 0.0075)$ in the $x - t$ plane.

Solution

$$\frac{\partial V}{\partial t} = \frac{1}{U}\frac{\partial U}{\partial t}; \quad \frac{\partial V}{\partial x} = \frac{1}{U}\frac{\partial U}{\partial x}; \quad \frac{\partial^2 V}{\partial x^2} = -\frac{1}{U^2}\left(\frac{\partial U}{\partial x}\right)^2 + \frac{1}{U}\frac{\partial^2 U}{\partial x^2}.$$

By $V = \log U$, $U = 1$, when $V = 0$. As $\partial V/\partial x = (1/U)(\partial U/\partial x)$ and $\partial V/\partial x = 1$ at $x = 0$, $t > 0$, therefore $\partial U/\partial x = U$ at $x = 0$, $t > 0$. The approximation equation is

$$u_{i,j+1} = \tfrac{1}{4}(u_{i-1,j} + 2u_{i,j} + u_{i+1,j}), \quad i = 1(1)9.$$

Putting $i = 0$ and eliminating $u_{-1,j}$ by means of the derivative boundary equation $(u_{1,j} - u_{-1,j})/0.2 = u_{0,j}$, leads to $u_{0,j+1} = 0.45u_{0,j} + 0.5u_{1,j}$.

$x =$	0	0.1	0.2
$t = 0$	1.0000	1.0000	1.0000
$t = 0.0025$	0.9500	1.0000	1.0000
$t = 0.0050$	0.9275	0.9875	1.0000
$t = 0.0075$	0.9111		

$V(0, 0.005) = \log_e 0.9275 = -0.0753$. $V(0, 0.0075) = \log_e 0.9111 = -0.0931$.

10. The equation

$$\frac{\partial U}{\partial t} - \frac{\partial^2 U}{\partial x^2} = 0$$

is approximated at the point (ih, jk) by the difference equation

$$\theta\left(\frac{u_{i,j+1} - u_{i,j-1}}{2k}\right) + (1 - \theta)\left(\frac{u_{i,j} - u_{i,j-1}}{k}\right) - \frac{1}{h^2}\delta_x^2 u_{i,j} = 0.$$

Show that the truncation error at this point is given by

$$-\tfrac{1}{2}k(1 - \theta)\frac{\partial^2 U}{\partial t^2} - \tfrac{1}{12}h^2\frac{\partial^4 U}{\partial x^4} + O(k^2, h^4).$$

Hence find the value of θ that will reduce this error to one of order k^2 and h^4.

Solution

$$T_{i,j} = \theta(U_{i,j+1} - U_{i,j-1})/2k + (1 - \theta)(U_{i,j} - U_{i,j-1})/k$$
$$- (U_{i-1,j} - 2U_{i,j} + U_{i+1,j})/h^2.$$

Expand each term by Taylor's series about (ih, jk) to get

$$T_{i,j} = -\tfrac{1}{2}k(1 - \theta)\frac{\partial^2 U_{i,j}}{\partial t^2} - \tfrac{1}{12}h^2\frac{\partial^4 U}{\partial x^4} + O(k^2) + O(h^4),$$

where

$$\frac{\partial U}{\partial t} = \frac{\partial^2 U}{\partial x^2},$$

so that

$$\frac{\partial^2 U}{\partial t^2} = \frac{\partial^4 U}{\partial x^4}.$$

Hence,

$$T_{i,j} = \{-\tfrac{1}{2}k(1 - \theta) - \tfrac{1}{12}h^2\}\frac{\partial^4 U_{i,j}}{\partial x^4} + O(k^2) + O(h^4).$$

Therefore,

$$\tfrac{1}{2}k(1 - \theta) = -\tfrac{1}{12}h^2, \text{ i.e. } \theta = 1 + \frac{h^2}{6k}$$

gives the required result.

11. Show that the local truncation error at the point (ih, jk) of the Crank–Nicolson approximation to $\partial U/\partial t = \partial^2 U/\partial x^2$ is $O(h^2) + O(k^2)$.

Solution

$$T_{i,j} = \frac{1}{k}(U_{i,j+1} - U_{i,j}) - \frac{1}{2h^2}(\delta_x^2 U_{i,j+1} + \delta_x^2 U_{i,j}),$$

where, by Taylor's expansion about the point (ih, jk),

$$U_{i+1,j} = \left[U + h\frac{\partial U}{\partial x} + \frac{1}{2}h^2\frac{\partial^2 U}{\partial x^2} + \frac{1}{6}h^3\frac{\partial^3 U}{\partial x^3} + \frac{1}{24}h^4\frac{\partial^4 U}{\partial x^4} + O(h^6) \right]_{i,j}$$

etc., giving

$$\delta_x^2 U_{i,j} = \left[h^2\frac{\partial^2 U}{\partial x^2} + \frac{1}{12}h^4\frac{\partial^4 U}{\partial x^4} + O(h^6) \right]_{i,j}.$$

Again, by Taylor's expansion,

$$\delta_x^2 U_{i,j+1} = \delta_x^2 U_{i,j} + k\frac{\partial}{\partial t}\delta_x^2 U_{i,j} + O(k^2).$$

Therefore,

$$T_{i,j} = \left[\frac{\partial U}{\partial t} - \frac{\partial^2 U}{\partial x^2} \right]_{i,j} + \frac{1}{2}k\frac{\partial}{\partial t}\left[\frac{\partial U}{\partial t} - \frac{\partial^2 U}{\partial x^2} \right]_{i,j}$$

$$-\frac{1}{12}h^2\left(\frac{\partial^4 U}{\partial x^4} \right)_{i,j} + O(k^2) + O(h^4) + O(kh^2),$$

where

$$\frac{\partial U}{\partial t} - \frac{\partial^2 U}{\partial x^2} = 0.$$

Hence the result.

12. The equation

$$\alpha\frac{\partial U}{\partial t} + \frac{\partial U}{\partial x} - f(x, t) = 0, \quad \alpha \text{ constant,}$$

is approximated at the point (ih, jk) in the x–t plane by the difference scheme

$$\frac{\alpha}{k}\{u_{i,j+1} - \frac{1}{2}(u_{i+1,j} + u_{i-1,j})\} + \frac{1}{2h}(u_{i+1,j} - u_{i-1,j}) - f_{i,j} = 0.$$

Investigate the consistency of this scheme for (a) $k = rh$, and (b) $k = rh^2$, r a positive constant, where it is assumed that U is sufficiently smooth for third-order derivatives in x and second-order derivatives in t to exist.

If either is inconsistent with the differential equation obtain the equation it does approximate.

Solution

$$U_{i,j+1} = U_{i,j} + k\frac{\partial U_{i,j}}{\partial t} + \tfrac{1}{2}k^2\frac{\partial^2 U_{i,j+\theta_1}}{\partial t^2},$$

$$U_{i+1,j} = U_{i,j} + h\frac{\partial U_{ij}}{\partial x} + \tfrac{1}{2}h^2\frac{\partial^2 U_{i,j}}{\partial x^2} + \tfrac{1}{6}h^3\frac{\partial^3 U_{i+\theta_2,j}}{\partial x^3},$$

$$U_{i-1,j} = U_{i,j} - h\frac{\partial U_{i,j}}{\partial x} + \tfrac{1}{2}h^2\frac{\partial^2 U_{i,j}}{\partial x^2} - \tfrac{1}{6}h^3\frac{\partial^3 U_{i-\theta_3,j}}{\partial x^3},$$

where $0 < \theta_1, \theta_2, \theta_3 < 1$. Hence

$$T_{i,j} = F_{i,j}(U) = \left(\alpha\frac{\partial U}{\partial t} + \frac{\partial U}{\partial x} - f\right)_{i,j} + \tfrac{1}{2}\alpha k\frac{\partial^2 U_{i,j+\theta_1}}{\partial t^2}$$
$$- \frac{\alpha h^2}{2k}\frac{\partial^2 U_{i,j}}{\partial x^2} - \frac{\alpha h^3}{12k}\left(\frac{\partial^3 U_{i+\theta_2,j}}{\partial x^3} - \frac{\partial^3 U_{i-\theta_3,j}}{\partial x^3}\right) + O(h^2).$$

(a) $k = rh$. Scheme is consistent.
(b) $k = rh^2$. The difference equation approximates

$$\alpha\frac{\partial U}{\partial t} + \frac{\partial U}{\partial x} - \frac{\alpha}{2r}\frac{\partial^2 U}{\partial x^2} - f = 0$$

as $h \to 0$.

13. Prove that when the explicit finite-difference scheme

$$u_{i,j+1} = ru_{i-1,j} + (1-2r)u_{i,j} + ru_{i+1,j}, \quad \text{where } r = \frac{k}{h^2},$$

is used to approximate the equation $\partial U/\partial t = \partial^2 U/\partial x^2$, and it is assumed that U possesses continuous and finite derivatives up to order three in t and order six in x, then the discretization error is the solution of the difference equation

$$e_{i,j+1} = re_{i-1,j} + (1-2r)e_{i,j} + re_{i+1,j} + k\omega(x, t),$$

where

$$\omega(x, t) = \frac{h^2}{12}\left(6r\frac{\partial^2 U}{\partial t^2} - \frac{\partial^4 U}{\partial x^4}\right)_{i,j} + \frac{k^2}{6}\frac{\partial^3 U(x_i, t_j + \theta_j k)}{\partial t^3}$$
$$- \frac{h^4}{360}\frac{\partial^6 U(x_i + \theta_i h, t_j)}{\partial x^6}, \quad -1 < \theta_i < 1 \quad \text{and} \quad 0 < \theta_j < 1.$$

If the maximum value of $|\omega|$ is M, deduce, for $0 < r \leq \frac{1}{2}$, that $|e_{i,j}| \leq tM$. Hence show that the discretization error is of order h^2, except when $r = \frac{1}{6}$ in which case it is of order h^4.

Solution

As in the text, p. 46.

14. (a) If the $n \times n$ matrices $\mathbf{A}$ and $\mathbf{B}$ are symmetric, prove that $\mathbf{AB}$ and $\mathbf{BA}$ are symmetric if and only if $\mathbf{A}$ and $\mathbf{B}$ commute.

(b) If the non-singular matrix $\mathbf{A}$ is symmetric, prove that $\mathbf{A}^{-1}$ is also symmetric.

(c) If the non-singular matrices $\mathbf{A}$ and $\mathbf{B}$ commute, prove that $\mathbf{A}^{-1}$ and $\mathbf{B}$, $\mathbf{A}$ and $\mathbf{B}^{-1}$, and also $\mathbf{A}^{-1}$ and $\mathbf{B}^{-1}$ commute.

(d) If the $n \times n$ non-singular and symmetric matrices $\mathbf{A}$ and $\mathbf{B}$ commute, prove that $\mathbf{A}^{-1}\mathbf{B}$, $\mathbf{AB}^{-1}$ and $\mathbf{A}^{-1}\mathbf{B}^{-1}$ are symmetric.

Solution

(a) Assume $\mathbf{AB} = \mathbf{BA}$. Since $\mathbf{A}$ and $\mathbf{B}$ are symmetric, $(\mathbf{AB})^T = \mathbf{B}^T\mathbf{A}^T = \mathbf{BA} = \mathbf{AB}$ by hypothesis. Hence $\mathbf{AB}$ is symmetric. Similarly for $\mathbf{BA}$. Assume $\mathbf{AB}$ is symmetric. Since $\mathbf{A}$ and $\mathbf{B}$ are symmetric, $(\mathbf{AB})^T = \mathbf{B}^T\mathbf{A}^T = \mathbf{BA}$. But $(\mathbf{AB})^T = \mathbf{AB}$ by hypothesis. Hence $\mathbf{AB} = \mathbf{BA}$, proving $\mathbf{A}$ and $\mathbf{B}$ commute.

(b) $\mathbf{I} = \mathbf{A}^{-1}\mathbf{A} = (\mathbf{AA}^{-1})^T = (\mathbf{A}^{-1})^T\mathbf{A}^T = (\mathbf{A}^{-1})^T\mathbf{A}$, since $\mathbf{A}^T = \mathbf{A}$. Hence $(\mathbf{A}^{-1})^T = \mathbf{A}^{-1}$, proving $\mathbf{A}^{-1}$ is symmetric.

(c) $\mathbf{AB} = \mathbf{BA}$, so $\mathbf{A}^{-1}(\mathbf{AB})\mathbf{A}^{-1} = \mathbf{A}^{-1}(\mathbf{BA})\mathbf{A}^{-1}$. Therefore $(\mathbf{A}^{-1}\mathbf{A})\mathbf{BA}^{-1} = \mathbf{A}^{-1}\mathbf{B}(\mathbf{AA}^{-1})$, proving that $\mathbf{BA}^{-1} = \mathbf{A}^{-1}\mathbf{B}$. Similarly for the other pairs.

(d) By (c), $\mathbf{A}^{-1}\mathbf{B} = \mathbf{BA}^{-1}$. Therefore $(\mathbf{A}^{-1}\mathbf{B})^T = (\mathbf{BA}^{-1})^T = (\mathbf{A}^{-1})^T\mathbf{B}^T$. But $\mathbf{A}^{-1}$ is symmetric by (b) and $\mathbf{B}$ is symmetric. Hence $(\mathbf{A}^{-1})^T\mathbf{B}^T = \mathbf{A}^{-1}\mathbf{B}$, proving that $\mathbf{A}^{-1}\mathbf{B}$ is symmetric. Similarly for the other pairs.

15. (a) Prove that the infinity norm of matrix $\mathbf{A}$ is equal to the maximum row sum of the moduli of the elements of $\mathbf{A}$, assuming $\|\mathbf{x}\|_\infty = 1$.

(b) Assuming that $\|\mathbf{A}\|_2^2 = \rho(\mathbf{A}^H\mathbf{A})$, prove that $\|\mathbf{A}\|_2^2 \leq \|\mathbf{A}\|_1 \|\mathbf{A}\|_\infty$. (For derivative boundary condition problems this can sometimes be used to establish stability in the 2-norm.)

Solution

(a) $\|\mathbf{A}\|_\infty = \max\limits_{\|\mathbf{x}\|_\infty=1} \|\mathbf{Ax}\|_\infty$ where

$$\|\mathbf{Ax}\|_\infty = \max_i \left| \sum_{j=1}^n a_{ij} x_j \right| \leq \max_i \sum_{j=1}^n |a_{ij}| \, |x_j| \leq \max_i \sum_{j=1}^n \left\{ |a_{ij}| \max_j |x_j| \right\}.$$

But $\|\mathbf{x}\|_\infty = \max\limits_j |x_j| = 1$. Hence, $\|\mathbf{Ax}\|_\infty \leq \max\limits_i \sum_{j=1}^n |a_{ij}| = \max$ row sum of the moduli of the elements of $\mathbf{A}$. If the maximum row sum is given by $i = k$ then $\|\mathbf{Ax}\|_\infty$ will attain this max value by choosing $x_j = 1$, when $a_{kj} \geq 0$ and $x_j = -1$ when $a_{kj} < 0$.

(b) $\|\mathbf{A}\|^2 = \rho(\mathbf{A}^H \mathbf{A})$. By Gerschgorin's first theorem, $\rho(\mathbf{A}^H \mathbf{A}) \leq \|\mathbf{A}^H \mathbf{A}\|_1 \leq \|\mathbf{A}^H\|_1 \|\mathbf{A}\|_1 = \|(\bar{\mathbf{A}})^T\|_1 \|\mathbf{A}\|_1 = \|\mathbf{A}\|_\infty \|\mathbf{A}\|_1$.

16. Prove that a real tridiagonal matrix with either all its off-diagonal elements positive or all its off-diagonal elements negative is similar to a real symmetric tridiagonal matrix with non-zero off-diagonal elements. Deduce that the eigenvalues of such a matrix are real.

Solution

$$\mathbf{A} = \begin{bmatrix} a_1 & c_2 & & & & \\ b_2 & a_2 & c_3 & & & \\ \cdot & b_3 & a_3 & c_4 & & \\ & & \cdot & & \cdot & \\ & & & & \cdot & \\ & & & & b_n & a_n \end{bmatrix}$$

Let $\mathbf{D}$ be a real diagonal matrix with elements $d_1, d_2, \ldots, d_n$, i.e. $\mathbf{D} = \text{diag}(d_1, d_2, \ldots, d_n)$. Calculate $\mathbf{DAD}^{-1}$. This matrix will be symmetric if

$$\frac{d_1^2}{d_2^2} = \frac{b_2}{c_2}, \quad \frac{d_2^2}{d_3^2} = \frac{b_3}{c_3}, \ldots, \quad \frac{d_{n-1}^2}{d_n^2} = \frac{b_n}{c_n}.$$

Each right-hand side is positive. Assign a real value to d_1, then $d_2, d_3, \ldots, d_n$ are determined. Matrix $\mathbf{A}$ and the real symmetric $\mathbf{DAD}^{-1}$ are similar so they have the same eigenvalues. Hence the eigenvalues of $\mathbf{A}$ are real because the eigenvalues of a real symmetric matrix are real.

17. If the real non-symmetric matrix **A** is similar to the real symmetric matrix **B**, prove that the equations $\mathbf{u}_{j+1} = \mathbf{A}\mathbf{u}_j + \mathbf{b}_j$ are stable in the Lax–Richtmyer sense if $\|\mathbf{B}^j\| \leq M$, $j = 1, 2, \ldots$, where M is a positive number independent of h, k, and j. Show, in particular, that the equations are stable if $\rho(\mathbf{B}) \leq 1$, but that the errors could be large. (See also Chapter 2, Exercise 16.)

Solution

Since **A** is similar to **B**, there exists a non-singular matrix **P** such that $\mathbf{B} = \mathbf{P}^{-1}\mathbf{AP}$. Hence, $\mathbf{A} = \mathbf{PBP}^{-1}$, $\mathbf{A}^2 = (\mathbf{PBP}^{-1})(\mathbf{PBP}^{-1}) = \mathbf{PB}^2\mathbf{P}^{-1}$ and $\mathbf{A}^j = \mathbf{PB}^j\mathbf{P}^{-1}$. Therefore, $\|\mathbf{A}^j\| \leq \|\mathbf{P}\| \|\mathbf{B}^j\| \|\mathbf{P}^{-1}\|$. As $\|\mathbf{P}\|$ and $\|\mathbf{P}^{-1}\|$ are finite and $\|\mathbf{B}^j\| \leq M$, therefore $\|\mathbf{A}^j\|$ is bounded for all j, h, and k, proving stability. Using the 2-norm, $\|\mathbf{B}^j\|_2 \leq \|\mathbf{B}\|_2^j = [\rho(\mathbf{B})]^j \leq 1$ by hypothesis and because **B** is real and symmetric. Therefore, $\|\mathbf{A}^j\|_2 \leq \|\mathbf{P}\|_2 \|\mathbf{P}^{-1}\|_2$ which is a finite number that could be large as j increases.

18. The equation

$$\frac{\partial U}{\partial t} = \alpha \frac{\partial^2 U}{\partial x^2} - \beta U, \quad 0 < x < 1, \quad t > 0,$$

where α and β are real positive constants, is approximated at the point (ih, jk) by the explicit difference scheme

$$\frac{1}{k} \Delta_t u_{i,j} = \frac{\alpha}{h^2} \delta_x^2 u_{i,j} - \beta u_{i,j}.$$

Given that U has known continuous initial values throughout the interval $0 \leq x \leq 1$, $t = 0$, known boundary-values at $x = 0$ and 1, $t > 0$, and that $Nh = 1$, find an upper bound for $r = k/h^2$ that will be sufficient to keep the difference equations stable, assuming that U does not increase with t.

Verify that Gerschgorin's circle theorem establishes the same sufficient condition for stability.

Solution

$$u_{i,j+1} - u_{i,j} = r\alpha(u_{i-1,j} - 2u_{i,j} + u_{i+1,j}) - k\beta u_{i,j}.$$

Hence $\mathbf{u}_{j+1} = \mathbf{A}\mathbf{u}_j + \mathbf{b}_j$, where $r = k/h^2$,

$$
\mathbf{A} = \begin{bmatrix} (1-2r\alpha - k\beta) & r\alpha & & \\ r\alpha & (1-2r\alpha - k\beta) & r\alpha & \\ & & \cdot & \\ & r\alpha & (1-2r\alpha - k\beta) \end{bmatrix}
$$

and $\mathbf{b}_j$ is a column vector of known values. The equations will be stable when $\|\mathbf{A}\| \le 1$.

Since $\mathbf{A}$ is real and symmetric, $\|\mathbf{A}\|_2 = \rho(\mathbf{A}) = \max_s |1 - k\beta - 4r\alpha \sin^2 s\pi/2N|$, $s = 1(1)N-1$. The largest value for r is given by $-1 \le 1 - rh^2\beta - 4r\alpha \sin^2(N-1)\pi/2N$, which implies $r \le 2/(4\alpha + \beta h^2)$. The 1-norm leads to the same result if $1 - 2r\alpha - k\beta \le 0$, for then $\|\mathbf{A}\|_1 = 2r\alpha + k\beta - 1 + 2r\alpha = 4r\alpha + rh^2\beta - 1 \le 1$ for stability, etc. If $1 - 2r\alpha - k\beta > 0$, then $0 < \|\mathbf{A}\|_1 = 1 - 2r\alpha - k\beta + 2r\alpha = 1 - rh^2\beta \le 1$, which implies that $r < 1/h^2\beta > 2/(4\alpha + h^2\beta)$.

19. The equation

$$
\frac{\partial U}{\partial t} = a\frac{\partial^2 U}{\partial x^2}, \quad 0 < x < 1, \quad t > 0
$$

where a is a positive constant, is approximated at the point (ih, jk) by the fully implicit backward-difference (or backward Euler) scheme

$$
u_{i,j+1} = ra(u_{i-1,j+1} - 2u_{i,j+1} + u_{i+1,j+1}),
$$

where $r = k/h^2$ and $Nh = 1$. Assuming that the boundary and initial values are known, prove that:

(a) The scheme is unconditionally stable in the Lax–Richtmyer sense.

(b) The local truncation error is $O(k) + O(h^2)$.

Comment

This is called a backward difference scheme because the time-difference relative to the time of the space-difference is a backward one. It is not as accurate as the Crank–Nicolson method because the truncation error for the latter is $O(k^2) + O(h^2)$.

Solution

(a) Show that $\mathbf{u}_{j+1} = (\mathbf{I} - ra\mathbf{T}_{N-1})^{-1}\mathbf{u}_j + \mathbf{b}_j$. The matrix is real and symmetric so

$$\|\mathbf{A}\|_2 = \rho(\mathbf{A}) = 1/\{1 + 4ra\ \sin^2 \pi/2N\} < 1$$

for all $r > 0$.

(b) $T_{i,j} = k\left(\dfrac{1}{2}\dfrac{\partial^2 U}{\partial t^2} - a\dfrac{\partial^3 U}{\partial x^2\,\partial t}\right)_{i,j} - \dfrac{1}{12}ah^2\left(\dfrac{\partial^4 U}{\partial x^4}\right)_{i,j}$

$$+ k^2\left(\dfrac{1}{6}\dfrac{\partial^3 U}{\partial t^3} - \dfrac{1}{2}a\dfrac{\partial^4 U}{\partial x^2\,\partial t^2}\right)_{i,j} + \ldots$$

20. The equation $\partial U/\partial t = \partial^2 U/\partial x^2$, $0 < x < 1$, $t > 0$, is approximated by the difference scheme

$$u_{i,j+1} - u_{i,j} = r\{\theta(u_{i-1,j+1} - 2u_{i,j+1} + u_{i+1,j+1})$$

$$+ (1 - \theta)(u_{i-1,j} - 2u_{i,j} + u_{i+1,j})\},$$

where $0 \leqslant \theta \leqslant 1$, $r = k/h^2$ and $Nh = 1$. Assuming that the initial values and boundary values are known, show that:

(a) The scheme is unconditionally stable in the Lax–Richtmyer sense for $\frac{1}{2} \leqslant \theta \leqslant 1$ and stable for $0 \leqslant \theta < \frac{1}{2}$ when $r \leqslant 1/2(1 - 2\theta)$.

(b) The von Neumann method gives exactly the same results as (a).

Solution

(a) The matrix $\mathbf{A}$ of the equations is

$(\mathbf{I} - r\theta\mathbf{T}_{N-1})^{-1}\{\mathbf{I} + r(1 - \theta)\mathbf{T}_{N-1}\}$. Since $(\mathbf{I} - r\theta\mathbf{T}_{N-1})$ and $\{\mathbf{I} + r(1 - \theta\mathbf{T}_{N+1}\}$ are both symmetric and commute, matrix $\mathbf{A}$ is symmetric. (Exercise 14(d)). Hence its 2-norm is equal to its spectral radius. The equations will be stable when

$$-1 \leqslant \max_s \frac{1 - 4r(1 - \theta)\sin^2(s\pi/2N)}{1 + 4r\theta\ \sin^2(s\pi/2N)} \leqslant 1.$$

The right-hand inequality is automatically satisfied for $r > 0$, $0 \leqslant \theta \leqslant 1$. The left-hand inequality gives $2r(1 - 2\theta) \leqslant 1$. Hence the result.

(b) Replacing $u_{p,q}$ by $e^{i\beta ph}\xi^q$ leads to

$$\xi = \{1 - 4r(1 - \theta)\sin^2\beta h/2\}/\{1 + 4r\theta\ \sin^2\beta h/2\}.$$

Because $u_{p,q}$ does not increase exponentially with q, the condition for stability is that $-1 \leq \xi \leq 1$, etc.

21. A set of linear difference equations is given by $\mathbf{u}_{j+1} = \mathbf{A}\mathbf{u}_j + \mathbf{b}_j$, where the non-singular $N \times N$ matrix $\mathbf{A}$ has N distinct non-zero eigenvalues λ_i, $i = 1(1)N$, and a corresponding non-singular matrix of eigenvectors $\mathbf{X} = [\mathbf{x}_1, \mathbf{x}_2, \dots, \mathbf{x}_N]$. Show that for the matrix method of analysis in which h, k, and N are kept constant, the stability condition $\|\mathbf{A}^j\| \leq M$, where M is a positive number independent only of j, is equivalent to $|\lambda_i| \leq 1$, $i = 1(1)N$. (N.B. For the Lax–Richtmyer definition of stability M must be independent of j, h, and k.)

Solution

Any perturbation $\mathbf{e} = \mathbf{u} - \mathbf{u}^*$ satisfies $\mathbf{e}_j = \mathbf{A}^j \mathbf{e}_0$. Hence $\|\mathbf{e}_j\| \leq \|\mathbf{A}^j\| \|\mathbf{e}_0\|$ and will be bounded if $\|\mathbf{A}^j\| \leq M$, a constant that is independent of j, since h and k are fixed.

The set of all equations $\mathbf{A}\mathbf{x}_i = \lambda_i \mathbf{x}_i$, $i = 1(1)N$, can be written as $\mathbf{A}\mathbf{X} = \mathbf{X}\mathbf{D}$, where $\mathbf{D} = \text{diag}(\lambda_1, \lambda_2, \dots, \lambda_N)$. (See Chapter 3, Eigenvector–eigenvalue solution). Hence $\mathbf{A} = \mathbf{X}\mathbf{D}\mathbf{X}^{-1}$, $\mathbf{A}^2 = \mathbf{X}\mathbf{D}\mathbf{X}^{-1}\mathbf{X}\mathbf{D}\mathbf{X}^{-1} = \mathbf{X}\mathbf{D}^2\mathbf{X}^{-1}$ and $\mathbf{A}^j = \mathbf{X}\mathbf{D}^j\mathbf{X}^{-1}$. Therefore, $\|\mathbf{A}^j\| \leq \|\mathbf{X}\| \|\mathbf{D}^j\| \|\mathbf{X}^{-1}\|$. But the 1, 2, and ∞ norms of the diagonal matrix $\mathbf{D}^j = \text{diag}(\lambda_1^j, \lambda_2^j, \dots, \lambda_N^j)$ are $\max_i |\lambda_i|^j$. Therefore, a sufficient condition for $\|\mathbf{A}^j\| \leq M$ is that $\max_i |\lambda_i| \leq 1$.

22. (a) Show that one explicit finite-difference scheme approximating the equation $\partial U/\partial t = \partial^2 U/\partial x^2$ is

$$u_{i,j+1} = ru_{i-1,j} + (1 - 2r)u_{i,j} + ru_{i+1,j}.$$

Given that the initial values and boundary values are known, show that Gerschgorin's first and second theorems establish stability of this scheme for $r \leq \frac{1}{2}$, but give no useful result for $r > \frac{1}{2}$.

(b) Show that Gerschgorin's first theorem is inadequate for establishing the unconditional stability of the Crank–Nicolson equations approximating the equation $\partial U/\partial t = \partial^2 U/\partial x^2$, the boundary and initial values being assumed known.

Solution

(a) Gerschgorin's theorem gives $|\lambda| \leq 2|r| + |1 - 2r|$. Consider $0 \leq r \leq \frac{1}{2}$. Then $|\lambda| \leq 2r + (1 - 2r) = 1$. Hence the finite-difference equations are stable. $r > \frac{1}{2}$ leads to $|\lambda| \leq 4r - 1 > 1$, which is inconclusive. The circle theorem gives $|\lambda - 1 + 2r| \leq 2r$, since $r > 0$. Hence $-2r \leq \lambda - 1 + 2r \leq 2r$, giving $1 - 4r \leq \lambda \leq 1$. The equations will be stable when $-1 \leq 1 - 4r$, i.e., $r \leq \frac{1}{2}$. When $r > \frac{1}{2}$, $\lambda \geq 1 - 4r < -1$, which is inconclusive.

(b) As shown in the text, the necessary condition for stability is $\lambda \geq 2$, where λ is an eigenvalue of **B**. Gerschgorin's first theorem leads to $|\lambda| \leq 2 + 4r$, where $r > 0$, which is insufficient to establish $\lambda \geq 2$.

23. The function U satisfies the equation

$$\frac{\partial U}{\partial t} = \frac{\partial^2 U}{\partial x^2} + \frac{2}{x}\frac{\partial U}{\partial x}, \quad 0 < x < 1, \quad t > 0,$$

the initial condition $U = 1 - x^2$ when $t = 0$, $0 \leq x \leq 1$, and the boundary conditions

$$\frac{\partial U}{\partial x} = 0 \text{ at } x = 0, t > 0; \quad U = 0 \text{ at } x = 1, t > 0.$$

The partial differential equation is approximated at the point (ih, jk) in the x–t plane by the explicit difference equation

$$\frac{1}{k}\Delta_t u_{i,j} = \frac{1}{h^2}\delta_x^2 u_{i,j} + \frac{1}{xh}(\Delta_x u_{i,j} + \nabla_x u_{i,j}),$$

and the limiting form of the partial differential equation at $x = 0$ is approximated in a similar manner. The derivative boundary condition is subsequently approximated by the usual central-difference formula.

Given that $Nh = 1$ and that $k = rh^2$, $r > 0$, show that the matrix A of the difference equations $u_{j+1} = Au_j$ is

$$A = \begin{bmatrix} (1-6r) & 6r & & & & \\ 0 & (1-2r) & 2r & & & \\ & r(1-\tfrac{1}{2}) & (1-2r) & r(1+\tfrac{1}{2}) & & \\ & & \ddots & \ddots & \ddots & \\ & & & r\left(1-\frac{1}{i}\right) & (1-2r) & r\left(1+\frac{1}{i}\right) \\ & & & & \ddots & \ddots \\ & & & & r\left(1-\frac{1}{N-1}\right) & (1-2r) \end{bmatrix}$$

Deduce that one eigenvalue of A is $(1-6r)$. Derive from this result a necessary upper bound M on r for possible stability of the equations. Deduce that for $r \leq M$ the infinity norm of matrix A is 1. Using the largest possible value for r and taking $h = 0.1$, calculate the solution values denoted by $u_{0,1}$, $u_{2,1}$, and $u_{9,1}$.

Solution

$$\frac{1}{k}(u_{i,j+1} - u_{i,j}) = \frac{1}{h^2}(u_{i-1,j} - 2u_{i,j} + u_{i+1,j}) + \frac{1}{ih^2}(u_{i+1,j} - u_{i-1,j}),$$

giving

$$u_{i,j+1} = r\left(1 - \frac{1}{i}\right)u_{i-1,j} + (1-2r)u_{i,j} + r\left(1 + \frac{1}{i}\right)u_{i+1,j}.$$

At $x = 0$, $\partial U/\partial t = 3\partial^2 U/\partial x^2$ which can be approximated by $(u_{0,j+1} - u_{0,j})/k = 3(u_{-1,j} - 2u_{0,j} + u_{1,j})/h^2$ where $u_{-1,j} = u_{1,j}$ because $\partial U/\partial x = 0$ at $x = 0$. In matrix form,

$$\mathbf{u}_{j+1} = \begin{bmatrix} (1-6r) & 6r & & & \\ 0 & (1-2r) & 2r & & \\ & r\left(1-\frac{1}{i}\right) & (1-2r)r\left(1+\frac{1}{i}\right) & & \\ & & & & \\ & & & r\left(1-\frac{1}{N-1}\right)(1-2r) \end{bmatrix} \mathbf{u}_j, \text{ i.e.,}$$

$\mathbf{u}_{j+1} = \mathbf{A}\mathbf{u}_j$, where $\mathbf{u}_j^T = (u_{0,j}, u_{1,j}, \ldots, u_{N-1,j})$. Expansion of the determinant of $(\mathbf{A} - \lambda\mathbf{I})$ by the first column gives $(1-6r)$ as an eigenvalue. For this eigenvalue, $-1 \leq 1-6r \leq 1$ for stability. Hence $r \leq \frac{1}{3}$. When $0 < r \leq \frac{1}{3}$ every term in the matrix is positive and the sum of the moduli of the terms along every row except the last is 1. For the last row,

$$0 < \sum_j |a_{N-1,j}| = 1 - r\left(1 + \frac{1}{N-1}\right) < 1.$$

Taking $r = \frac{1}{3}$, $h = 0.1$, $u_{0,j+1} = -u_{0,j} + 2u_{1,j}$ and $u_{i,j+1} =$

$\frac{1}{3}\{(1-(1/i))u_{i-1,j}+u_{i,j}+(1+(1/i))u_{i+1,j}\}$, $i=1(1)9$. Hence $u_{0,1}=$ 0.98, $u_{2,1}=0.94$, and $u_{9,1}=0.17$.

24. Prove that the equations of Exercise 6(b), Chapter 2, are stable in the Lax–Richtmyer sense for $0<r\leqslant\frac{1}{2}$.

Solution

$u_{0,j+1}$ and $u_{N,j+1}$ are given in terms of $u_{1,j+1}$ and $u_{N-1,j+1}$ respectively, so any errors in their solution are not propagated forward in time. Hence we need consider only the remaining equations, namely,

$$
\begin{bmatrix} u_{1,j+1} \\ u_{2,j+1} \\ \\ \\ u_{N-1,j+1} \end{bmatrix} =
$$

$$
\begin{bmatrix} \left\{1-2r+\dfrac{r}{1+h_1\delta x}\right\} & r & & & \\ r & (1-2r) & r & & \\ & & \ddots & & \\ & & r & (1-2r) & r \\ & & & r & \left\{1-2r+\dfrac{r}{1+h_2\delta x}\right\} \end{bmatrix} \begin{bmatrix} u_{1,j} \\ u_{2,j} \\ \\ \\ u_{N-1,j} \end{bmatrix} + \mathbf{c}
$$

where $\mathbf{c}$ is a column vector of constants.

If $1-2r\geqslant0$, i.e. $0<r\leqslant\frac{1}{2}$, the sum of the moduli of the elements along rows 2 to $N-2$ is 1 and along row $1=$ $1-r+r/(1+h_1\delta x)=1-rh_1\delta x/(1+h_1\delta x)<1$ for $r>0$, $h_1>0$. Similarly for row $N-1$. Hence $\|\mathbf{A}\|_\infty=1$, proving that the equations are stable for $0<r\leqslant\frac{1}{2}$. When $r>\frac{1}{2}$, the sum of the moduli of the terms along rows 2 to $N-2=4r-1>1$. Therefore $\|\mathbf{A}\|_\infty>1$ and the equations will be unstable.

Comment

As mentioned in the text, Gerschgorin's circle theorem combined with $|\lambda| \leqslant 1$ is equivalent to $\|A\|_\infty \leqslant 1$, and this method can be used to establish stability for $0 < r \leqslant \frac{1}{2}$ as in the next exercise.

25. Prove that the solution and rounding errors of the equations in Exercise 8, Chapter 2, will not increase exponentially with increasing time if $r \leqslant 2/(2 + \frac{1}{2}h)$.

Solution

The $N \times N$ matrix A of the equations is

$$\begin{bmatrix} (1-2rh) & rh & & & \\ 2rh & (1-4rh) & 2rh & & \\ & & \ddots & & \\ & & irh & (1-2irh) & irh \\ & & & r & (1-r-\frac{1}{2}rh) \end{bmatrix}$$

We can proceed either as in Exercise 24, or use Gerschgorin's circle theorem together with $|\lambda| \leqslant 1$, which is equivalent to $\|A\|_\infty \leqslant 1$, as follows.

The circle theorem applied to row N gives that $|\lambda - (1-r-\frac{1}{2}rh)| \leqslant r$, i.e. $1-2r-\frac{1}{2}rh \leqslant \lambda \leqslant 1-\frac{1}{2}rh$. The right-hand inequality merely shows that $\lambda < 1$ for r, $h > 0$. The condition $-1 \leqslant \lambda$ will be satisfied by $-1 \leqslant 1-2r-\frac{1}{2}rh$, giving $r \leqslant 2/(2+\frac{1}{2}h)$. Similarly, row 1 gives $r \leqslant 2/3h$ and rows 2 to $N-1$ that $r \leqslant 2/4ih$, $i = 2(1)N-1$, where $Nh = \frac{1}{2}$. The smallest range is $0 < r \leqslant 2/(2+\frac{1}{2}h)$.

26. Use the Fourier series method to prove that

(a) $$u_{p,q+1} - u_{p,q} = r(u_{p-1,q} - 2u_{p,q} + u_{p+1,q})$$

is stable for $0 < r \leqslant \frac{1}{2}$. (The forward-difference explicit approximation to $\partial U/\partial t = \partial^2 U/\partial x^2$.)

(b) $$u_{p,q+1} - u_{p,q-1} = 2r(u_{p-1,q} - 2u_{p,q} + u_{p+1,q})$$

is *unstable* for all positive values of r. (The central-difference explicit approximation to $\partial U/\partial t = \partial^2 U/\partial x^2$, often called Richardson's method.)

(c) $u_{p,q+1} - 2u_{p,q} + u_{p,q-1}$
$$= \tfrac{1}{2}r^2\{(u_{p+1,q+1} - 2u_{p,q+1} + u_{p-1,q+1})$$
$$+ (u_{p+1,q-1} - 2u_{p,q-1} + u_{p-1,q-1})\}$$

is stable for all positive values of $r = \delta t/\delta x$. (As implicit approximation to the hyperbolic equation $\partial^2 U/\partial t^2 = \partial^2 U/\partial x^2$.)

Solution

(a) $\xi = 1 - 2r(1 - \cos \beta h)$ where $|\xi| \leqslant 1$. This gives $r \leqslant$ $1/(1 - \cos \beta h)$,
The least value of the right side is $\tfrac{1}{2}$.

(b) $\xi^2 + 8r\xi \sin^2(\beta h/2) - 1 = 0$, so $\xi_1 = -1/\xi_2$ and $\xi_1 + \xi_2 = -8r \sin^2(\beta h/2)$. For stability $|\xi_1| \leqslant 1$ and $|\xi_2| \leqslant 1$. When $|\xi_1| < 1$, $|\xi_2| > 1$, giving instability. When $\xi_1 = 1$, $\xi_2 = -1$, and $\xi_1 + \xi_2 = 0$ giving $r = 0$.

(c) $\xi^2 - 2A\xi + 1 = 0$ where $A = 1/\{1 + 2r^2 \sin^2(\beta h/2)\} > 1$. Hence $\xi = A \pm i(1 - A^2)^{\frac{1}{2}}$, giving $|\xi| = 1$ for all real values of r.

27. The equation

$$\frac{\partial U}{\partial t} = \frac{\partial^2 U}{\partial x^2} + \frac{1}{x}\frac{\partial U}{\partial x}, \quad 0 < x < 1, \quad t > 0,$$

is approximated at the point (ph, qk) by the difference equation

$$\frac{1}{k}\Delta_t u_{p,q} = \frac{1}{h^2}\delta_x^2 u_{p,q} + \frac{1}{2xh}(\Delta_x u_{p,q} + \nabla_x u_{p,q}).$$

Use the von Neumann method of analysis to show that the difference equations are stable for $x > 0$ when

$$\frac{k}{h^2} \leqslant \frac{2}{4 + p^{-2}}.$$

Evaluate the form of the differential equation at $x = 0$ given that $\partial U/\partial x = 0$ at $x = 0$, $t > 0$.

Given also that U is constant at $x = 1$, that $Nh = 1$, and that the derivative boundary condition at $x = 0$ is approximated by a central-difference formula, write out in matrix form the corresponding explicit difference equations approximating the partial differential equation and associated boundary conditions.

Given that $k/h^2 \leqslant \frac{1}{4}$, deduce that the errors will not increase exponentially with increasing q.

Solution

$$\frac{1}{k}(u_{p,q+1} - u_{p,q}) = \frac{1}{h^2}(u_{p+1,q} - 2u_{p,q} + u_{p-1,q}) + \frac{1}{2ph^2}(u_{p+1,q} - u_{p-1,q}).$$

The error function $E_{p,q} = e^{i\beta ph}\xi^q$, $i = \sqrt{(-1)}$, satisfies

$$\xi = (1 - 4r\sin^2\tfrac{1}{2}\beta h) + i\frac{r}{p}\sin\beta h, \quad r = \frac{k}{h^2}, \quad p \geqslant 1.$$

$$|\xi|^2 = 1 + 16r^2\sin^4\tfrac{1}{2}\beta h - 4r\sin^2\tfrac{1}{2}\beta h\left(2 - \frac{r}{p^2}\cos^2\tfrac{1}{2}\beta h\right).$$

$$|\xi| \leqslant 1 \text{ if } 4r\sin^2\tfrac{1}{2}\beta h\left(2 - \frac{r}{p^2}\cos^2\tfrac{1}{2}\beta h\right) \geqslant 16r^2\sin^4\tfrac{1}{2}\beta h,$$

giving

$$r \leqslant \frac{2}{4\sin^2\tfrac{1}{2}\beta h + \dfrac{1}{p^2}\cos^2\tfrac{1}{2}\beta h} > \frac{2}{4 + p^{-2}}.$$

At $x = 0$, $\partial U/\partial t = 2\partial^2 U/\partial x^2$ and this can be approximated by $(u_{0,q+1} - u_{0,q})/k = 2(u_{-1,q} - 2u_{0,q} + u_{1,q})/h^2$. As $\partial U/\partial x = 0$ at $x = 0$, $u_{1,q} = u_{-1,q}$. Hence $u_{0,q+1} = (1 - 4r)u_{0,q} + 4ru_{1,q}$. In matrix form $\mathbf{u}_{q+1} = \mathbf{A}\mathbf{u}_q + \mathbf{c}_q$, where $\mathbf{c}_q$ is a column of constants and

$$\mathbf{A} = \begin{bmatrix} (1-4r) & 4r & & & \\ (1-\tfrac{1}{2})r & (1-2r) & (1+\tfrac{1}{2})r & & \\ & & \cdot & & \\ \left(1-\dfrac{1}{2p}\right)r & (1-2r) & \left(1+\dfrac{1}{2p}\right)r & & \\ & & & \cdot & \\ & & & \left(1-\dfrac{1}{2(N-1)}\right)r & (1-2r) \end{bmatrix}.$$

When $r \leqslant \frac{1}{4}$ the sum of the moduli of the terms along each row $\leqslant 1$. Hence $\|\mathbf{A}\|_\infty \leqslant 1$.

28. The function U satisfies the diffusion–convection equation

$$\frac{\partial U}{\partial t} = \frac{\partial^2 U}{\partial x^2} - \lambda \frac{\partial U}{\partial x}, \quad 0 < x < 1, t > 0, \lambda > 0,$$

and the conditions

$$U(x, 0) = 0, \quad 0 < x < 1,$$

$$U(0, t) = 1, \quad \frac{\partial U(1, t)}{\partial x} = 0, \quad t > 0.$$

If the equation is approximated at (ph, qk) by the central-difference scheme

(a) $\quad \dfrac{u_{p,q+1} - u_{p,q}}{k} = \dfrac{u_{p+1,q} - 2u_{p,q} + u_{p-1,q}}{h^2} - \lambda \dfrac{u_{p+1,q} - u_{p-1,q}}{2h}$

and by the 'upwind' scheme

(b) $\quad \dfrac{u_{p,q+1} - u_{p,q}}{k} = \dfrac{u_{p+1,q} - 2u_{p,q} + u_{p-1,q}}{h^2} - \lambda \dfrac{u_{p,q} - u_{p-1,q}}{h},$

use the von Neumann method to show that the Lax–Richtmyer definition of stability is satisfied by $0 < r \leqslant \frac{1}{2}$ for Scheme (a) and by $0 < r \leqslant 1/\{2 + (\lambda^2/2K)\} < \frac{1}{2}$ for Scheme (b), K a positive number and $r = k/h^2$. (It must be assumed, with the usual notation, that the Fourier solution of the difference equations and boundary conditions at (ph, qk), $p = 1(1)N$, $Nh = 1$, is such that $\beta h = \pm \pi/N, \pm 2\pi/N, \ldots, \pm \pi$.)

Demonstrate that the upwind scheme is unstable for $r = \frac{1}{2}$ by showing that the initial values $u_{p,0} = (-1)^p$ are amplified by a factor of $(1 + \lambda h)$ at each time-step.

If the growth of errors and of $u_{p,q} = e^{i\beta ph}\xi^q$ are restricted by the severer condition $|\xi| \leqslant 1$, show that:

(i) Scheme (a) is stable when $0 < r \leqslant 2/\{\lambda^2 h^2 + (4 - \lambda^2 h^2)\sin^2 \beta h/2\}$. Deduce that $r \leqslant \min\{\frac{1}{2}, 2/\lambda^2 h^2\}$.

(ii) Scheme (b) is stable when $0 < r \leqslant (2 + \lambda h)/\{(2 + \lambda h)^2 \sin^2(\beta h/2) + \lambda^2 h^2 \cos^2 \beta h/2\}$. Deduce that $0 < r \leqslant 1/(2 + \lambda h)$. (See reference 19).

Solution

Substitution of $u_{p,q} = e^{i\beta ph}\xi^q$ into (a) leads to

$$\xi = 1 - 4r \sin^2 \beta h/2 - i(\lambda k/h)\sin \beta h, \quad i = \sqrt{-1}.$$

The stability condition $|\xi| \le 1 + Kk$ for all h, k, and β, where $K > 0$ is a positive number independent of h, k, and β, gives $|\xi|^2 = (1 - 4r \sin^2 \beta h/2)^2 + \lambda^2 rk \sin^2 \beta h \le 1 + 2Kk + K^2k^2 > 1 + 2Kk$. This is satisfied by $|\xi|^2 \le 1 + 2Kk$. When $0 < r \le \frac{1}{2}$, $(1 - 4r \sin^2 \beta h/2)^2 \le 1$ for all βh, and $\lambda^2 rk \sin^2 \beta h \le 2Kk$ for sufficiently large K.

Substitution of $u_{p,q} = e^{i\beta ph} \xi^q$ into (b) leads to $\xi = 1 - 2r(1 - \cos \beta h) - \lambda rh(1 - e^{-i\beta h}) = [1 - (4r + 2\lambda rh) \sin^2 \beta h/2] - i\lambda rh \sin \beta h$. As in (a) this gives that

$$|\xi|^2 = [1 - (4r + 2\lambda rh) \sin^2 \beta h/2]^2 + \lambda^2 rk \sin^2 \beta h \le 1 + 2Kk.$$

When $-1 \le 1 - (4r + 2\lambda rh) \sin^2 \beta h/2$, the inequality is automatically satisfied for $|\beta h| < \pi$ and for sufficiently large K. This gives $r < 1/(2 + \lambda h) \sin^2 \beta h/2$. When $\beta h = \pm \pi$, $\sin^2 \beta h = 0$ and the inequality is satisfied by $-1 - Kk \le 1 - (4r + 2\lambda rh)$, giving $r(2 + \lambda h - \frac{1}{2} Kh^2) \le 1$. Completing the square, $2 + \lambda h - \frac{1}{2} Kh^2 = -\frac{1}{2} K(h - \lambda/K)^2 + 2 + (\lambda^2/2K)$ is seen to have a maximum value of $2 + (\lambda^2/2K)$ when $h = \lambda/K$. Hence $r \le 1/(2 + \lambda^2/2K)$.

When $|\xi|^2 \le 1$, Scheme (a) gives

$$0 \le (1 - 4r \sin^2 \beta h/2)^2 + \lambda^2 rk \sin^2 \beta h \le 1.$$

The left-hand inequality is automatically satisfied. The right-hand inequality and $\sin \beta h = 2 \sin \beta h/2 \cos \beta h/2$ lead to $r\{\lambda^2 h^2 + (4 - \lambda^2 h^2) \sin^2 \beta h/2\} \le 2$. The range $0 < \beta h \le \pi$ gives $r \le \min\{\frac{1}{2}, 2/\lambda^2 h^2\}$. (For the latter, let $\beta h \to 0$.)

Scheme (b) gives

$$[1 - (4r + 2\lambda rh) \sin^2 \beta h/2]^2 + \lambda^2 r^2 h^2 \sin^2 \beta h \le 1.$$

The use of $\sin \beta h = 2 \sin \beta h/2 \cos \beta h/2$ leads to $r\{(2 + \lambda h)^2 \sin^2 \beta h/2 + \lambda^2 h^2 \cos^2 \beta h/2\} \le 2 + \lambda h$. Hence $r \le (2 + \lambda h)/\{(4 + 4\lambda h) \sin^2 \beta h/2 + \lambda^2 h^2\}$ and its upper bound is given by $\beta h = \pi$, etc.

29. The function U is a solution of the equation

$$\frac{\partial U}{\partial t} = \frac{\partial^2 U}{\partial r^2} + \frac{1}{r} \frac{\partial U}{\partial r} + \frac{1}{r^2} \frac{\partial^2 U}{\partial \theta^2}, \quad 0 < r < 1, \quad t > 0,$$

at every point $P(r, \theta, t)$ of the open-bounded domain $0 < r < 1$, $t > 0$, and satisfies the initial condition $U = r \sin \frac{1}{2}\theta$, $0 \le r \le 1$,

$t = 0$, and the boundary condition $\partial U/\partial r = -U$ at $r = 1$, $t > 0$, where (r, θ, t) are the cylindrical polar co-ordinates of P.

Take $\delta r = 0.1$, $\delta \theta = \pi/16$, and $\delta t = 0.01(\delta r)^2$, and use an explicit method to calculate the numerical values of $u(o, \pi, \delta t)$ and $u(1, \pi, \delta t)$, where u is an approximation to U.

Solution

As shown on p. 75, the equation may be approximated at $(0, \theta_j, t_n)$ by $(u_{o,j,n+1} - u_{o,j,n})/\delta t = 4(u_M - u_{0,j,n})/(\delta r)^2$, where u_M is the mean value of u round the circle $r = 0.1$ at time $t_n = n\delta t$. Initially, $u(r, \theta, 0) = U(r, \theta, 0)$. Hence

$$u_M = \frac{1}{2\pi} \int_0^{2\pi} r \sin\frac{\theta}{2}\, d\theta = \frac{0.1}{2\pi} \int_0^{2\pi} \sin\frac{\theta}{2}\, d\theta = 0.0637.$$

As $u_{0,j,0} = 0$ for all j,

$$u(0, \pi, \delta t) = u_{0,16,1} = 4\delta t u_M/(\delta r)^2 = 0.04 u_M = 0.0025.$$

As shown on p. 246, the equation can be approximated at the point (r_i, θ_j, t_n) by

$$\frac{u_{i,j,n+1} - u_{i,j,n}}{\delta t} = \frac{1}{(\delta r)^2} \left\{ \left(1 - \frac{1}{2i}\right) u_{i-1,j,n} + \left(1 + \frac{1}{2i}\right) u_{i+1,j,n} \right.$$

$$\left. - 2\left(1 + \frac{1}{i^2(\delta\theta)^2}\right) u_{i,j,n} + \frac{1}{i^2(\delta\theta)^2} (u_{i,j-1,n} + u_{i,j+1,n}) \right\}.$$

By the boundary condition, $(u_{11,j,n} - u_{9,j,n})/2(\delta r) = -u_{10,j,n}$. Hence $u_{11,j,n} = u_{9,j,n} - 0.2 u_{10,j,n}$. Putting $i = 10$, $j = 16$, and $n = 0$ in the difference equation leads to

$$u_{10,16,1} = u_{10,16,0} + 0.01 \left\{ \tfrac{19}{20} u_{9,16,0} + \tfrac{21}{20}(u_{9,16,0} - 0.2 u_{10,16,0}) \right.$$

$$\left. - 2\left(1 + \frac{16^2}{100\pi^2}\right) u_{10,16,0} + \frac{16^2}{100\pi^2} (u_{10,15,0} + u_{10,17,0}) \right\},$$

where $u_{10,16,0} = \sin(\pi/2) = 1$, $u_{9,16,0} = 0.9$, and $u_{10,15,0} = u_{10,17,0} = 0.9952$. Therefore $u_{10,16,1} = 1 - 0.01882 = 0.9812$ to $4D$.

3 Parabolic equations: alternative derivation of difference equations and miscellaneous topics

This chapter is not necessary for any reader who would prefer to study, at this stage, the numerical solution of hyperbolic and/or elliptic equations.

Reduction to a system of ordinary differential equations

Consider the equation

$$\frac{\partial U}{\partial t} = \frac{\partial^2 U}{\partial x^2}, \quad 0 < x < X, \quad t > 0, \tag{3.1}$$

where U satisfies the initial condition $U(x, 0) = g(x)$, $0 \leqslant x \leqslant X$, and has known boundary values at $x = 0$ and X, $t > 0$.

If the x derivative at (x, t) is replaced by

$$\frac{1}{h^2} \{ U(x - h, t) - 2U(x, t) + U(x + h, t) \} + O(h^2)$$

and x is considered as a constant, eqn (3.1) can be written as the ordinary differential equation

$$\frac{dU(t)}{dt} = \frac{1}{h^2} \{ U(x - h, t) - 2U(x, t) + U(x + h, t) \} + O(h^2).$$

$$\tag{3.2}$$

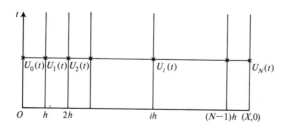

Fig. 3.1

Subdivide the interval $0 \le x \le X$ into N equal subintervals by the grid lines $x_i = ih$, $i = 0(1)N$, where $Nh = X$, and write down eqn (3.2) at every mesh point $x_i = ih$, $i = 1(1)N - 1$, along time-level t. It then follows that the values $V_i(t)$ approximating $U_i(t)$ will be the exact solution values of the system of $(N - 1)$ ordinary differential equations

$$\frac{dV_1(t)}{dt} = \frac{1}{h^2}(V_0 - 2V_1 + V_2)$$

$$\frac{dV_2(t)}{dt} = \frac{1}{h^2}(V_1 - 2V_2 + V_3)$$

$$\vdots$$

$$\frac{dV_{N-1}}{dt} = \frac{1}{h^2}(V_{N-2} - 2V_{N-1} + V_N),$$

where V_0 and V_N are known boundary-values. These can be written in matrix form as

$$\frac{d}{dt}\begin{bmatrix} V_1 \\ V_2 \\ \vdots \\ V_{N-2} \\ V_{N-1} \end{bmatrix} = \frac{1}{h^2}\begin{bmatrix} -2 & 1 & & & \\ 1 & -2 & 1 & & \\ & \cdot & \cdot & \cdot & \\ & & 1 & -2 & 1 \\ & & & 1 & -2 \end{bmatrix}\begin{bmatrix} V_1 \\ V_2 \\ \vdots \\ V_{N-2} \\ V_{N-1} \end{bmatrix} + \frac{1}{h^2}\begin{bmatrix} V_0 \\ 0 \\ \vdots \\ 0 \\ V_N \end{bmatrix},$$

i.e. as

$$\frac{d\mathbf{V}(t)}{dt} = \mathbf{A}\mathbf{V}(t) + \mathbf{b}, \tag{3.3}$$

where $\mathbf{V}(t) = [V_1, V_2, \ldots, V_{N-1}]^T$, $\mathbf{b}$ is a column vector of zeros and known boundary-values and matrix $\mathbf{A}$ of order $(N - 1)$ is given by

$$\mathbf{A} = \frac{1}{h^2}\begin{bmatrix} -2 & 1 & & & \\ 1 & -2 & 1 & & \\ & & \cdot & & \\ & & & -2 & 1 \\ & & & 1 & -2 \end{bmatrix}. \tag{3.4}$$

The solution of the ordinary scalar differential equation

$$\frac{dV}{dt} = AV + b,$$

where A and b are independent of t and $V(t)$ satisfies the initial condition $V(0) = g$, is easily shown, by the method of separation of the variables, to be

$$V(t) = -\frac{b}{A} + \left(g + \frac{b}{A}\right)\exp(At).$$

At the end of this section it is shown that the solution of (3.3) satisfying the initial condition $\mathbf{V}(0) = [g_1, g_2, \ldots, g_{N-1}]^T = \mathbf{g}$, and where $\mathbf{b}$ is independent of t, is

$$\mathbf{V}(t) = -\mathbf{A}^{-1}\mathbf{b} + \{\exp(t\mathbf{A})\}(\mathbf{g} + \mathbf{A}^{-1}\mathbf{b}). \tag{3.5}$$

Hence,

$$\mathbf{V}(t + k) = -\mathbf{A}^{-1}\mathbf{b} + \{\exp(t + k)\mathbf{A}\}(\mathbf{g} + \mathbf{A}^{-1}\mathbf{b})$$
$$= -\mathbf{A}^{-1}\mathbf{b} + \{\exp(k\mathbf{A})\}\{\exp(t\mathbf{A})\}(\mathbf{g} + \mathbf{A}^{-1}\mathbf{b}).$$

By eqn (3.5) this leads to

$$\mathbf{V}(t + k) = -\mathbf{A}^{-1}\mathbf{b} + \{\exp(k\mathbf{A})\}(\mathbf{V}(t) + \mathbf{A}^{-1}\mathbf{b}). \tag{3.6}$$

If all boundary values are zero,

$$\mathbf{V}(t + k) = \{\exp(k\mathbf{A})\}\mathbf{V}(t). \tag{3.7}$$

The boundary values can always be eliminated if we are concerned more, say, with stability than with a particular numerical solution. Perturb the vector of initial values from $\mathbf{g}$ to $\mathbf{g}^*$. By eqn (3.5), the solution $\mathbf{V}^*(t)$ is

$$\mathbf{V}^*(t) = -\mathbf{A}^{-1}\mathbf{b} + \{\exp(t\mathbf{A})\}(\mathbf{g}^* + \mathbf{A}^{-1}\mathbf{b}). \tag{3.8}$$

Equations (3.5) and (3.8) then show that

$$\mathbf{V}^*(t) - \mathbf{V}(t) = \{\exp(t\mathbf{A})\}(\mathbf{g}^* - \mathbf{g}).$$

Hence the perturbation vector $\mathbf{e}(t) = \mathbf{V}^*(t) - \mathbf{V}(t)$ at time t is related to the initial perturbation vector $\mathbf{e}(0) = \mathbf{g}^* - \mathbf{g}$ by

$$\mathbf{e}(t) = \{\exp(t\mathbf{A})\}\mathbf{e}(0).$$

As before, $\mathbf{e}(t + k) = \{\exp(k\mathbf{A})\}\mathbf{e}(t)$.

A note on the solution of $d\mathbf{V}/dt = \mathbf{A}\mathbf{V} + \mathbf{b}$

Define the exponential matrix of the real $n \times n$ matrix $\mathbf{P}$ by

$$\exp \mathbf{P} = e^{\mathbf{P}} = \mathbf{I}_n + \mathbf{P} + \frac{\mathbf{P}^2}{2!} + \frac{\mathbf{P}^3}{3!} + \cdots \sum_{m=0}^{\infty} \frac{\mathbf{P}^m}{m!}, \tag{3.9}$$

where $\mathbf{P}^0 = \mathbf{I}_n$ is the unit matrix of order n.

Parabolic equations

If $\mathbf{Q}$ is a real $n \times n$ matrix such that $\mathbf{PQ} = \mathbf{QP}$, it can be proved by (3.9) that

$$e^{\mathbf{P}}e^{\mathbf{Q}} = e^{\mathbf{Q}}e^{\mathbf{P}} = e^{\mathbf{P}+\mathbf{Q}}.$$

Hence

$$e^{\mathbf{P}}e^{-\mathbf{P}} = e^{-\mathbf{P}}e^{\mathbf{P}} = e^{\mathbf{O}}.$$

But, by (3.9),

$$e^{\mathbf{O}} = \mathbf{I}_n.$$

Therefore

$$e^{\mathbf{P}}e^{-\mathbf{P}} = \mathbf{I}_n. \tag{3.10}$$

Premultiplication of both sides of (3.10) by the inverse $(e^{\mathbf{P}})^{-1}$ of $e^{\mathbf{P}}$, defined by $(e^{\mathbf{P}})^{-1}e^{\mathbf{P}} = \mathbf{I}_n$, then shows

$$e^{-\mathbf{P}} = (e^{\mathbf{P}})^{-1}.$$

On putting $\mathbf{P} = \mathbf{A}t$ into (3.9), where matrix $\mathbf{A}$ is independent of t, and differentiating with respect to t, it follows that

$$\frac{\mathrm{d}}{\mathrm{d}t}(e^{\mathbf{A}t}) = \mathbf{A}e^{\mathbf{A}t} = e^{\mathbf{A}t}\mathbf{A}.$$

Now consider $\mathbf{V}(t) = e^{\mathbf{A}t}\mathbf{g}$, where $\mathbf{g}$ is independent of t. This clearly satisfies the initial condition $\mathbf{V}(0) = \mathbf{g}$. Differentiation with respect to t gives that

$$\frac{\mathrm{d}\mathbf{V}}{\mathrm{d}t} = \mathbf{A}e^{\mathbf{A}t}\mathbf{g} = \mathbf{A}\mathbf{V}.$$

In other words, the solution of

$$\frac{\mathrm{d}\mathbf{V}}{\mathrm{d}t} = \mathbf{A}\mathbf{V} \tag{3.11}$$

which satisfies $\mathbf{V}(0) = \mathbf{g}$, is

$$\mathbf{V}(t) = e^{\mathbf{A}t}\mathbf{g}.$$

Similarly, the vector function

$$\mathbf{V}(t) = -\mathbf{A}^{-1}\mathbf{b} + e^{t\mathbf{A}}(\mathbf{g} + \mathbf{A}^{-1}\mathbf{b}),$$

which obviously satisfies the initial condition $\mathbf{V}(0) = \mathbf{g}$, is the solution of

$$\frac{\mathrm{d}\mathbf{V}}{\mathrm{d}t} = \mathbf{A}\mathbf{V} + \mathbf{b},$$

provided vector **b** and matrix **A** are independent of *t*. The analytical solution of (3.11) in terms of the eigenvalues and eigenvectors of matrix **A** is given later.

Finite difference approximation via the ordinary differential equations

For simplicity, assume that the boundary values associated with

$$\frac{\partial U}{\partial t} = \frac{\partial^2 U}{\partial x^2}, \quad 0 < x < X, \quad t > 0,$$

are zero. By eqn (3.7) the recurrence relationship satisfied by the vector of values $\mathbf{V}(t)$ approximating $U(t)$ at the mesh points $x_i = ih$, $i = 1(1)N - 1$, along time-level t, is

$$\mathbf{V}(t + k) = \{\exp(k\mathbf{A})\}\mathbf{V}(t), \quad t = 0, k, 2k, \ldots, \tag{3.12}$$

where matrix **A** is defined by eqn (3.4).

In order to derive a set of finite difference equations from this it is necessary to approximate the exponential of $k\mathbf{A}$ by a finite algebraic function of $k\mathbf{A}$. Since $\exp(k\mathbf{A})$, by definition, is

$$\mathbf{I} + k\mathbf{A} + \tfrac{1}{2}k^2\mathbf{A}^2 + \tfrac{1}{6}k^3\mathbf{A}^3 + \ldots,$$

one obvious approximation is $\mathbf{I} + k\mathbf{A}$, with a leading error term of order k^2. The vector of values $\mathbf{u} = [u_1, u_2, \ldots, u_{N-1}]^T$ approximating $\mathbf{V}$ in eqn (3.12) will then be the solution of the finite-difference equations

$$\mathbf{u}(t + k) = (\mathbf{I} + k\mathbf{A})\mathbf{u}(t). \tag{3.13}$$

If $t = t_j = jk$ and $r = k/h^2$, these equations in detail, for zero boundary-values, are

$$
\begin{bmatrix} u_{1,j+1} \\ u_{2,j+1} \\ \vdots \\ u_{N-2,j+1} \\ u_{N-1,j+1} \end{bmatrix} =
\begin{bmatrix} (1-2r) & r & & & \\ r & (1-2r) & r & & \\ & & \ddots & & \\ & & r & (1-2r) & r \\ & & & r & (1-2r) \end{bmatrix}
\begin{bmatrix} u_{1,j} \\ u_{2,j} \\ \vdots \\ u_{N-2,j} \\ u_{N-1,j} \end{bmatrix}.
$$

The *i*th equation is the classical explicit approximation

$$u_{i,j+1} = ru_{i-1,j} + (1-2r)u_{i,j} + ru_{i+1,j}, \quad i = 1(1)N - 1.$$

Other higher-order approximations are given by the Padé approximants to the exponential function.

The Padé approximants to e^θ, θ real

Assume that e^θ is to be approximated by $(1+p_1\theta)/(1+q_1\theta)$, where p_1 and q_1 are constants. The determination of p_1 and q_1 requires two equations, which will come from the coefficients of θ and θ^2, so the leading error term will be of order θ^3. Hence

$$e^\theta \equiv \frac{1+p_1\theta}{1+q_1\theta} + c_3\theta^3 + c_4\theta^4 + \ldots .$$

Therefore,

$$(1+q_1\theta)(1+\theta+\tfrac{1}{2}\theta^2+\tfrac{1}{6}\theta^3+\ldots)$$
$$\equiv 1+p_1\theta+(1+q_1\theta)(c_3\theta^3+c_4\theta^4+\ldots).$$

Hence,

$$(1+q_1-p_1)\theta+(\tfrac{1}{2}+q_1)\theta^2+(\tfrac{1}{6}+\tfrac{1}{2}q_1-c_3)\theta^3$$
$$+ \text{higher order terms} \equiv 0.$$

This is satisfied uniquely *to terms of order three* by

$$p_1=\tfrac{1}{2}, \quad q_1=-\tfrac{1}{2} \quad \text{and} \quad c_3=-\tfrac{1}{12}.$$

The rational approximation $(1+\tfrac{1}{2}\theta)/(1-\tfrac{1}{2}\theta)$ is called the $(1, 1)$ Padé approximation of order 2 to $\exp\theta$ and has a leading error term of order 3.

In general, it is possible to approximate $\exp\theta$ by

$$e^\theta = \frac{1+p_1\theta+p_2\theta^2+\ldots+p_T\theta^T}{1+q_1\theta+q_2\theta^2+\ldots+q_S\theta^S} + c_{S+T+1}\theta^{S+T+1} + O(\theta^{S+T+2}),$$

where c_{S+T+1} is a constant. The rational function

$$R_{S,T}(\theta) = \frac{1+p_1\theta+\ldots+p_T\theta^T}{1+q_1\theta+\ldots+q_S\theta^S} = \frac{P_T(\theta)}{Q_S(\theta)}$$

is called the (S, T) Padé approximant of order $(S + T)$ to e^θ. The following table gives the first eight Padé approximants to $\exp\theta$ and their leading error terms.

TABLE 3.1

(S, T)	$R_{S,T}(\theta)$	Principal error term
$(0, 1)$	$1 + \theta$	$\frac{1}{2}\theta^2$
$(0, 2)$	$1 + \theta + \frac{1}{2}\theta^2$	$\frac{1}{6}\theta^3$
$(1, 0)$	$\dfrac{1}{1 - \theta}$	$-\frac{1}{2}\theta^2$
$(1, 1)$	$\dfrac{1 + \frac{1}{2}\theta}{1 - \frac{1}{2}\theta}$	$-\frac{1}{12}\theta^3$
$(1, 2)$	$\dfrac{1 + \frac{2}{3}\theta + \frac{1}{6}\theta^2}{1 - \frac{1}{3}\theta}$	$-\frac{1}{72}\theta^4$
$(2, 0)$	$\dfrac{1}{1 - \theta + \frac{1}{2}\theta^2}$	$\frac{1}{6}\theta^3$
$(2, 1)$	$\dfrac{1 + \frac{1}{3}\theta}{1 - \frac{2}{3}\theta + \frac{1}{6}\theta^2}$	$\frac{1}{72}\theta^4$
$(2, 2)$	$\dfrac{1 + \frac{1}{2}\theta + \frac{1}{12}\theta^2}{1 - \frac{1}{2}\theta + \frac{1}{12}\theta^2}$	$\frac{1}{720}\theta^5$

Standard finite difference equations via the Padé approximants

The classical explicit approximation

It is seen by Table 3.1 and eqn (3.13) that the classical explicit approximation to $\partial U/\partial t = \partial^2 U/\partial x^2$ is given by approximating $\exp(k\mathbf{A})$ by its $(0, 1)$ Padé approximant.

The classical implicit approximation

The $(1, 0)$ Padé approximant approximates

$$\mathbf{V}(t + k) = \{\exp(k\mathbf{A})\}\mathbf{V}(t)$$

by

$$\mathbf{u}(t+k) = (\mathbf{I} - k\mathbf{A})^{-1}\mathbf{u}(t).$$

Premultiplication of both sides by the matrix $(\mathbf{I} - k\mathbf{A})$ yields

$$(\mathbf{I} - k\mathbf{A})\mathbf{u}(t_j + k) = \mathbf{u}(t_j), \quad j = 0, 1, 2, \ldots,$$

where $\mathbf{u}(t_j + k) = [u_{1,j+1}, u_{2,j+1}, \ldots, u_{N-1,j+1}]^T$ and matrix $\mathbf{A}$ is defined by (3.4). In detail, for zero boundary-conditions,

$$\begin{bmatrix} (1+2r) & -r & & & \\ -r & (1+2r) & -r & & \\ & & \ddots & & \\ & & -r & (1+2r) & -r \\ & & & -r & (1+2r) \end{bmatrix}$$

$$\times \begin{bmatrix} u_{1,j+1} \\ u_{2,j+1} \\ \vdots \\ u_{N-2,j+1} \\ u_{N-1,j+1} \end{bmatrix} = \begin{bmatrix} u_{1,j} \\ u_{2,j} \\ \vdots \\ u_{N-2,j} \\ u_{N-1,j} \end{bmatrix}.$$

The ith equation gives the implicit or backward-difference scheme

$$-ru_{i-1,j+1} + (1-2r)u_{i,j+1} - ru_{i+1,j+1} = u_{i,j}, \quad i = 1(1)N - 1.$$

This is unconditionally stable for all $r = k/h^2 > 0$. The leading error terms are of order h^2 in x because of the central-difference approximation to $\partial^2 U/\partial x^2$ and of order k in t. (The leading error term of the $(1, 0)$ Padé approximant to $\exp(k\mathbf{A})$ is $O(k^2)$ but eqn (3.12) was derived by integrating the ordinary differential for $\mathbf{V}(t)$ with respect to t.) The method is said to be first order accurate in t.

The Crank–Nicolson equations

Table 3.1 shows that the $(1, 1)$ Padé approximant replaces

$$\mathbf{V}(t+k) = \{\exp(k\mathbf{A})\}\mathbf{V}(t)$$

by

$$\mathbf{u}(t+k) = (\mathbf{I} - \tfrac{1}{2}k\mathbf{A})^{-1}(\mathbf{I} + \tfrac{1}{2}k\mathbf{A})\mathbf{u}(t).$$

For numerical calculations this needs to be written as

$$(\mathbf{I}-\tfrac{1}{2}k\mathbf{A})\mathbf{u}(t+k)=(\mathbf{I}+\tfrac{1}{2}k\mathbf{A})\mathbf{u}(t).$$

This gives the Crank–Nicolson scheme

$$-ru_{i-1,j+1}+2(1+r)u_{i,j+1}-ru_{i+1,j+1}$$
$$=ru_{i-1,j}+2(1-r)u_{i,j}+ru_{i+1,j}, \quad i=1(1)N-1.$$

It is second-order accurate in t, having an error term via its Padé approximant of order k^3, and may be used with larger time-intervals than the backward difference method. As shown later, it can, however, produce unwanted finite oscillations near points of discontinuity if $k/h > X/\pi$. (See Fig. 3.2.)

A_0-stability, L_0-stability and the symbol of the method

Assume that the boundary values associated with the equation

$$\frac{\partial U}{\partial t}=\frac{\partial^2 U}{\partial x^2}, \quad 0<x<X, \quad t>0,$$

are zero and that the vector of initial values is

$$\mathbf{U}(0)=\mathbf{g}=[g_1, g_2, \ldots, g_{N-1}]^T, \quad Nh=X.$$

Then the vector of values $\mathbf{V}(t)$ approximating $\mathbf{U}(t)$ at the mesh points $x_i = ih$, $i=1(1)N-1$, along time-level $t_j = jk$ satisfies the recurrence relationship

$$\mathbf{V}(t_j + k)=\{\exp(k\mathbf{A})\}\mathbf{V}(t_j), \quad j=0, 1, 2, \ldots.$$

If the exponential of $k\mathbf{A}$ is approximated by its (S, T) Padé approximant $R_{S,T}(k\mathbf{A})$, the resulting set of finite difference equations is

$$\mathbf{u}(t_j + k)=R_{S,T}(k\mathbf{A})\mathbf{u}(t_j),$$

which can be written equivalently as

$$\mathbf{u}(t_j)=R_{S,T}(k\mathbf{A})\mathbf{u}(t_{j-1}).$$

Applied recursively this leads to

$$\mathbf{u}(t_j)=[R_{S,T}(k\mathbf{A})]^j\mathbf{u}(0), \tag{3.14}$$

where $\mathbf{u}(0) = \mathbf{U}(0) = \mathbf{g}$. Now the eigenvalues of matrix $\mathbf{A}$ are

$$\lambda_s = -\frac{4}{h^2}\sin^2\frac{s\pi}{2N}, \quad s = 1(1)N-1,$$

and are all different. Hence the $(N-1)$ eigenvectors $\mathbf{v}_s$ of $\mathbf{A}$ are linearly independent and can be used as a basis for the $(N-1)$ dimensional space of the vector $\mathbf{g}$ of initial values. In other words, $\mathbf{g}$ can be expressed as

$$\mathbf{g} = \sum_{s=1}^{N-1} c_s \mathbf{v}_s,$$

where the c_s are constants. Equation (3.14) can therefore be written as

$$\mathbf{u}(t_j) = [R_{S,T}(k\mathbf{A})]^j \sum_{s=1}^{N-1} c_s \mathbf{v}_s$$

$$= \sum_{s=1}^{N-1} c_s [R_{S,T}(k\mathbf{A})]^j \mathbf{v}_s. \tag{3.15}$$

Remembering that $\mathbf{A}\mathbf{v}_s = \lambda_s \mathbf{v}_s$ and that $f(\mathbf{A})\mathbf{v}_s = f(\lambda_s)\mathbf{v}_s$, it follows by eqn (3.15) that the solution of the finite-difference equations can be expressed as

$$\mathbf{u}(t_j) = \sum_{s=1}^{N-1} c_s [R_{S,T}(k\lambda_s)]^j \mathbf{v}_s. \tag{3.16}$$

Equation (3.16) shows that $\mathbf{u}(t_j)$ will tend to the null vector as $j \to \infty$ if and only if $|R_{S,T}(k\lambda_s)| < 1$, $s = 1(1)N-1$. All rounding errors will also tend to zero because they are subject to the same arithmetic operations as the components of $\mathbf{u}(t_j)$, $j = 1, 2, \ldots$. If this condition depends on the value of $r = k/h^2$, the equations are conditionally stable. This definition corresponds exactly to conditional stability defined by the matrix method for fixed h and k, when the solution tends to zero with increasing t.

When $|R_{S,T}(k\lambda_s)| < 1$ *for all $r > 0$, the equations are said to be* A_0-*stable.* This definition corresponds exactly to unconditional stability defined by the matrix method in the sense that $\mathbf{u}(t_j)$ tends to the null vector as $j \to \infty$ and is not merely bounded for fixed h and k.

The suffix '0' refers to the fact that all the eigenvalues of matrix $\mathbf{A}$ are real, negative and non-zero, so that λ_s, considered as a complex number, is such that $\pi + 0 = \arg \lambda_s = \pi - 0$. (If the eigenvalues are within the wedge defined by $\pi + \alpha > \arg \lambda_s > \pi - \alpha$, $0 < \alpha < \pi/2$, then the method is $A(\alpha)$-stable when, within this wedge, $|R_{S,T}(k\lambda_s)| < 1$, $s = 1(1)N - 1$, for all $r > 0$.)

Although A_0-stability implies that $-1 < R_{S,T}(k\lambda_s) < 1$ for real $R_{S,T}(k\lambda_s)$, it is possible that some values of $R_{S,T}(k\lambda_s)$ might be close to -1 for particular values of $k\lambda_s$. The corresponding values of $[R_{S,T}(k\lambda_s)]^j$ in eqn (3.16) will then alternate in sign as j increases and diminish in magnitude only very slowly. If the corresponding terms on the right-hand side of (3.16) are large in comparison with the remaining terms, the numerical solution could then oscillate finitely as j increases. This phenomenon is particularly pronounced in the x-neighbourhoods of points of discontinuity either in the initial values or between boundary values and initial values, as is illustrated in the next section.

The real coefficients $R_{S,T}(k\lambda_s)$ in (3.16) would clearly give a solution that is stable and free of unwanted oscillations if $0 < R_{S,T}(k\lambda_s) < 1$, $s = 1(1)N - 1$, and $R_{S,T}(k\lambda_s) \to 0$ monotonically as $k\lambda_s$ increases in magnitude. The $(1, 0)$ Padé approximant, for which $R_{1,0}(k\lambda_s) = \dfrac{1}{1 - k\lambda_s}$, λ_s real and negative, is of this character. In practice, this is an unnecessarily severe set of conditions and it is sufficient for $R_{S,T}(k\lambda_s)$ to tend to zero through positive and/or negative values as $k\lambda_s \to -\infty$, $\lambda_s < 0$ and real.

As a consequence, a set of difference equations is said to be L_0-stable if $|R_{S,T}(k\lambda_s)| < 1$, $s = 1(1)N - 1$, and $R_{S,T}(k\lambda_s) \to 0$ as $k\lambda_s \to -\infty$, where λ_s is real, negative, and non-zero.

Following Gourlay and Morris 1980, reference 10, it is usual to put $k\lambda_s = -z$, which makes z positive since $\lambda_s = -(4/h^2)\sin^2 s\pi/2N$, and to call $R_{S,T}(-z)$ the symbol of the method.

A set of difference equations would then be L_0-stable if $|R_{S,T}(-z)| < 1$ for all $z > 0$ and $R_{S,T}(-z) \to 0$ as $z \to \infty$. For the $(1, 0)$ Padé approximant, $R_{1,0}(-z) = 1/(1 + z)$ and obviously satisfies both conditions, showing that the backward-difference scheme is L_0-stable. (See Fig. 3.3.)

For the Crank–Nicolson scheme,

$$R_{1,1}(-z) = \frac{1 - \frac{1}{2}z}{1 + \frac{1}{2}z} = \frac{2/z - 1}{2/z + 1}. \tag{3.17}$$

Clearly, $|R_{1,1}(-z)| < 1$ for all $z > 0$, but $R_{1,1}(-z) \to -1$ as $z \to \infty$, showing that the scheme is A_0-stable.

In general, L_0-stable schemes are preferable to A_0-stable ones because possible unwanted finite oscillations in the numerical solution are rapidly dampened and this eliminates constraints on the time-step k in relation to the space-step h, as occurs with the Crank–Nicolson method. (See the next section.) With A_0-stable methods it is also common for unwanted finite oscillations to increase in magnitude when any attempt is made to improve accuracy by decreasing h. This occurs if the term $R_{S,T}(k\lambda_s)$ of eqn (3.16) approaches -1 more closely because of the decrease in h. The Crank–Nicolson equations exhibit this phenomenon because a decrease in h increases the magnitude of $\lambda_s = (-4/h^2)\sin^2 s\pi/2N$, i.e. increases z and brings $R_{1,1}(-z)$ of eqn (3.17) closer to -1. (See also the section on Stiff Equations.)

As a consequence, a great deal of research since 1978 has been directed towards the generation of L_0-stable methods of high-order accuracy in t. (Reference Lawson and Morris 1978, Gourlay and Morris 1980, Twizell and Khaliq 1982.)

The Padé approximation schemes for parabolic equations are L_0-stable when $S > T$ and A_0-stable when $S = T$.

A necessary constraint on the time-step for the Crank–Nicolson method

Numerical studies indicate that very slowly decaying finite oscillations can occur with the Crank–Nicolson method in the neighbourhood of discontinuities in the initial values or between initial values and boundary values.

Assume that the boundary values for

$$\frac{\partial U}{\partial t} = \frac{\partial^2 U}{\partial x^2}, \quad 0 < x < X, \quad t > 0,$$

are zero and that the vector of initial values is $\mathbf{U}_0 = \mathbf{g}$. Then the Crank–Nicolson or $(1, 1)$ Padé approximant solution corresponding to eqn (3.16) will be

$$\mathbf{u}_j = \sum_{s=1}^{N-1} c_s \mu_s^j \mathbf{v}_s, \tag{3.18}$$

where

$$\mu_s = \frac{1 + \frac{1}{2}k\lambda_s}{1 - \frac{1}{2}k\lambda_s}, \quad s = 1(1)N - 1, \tag{3.19}$$

are the eigenvalues of the amplification matrix $(\mathbf{I}-\frac{1}{2}k\mathbf{A})^{-1}(\mathbf{I}+\frac{1}{2}k\mathbf{A})$, λ_s and $\mathbf{v}_s$ are the eigenvalues and corresponding eigenvectors of matrix $\mathbf{A}$ of (3.4), $Nh = X$, $\lambda_s = -(4/h^2)\sin^2 s\pi/2N$, and

$$\mathbf{g} = \sum_{s=1}^{N-1} c_s \mathbf{v}_s. \tag{3.20}$$

The eigenvalue μ_s, often called the growth factor associated with $\mathbf{v}_s$, is less than one in modulus for all s because λ_s is negative. Hence eqn (3.18) shows that $\mathbf{u}_j$ tends to the null vector as $j \to \infty$. In this sense the equations are unconditionally stable. By eqn (3.19) it is seen, however, that μ_s will be close to -1 when $k\lambda_s = -4r\sin^2 s\pi/2N$ is large, where $r = k/h^2$. This will occur when r is large and $s\pi/2N \simeq \pi/2$, implying N and s both large, i.e. $s = N-1, N-2, \ldots$.

By eqns (3.20) and (3.18) it is seen that the high-frequency components $c_{N-1}\mathbf{v}_{N-1}$, $c_{N-2}\mathbf{v}_{N-2}$, ..., of the initial values have been transformed to $c_{N-1}\mu_{N-1}^j\mathbf{v}_{N-1}$, $c_{N-2}\mu_{N-2}^j\mathbf{v}_{N-2}$, ..., respectively, at the jth time-level. As j increases these components will alternate in sign and decay only very slowly. If $c_{N-1}, c_{N-2}, \ldots$, are large, as tends to occur when there are discontinuities between initial values and boundary values, the solution will oscillate finitely near the points of discontinuity. (See Fig. 3.2.)

Lawson and Morris, reference 14, have pointed out, however, that these oscillatory terms will not have disastrous consequences provided the highest-frequency component $c_{N-1}\mathbf{v}_{N-1}$ decays to zero faster than the lowest-frequency component $c_1\mathbf{v}_1$. The condition for this, by (3.18), is that $|\mu_{N-1}| < |\mu_1|$, i.e. $-\mu_1 < \mu_{N-1} < \mu_1$, giving

$$\frac{-1-\frac{1}{2}k\lambda_1}{1-\frac{1}{2}k\lambda_1} < \frac{1+\frac{1}{2}k\lambda_{N-1}}{1-\frac{1}{2}k\lambda_{N-1}} < \frac{1+\frac{1}{2}k\lambda_1}{1-\frac{1}{2}k\lambda_1}.$$

The right-hand inequality is automatically satisfied by $\lambda_s = -(4/h^2)\sin^2 s\pi/2N$. The left-hand inequality leads to

$$k^2\lambda_1\lambda_{N-1} < 4. \tag{3.21}$$

For large N, $\lambda_1 \simeq -4\pi^2/4h^2N^2 = -\pi^2/X^2$, and

$$\lambda_{N-1} \simeq -(4/h^2)\sin^2\pi/2 = -4/h^2.$$

Hence by (3.21),

$$k/h < X/\pi \text{ approximately, or } k < hX/\pi.$$

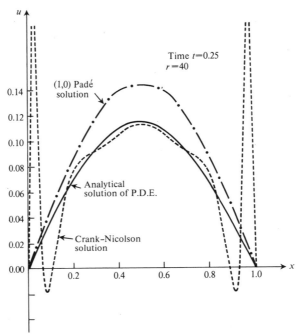

Fig. 3.2

Figure 3.2 exhibits the analytical solution at $t = 0.25$ of $\partial U/\partial t = \partial^2 U/\partial x^2$, $0 < x < 1$, $t > 0$, satisfying $U(0, t) = U(1, t) = 0$, $t > 0$, and $U(0, x) = 1$, $0 \leqslant x \leqslant 1$, $t = 0$, together with the Crank–Nicolson and backward-difference or $(1, 0)$ Padé approximant solutions for $h = 0.025$ and $r = 40$. For these values, $k = 0.025 > hX/\pi = 0.00796$.

The local truncation errors associated with the Padé approximants

When the exponential of matrix $k\mathbf{A}$, defined by (3.4), is approximated by its (S, T) Padé approximant, the difference equation approximating

$$\frac{\partial U}{\partial t} - \frac{\partial^2 U}{\partial x^2} = 0$$

at the point (i, j) is the ith row of the equations

$$\frac{1}{k}\{\mathbf{u}(t_j + k) - \mathbf{R}_{S,T}(k\mathbf{A})\mathbf{u}(t_j)\} = 0, \quad i = 1(1)N - 1,$$

where $\mathbf{R}_{S,T}(k\mathbf{A}) = \mathbf{Q}_S^{-1}(k\mathbf{A})\mathbf{P}_T(k\mathbf{A})$. Assuming that $\mathbf{Q}_s^{-1}$ is non-singular, it follows that the difference equations can be written as

$$\frac{1}{k}\{\mathbf{Q}_S(k\mathbf{A})\mathbf{u}_{j+1} - \mathbf{P}_T(k\mathbf{A})\mathbf{u}_j\} = 0, \quad i = 1(1)N - 1.$$

The definition of the local truncation error $T_{i,j}(U)$ then gives

$$T_{i,j}(U) = i\text{th row of } \frac{1}{k}\{\mathbf{Q}_S(k\mathbf{A})\mathbf{U}_{j+1} - \mathbf{P}_T(k\mathbf{A})\mathbf{U}_j\},$$

$$i = 2(1)N - 2.$$

If all the terms on the right-hand side are expanded about the point (i, j) by Taylor's series, (see Exercise 4), it can be shown that the principal part of the local truncation error at (i, j) is

$$\left[-\tfrac{1}{12}h^2\frac{\partial^4 U}{\partial x^4} + C_q k^{q-1}\frac{\partial^q U}{\partial t^q}\right]_{i,j},$$

where $q = S + T + 1$. Some of the constants C_q are given in Table 3.2. (Reference 13.) The component $-\tfrac{1}{12}h^2\,\partial^4 U/\partial x^4$ arises from

TABLE 3.2

(S, T)	q and r	C_q	E_r
$(0, 1)$	2	$\frac{1}{2}$	$\frac{4}{3}$
$(1, 1)$	3	$-\frac{1}{12}$	$\frac{1}{10}$
$(1, 0)$	2	$-\frac{1}{2}$	$\frac{4}{3}$
$(2, 0)$	3	$\frac{1}{6}$	$-\frac{1}{3}$
$(2, 1)$	4	$\frac{1}{72}$	$-\frac{8}{945}$
$(2, 2)$	5	$\frac{1}{720}$	$-\frac{1}{1890}$
$(1, 2)$	4	$-\frac{1}{72}$	$-\frac{8}{945}$

the approximation of $\partial^2 U/\partial x^2$ by $\{u(x-h, t)-2u(x, t)+u(x+h, t)\}/h^2$ and will appear in the principal part of the local truncation error of every difference scheme that approximates $\exp(k\mathbf{A})$ by a Padé approximant. The C_q term is determined by (S, T).

Stiff equations

The magnitude of the component $-h^2(\partial^4 U/\partial x^4)/12$ of the local truncation error can obviously be decreased by increasing N, since $Nh = X$, a fixed number. This inevitably increases the number and range of the eigenvalues $\lambda_s = -(4/h^2)\sin^2 s\pi/2N$, $s = 1(1)N-1$, of matrix $\mathbf{A}$. As a consequence, the analytical solution of the difference equations, namely

$$\mathbf{u}_j = \sum_{s=1}^{N-1} c_s[R_{S,T}(k\lambda_s)]^j \mathbf{v}_s,$$

may contain a large number of components with widely varying rates of decay, assuming $|R_{S,T}(k\lambda_s)| < 1$. Equations giving rise to this phenomenon are said to be 'stiff'. If, in general, the eigenvalues of matrix $\mathbf{A}$ are $\lambda_s = -\mu_s + i\nu_s$, where μ_s, ν_s are real and $\mu_s > 0$, the 'stiffness ratio' of the equations is measured by $\max \mu_s/\min \mu_s$. For $\mathbf{A}$ defined by (3.4), the stiffness ratio for large N is

$$\{\sin^2(N-1)\pi/2N\}/\sin^2\pi/2N \simeq 4N^2/\pi^2.$$

An extrapolation method for improving accuracy in t

Consider the equation

$$\frac{\partial U}{\partial t} = \frac{\partial^2 U}{\partial x^2}, \quad 0 < x < X, \quad t > 0,$$

satisfying the boundary conditions $U(0, t) = U(X, t) = 0$, $t > 0$. As shown previously, the vector of values $\mathbf{V}$ approximating U at the mesh points $x_i = ih$, $i = 1(1)N-1$, along time-levels t and $t+k$ satisfy

$$\mathbf{V}(t+k) = \{\exp(k\mathbf{A})\}\mathbf{V}(t), \quad t = 0, k, 2k, \ldots . \quad (3.22)$$

If the exponential is approximated by its $(1, 0)$ Padé approximant, the vector of values $\mathbf{u} = [u_1, u_2, \ldots, u_{N-1}]^T$ approximating $\mathbf{V}$ will be the solution of the implicit backward difference equations

$$\mathbf{u}(t+k) = (\mathbf{I} - k\mathbf{A})^{-1}\mathbf{u}(t), \tag{3.23}$$

where matrix $\mathbf{A}$ is defined by (3.4). Over a time-interval of $2k$ this gives

$$\mathbf{u}^{(1)}(t+2k) = (\mathbf{I} - 2k\mathbf{A})^{-1}\mathbf{u}(t). \tag{3.24}$$

Alternatively, the application of eqn (3.23) twice, each over a time-interval of k, leads to the implicit equations

$$\mathbf{u}^{(2)}(t+2k) = (\mathbf{I} - k\mathbf{A})^{-1}\mathbf{u}(t+k) = (\mathbf{I} - k\mathbf{A})^{-1}(\mathbf{I} - k\mathbf{A})^{-1}\mathbf{u}(t)$$

i.e.

$$\mathbf{u}^{(2)}(t+2k) = (\mathbf{I} - k\mathbf{A})^{-2}\mathbf{u}(t). \tag{3.25}$$

Equations (3.24) and (3.25) are two different backward-difference schemes for calculating approximations to $U_i(t+2k)$, $i = 1(1)N - 1$. The detail of the arithmetic is given later.

The binomial expansion of the matrix inverses of (3.24) and (3.25) show that

$$\mathbf{u}^{(1)}(t+2k) = (\mathbf{I} + 2k\mathbf{A} + 4k^2\mathbf{A}^2)\mathbf{u}(t) + O(k^3) \tag{3.26}$$

and

$$\mathbf{u}^{(2)}(t+2k) = (\mathbf{I} + 2k\mathbf{A} + 3k^2\mathbf{A}^2)\mathbf{u}(t) + O(k^3). \tag{3.27}$$

But the Maclaurin expansion of $\exp(2k\mathbf{A})$ in

$$\mathbf{V}(t+2k) = \{\exp(2k\mathbf{A})\}\mathbf{V}(t),$$

an equation giving a more accurate approximation to $U_i(t+2k)$ than either (3.26) or (3.27), $i = 1(1)N - 1$, results in

$$\mathbf{V}(t+2k) = (\mathbf{I} + 2k\mathbf{A} + 2k^2\mathbf{A}^2)\mathbf{V}(t) + O(k^3). \tag{3.28}$$

Comparison of eqns (3.26), (3.27), and (3.28) shows that neither (3.26) nor (3.27) is accurate to terms of order k^2. A simple linear combination of (3.26) and (3.27) will, however, produce an extrapolated vector $\mathbf{u}^{(E)}$ that is second-order accurate in t, i.e. with a leading error term $O(k^3)$, namely,

$$\mathbf{u}^{(E)}(t+2k) = 2\mathbf{u}^{(2)}(t+2k) - \mathbf{u}^{(1)}(t+2k)$$
$$= (\mathbf{I} + 2k\mathbf{A} + 2k^2\mathbf{A}^2)\mathbf{u}(t)$$

The algorithm for the extrapolation is therefore

$$(\mathbf{I} - 2k\mathbf{A})\mathbf{u}^{(1)}(t+2k) = \mathbf{u}(t), \tag{3.29}$$

$$(\mathbf{I} - k\mathbf{A})^2\mathbf{u}^{(2)}(t+2k) = \mathbf{u}(t) \tag{3.30}$$

and

$$\mathbf{u}^{(E)}(t+2k) = 2\mathbf{u}^{(2)} - \mathbf{u}^{(1)}. \tag{3.31}$$

Naturally, the extrapolated solution values are used as the starting values for the extrapolation procedure over the next two time-levels.

The symbol for the extrapolation method

If eqn (3.31) is written in the form

$$\mathbf{u}^{(E)}(t+2k) = S_{1,0}(k\mathbf{A})\mathbf{u}(t) = \{2(\mathbf{I}-k\mathbf{A})^{-2} - (\mathbf{I}-2k\mathbf{A})^{-1}\}\mathbf{u}(t),$$

then

$$S_{1,0}(k\mathbf{A}) = 2(\mathbf{I}-k\mathbf{A})^{-2} - (\mathbf{I}-2k\mathbf{A})^{-1}.$$

Therefore the symbol $S_{1,0}(-z)$ of the extrapolation method is

$$S_{1,0}(-z) = \frac{2}{(1+z)^2} - \frac{1}{1+2z} = \frac{1+2z-z^2}{1+4z+5z^2+2z^3}.$$

Division of the numerator and denominator by z^2 shows that $S_{1,0}(-z) \to 0$ as $z \to \infty$, whilst Fig. 3.3 demonstrates that $|S_{1,0}(-z)| < 1$ for all $z > 0$. Hence the scheme is L_0-stable. The symbol is small and negative for $z > 1+\sqrt{2}$, which implies that small oscillations or fluctuations could occur in the numerical solution for $z = -k\lambda_s > 1+\sqrt{2}$. They would, of course, be heavily

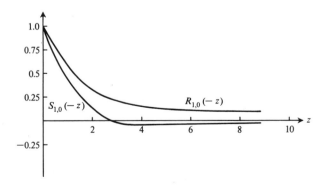

Fig. 3.3

damped in subsequent calculations because $S_{1,0}(-z)$ is very small for these values of z.

In practice, only L_0-stable methods are worth extrapolating.

The arithmetic of the extrapolation method

As

$$\mathbf{I} - 2k\mathbf{A} = \begin{bmatrix} (1+4r) & -2r & & \\ -2r & (1+4r) & -2r & \\ & & \ddots & \\ & & -2r & (1+4r) \end{bmatrix},$$

it follows that eqn (3.29), with zero boundary-values, yields the equations

$$(1+4r)u_{1,j+2}^{(1)} - 2ru_{2,j+2}^{(1)} = u_{1,j}$$
$$-2ru_{1,j+2}^{(1)} + (1+4r)u_{2,j+2}^{(1)} - 2ru_{3,j+2}^{(1)} = u_{2,j}$$
$$\vdots \qquad \vdots \qquad \vdots \qquad \vdots$$
$$-2ru_{N-2,j+2}^{(1)} + (1+4r)u_{N-1,j+2}^{(1)} = u_{N-1,j}.$$

These can be solved by the algorithm on p. 24. Similarly, eqn (3.30) can be treated as two tridiagonal systems, namely,

$$(\mathbf{I} - k\mathbf{A})\mathbf{u}^{(2)}(t+k) = \mathbf{u}(t) \tag{3.32}$$

and

$$(\mathbf{I} - k\mathbf{A})\mathbf{u}^{(2)}(t+2k) = \mathbf{u}^{(2)}(t+k), \tag{3.33}$$

where the equations for (3.32) are

$$(1+2r)u_{1,j+1}^{(2)} - ru_{2,j+1}^{(2)} = u_{1,j}$$
$$-ru_{1,j+1}^{(2)} + (1+2r)u_{2,j+1}^{(2)} - ru_{3,j+1}^{(2)} = u_{2,j}$$
$$-ru_{N-2,j+1}^{(2)} + (1+2r)u_{N-1,j+1}^{(2)} = u_{N-1,j+1}$$

Similarly for (3.33). Alternatively, eqn (3.30) can be written as

$$(\mathbf{I} - 2k\mathbf{A} + k^2\mathbf{A}^2)\mathbf{u}^{(2)}(t+2k) = \mathbf{u}(t), \tag{3.34}$$

where

$$\mathbf{I} - 2k\mathbf{A} + k^2\mathbf{A}^2$$

$$= \mathbf{I} - 2r \begin{bmatrix} -2 & 1 & & & \\ 1 & -2 & 1 & & \\ & 1 & -2 & 1 & \\ & & & \ddots & \\ & & & 1 & -2 \end{bmatrix} + r^2 \begin{bmatrix} 5 & -4 & 1 & & & \\ -4 & 6 & -4 & 1 & & \\ 1 & -4 & 6 & -4 & 1 & \\ & 1 & -4 & 6 & -4 & 1 \\ & & & \ddots & & \\ & & & 1 & -4 & 6 & -4 \\ & & & & 1 & -4 & 5 \end{bmatrix}$$

$$= \begin{bmatrix} (1+4r+5r^2) & (-2r-4r^2) & r^2 & & & \\ (-2r-4r^2) & (1+4r+6r^2) & (-2r-4r^2) & r^2 & & \\ r^2 & (-2r-4r^2) & (1+4r+6r^2) & (-2r-4r^2) & r^2 & \\ & r^2 & (-2r-4r^2) & (1+4r+6r^2) & (-2r-4r^2) & r^2 \\ & & & \ddots & & \\ & & r^2 & (-2r-4r^2) & (1+4r+6r^2) & (-2r-4r^2) \\ & & & r^2 & (-2r-4r^2) & (1+4r+5r^2) \end{bmatrix}$$

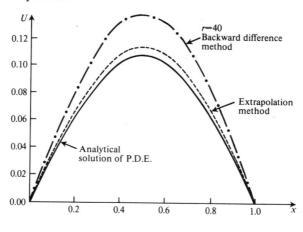

Fig. 3.4

Equation (3.34) can then be solved directly for the components of $\mathbf{u}^{(2)}(t+2k)$ by means of a quindiagonal solver.

For the heat-flow problem $\partial U/\partial t = \partial^2 U/\partial x^2$, $0 < x < 1$, $t > 0$, $U(0, t) = U(1, t) = 0$, $t > 0$, and $U(x, 0) = 1$, $0 \leqslant x \leqslant 1$, Fig. 3.4 shows the graphs of the solutions at $t = 0.25$ by the backward-difference method, eqn (3.23), and the extrapolation method, eqn (3.31) taking $h = k = 0.025$ and $r = 40$. The maximum errors, at $x = 0.5$, are 0.0324 (backward-difference) and 0.0061 (extrapolation).

Further comments

If the classical explicit method is extrapolated it is easily shown that $|S_{0,1}(-z)| \leqslant 1$ for $0 < r \leqslant 1/4$. (Exercise 2.) In general, there is little advantage in the extrapolation of conditionally stable and A_0-stable methods, especially if discontinuities exist in the initial values or between initial values and boundary values.

As one would expect, L_0-stable difference methods of third and fourth accuracy in t can be achieved by extrapolating over three or four time-levels respectively and two such schemes are considered in Gourlay and Morris, reference 10. The extrapolation of (2, 0), (2, 1), (2, 2), and (3, 0) Padé approximations are given in references 28 and 13.

The local truncation errors and amplification symbols of extrapolated schemes

In general, the extrapolation formula for the (S, T) Padé approximant is

$$\mathbf{u}^{(E)}(t+2k) = \alpha\mathbf{u}^{(2)} - (\alpha-1)\mathbf{u}^{(1)},$$

where $\alpha = 2^{S+T}/(2^{S+T}-1)$.

The principal part of the local truncation error of the extrapolated scheme at (i, j) is

$$\left[-\tfrac{1}{12}h^2\frac{\partial^4 U}{\partial x^4} + E_r k^{r-1}\frac{\partial^r U}{\partial t^r}\right]_{i,j}, \quad i = 2(1)N-2,$$

where $r = S + T + 2$. Some of the constants E_r are given in Table 3.2. The amplification symbol of the extrapolated method associated with the (S, T) Padé approximant is

$$S_{S,T}(-z) = \alpha[R_{S,T}(-z)]^2 - (\alpha-1)R_{S,T}(-2z).$$

As an example, consider the $(2, 0)$ Padé approximant to $\exp\theta$, namely $R_{2,0}(\theta) = 1/(1 - \theta + \tfrac{1}{2}\theta^2)$. For this,

$$\alpha = 2^2/(2^2 - 1) = 4/3, \quad \mathbf{u}^{(E)} = \tfrac{4}{3}\mathbf{u}^{(2)} - \tfrac{1}{3}\mathbf{u}^{(1)} \quad \text{and} \quad r = 4.$$

The principal part of the local truncation error is

$$\left[-\tfrac{1}{12}h^2\frac{\partial^4 U}{\partial x^4} - \tfrac{1}{3}k^3\frac{\partial^4 U}{\partial t^4}\right]_{i,j}$$

and

$$S_{2,0}(-z) = \frac{4}{3(1 + z + \tfrac{1}{2}z^2)} - \frac{1}{3(1 + 2z + 2z^2)},$$

which is easily shown to give L_0-stability.

The eigenvalue–eigenvector solution of a system of O.D.E.'s

Preliminary results

(i) If matrix $\mathbf{A}$ of order N has N linearly independent eigenvectors $\mathbf{x}_i$ corresponding to the N different eigenvalues λ_i, $i = 1(1)N$, then the N equations

$$\mathbf{A}\mathbf{x}_i = \lambda_i\mathbf{x}_i, \quad i = 1(1)N,$$

may be written in matrix form as

$$\mathbf{A}[\mathbf{x}_1, \mathbf{x}_2, \ldots, \mathbf{x}_N] = [\mathbf{A}\mathbf{x}_1, \mathbf{A}\mathbf{x}_2, \ldots, \mathbf{A}\mathbf{x}_N]$$
$$= [\lambda_1\mathbf{x}_1, \lambda_2\mathbf{x}_2, \ldots, \lambda_N\mathbf{x}_N]$$
$$= [\mathbf{x}_1, \mathbf{x}_2, \ldots, \mathbf{x}_N] \begin{bmatrix} \lambda_1 & & & \\ & \lambda_2 & & \\ & & \ddots & \\ & & & \lambda_N \end{bmatrix}.$$

Putting the matrix of eigenvectors $[\mathbf{x}_1, \mathbf{x}_2, \ldots, \mathbf{x}_N] = \mathbf{X}$, often called the modal matrix, and

$$\begin{bmatrix} \lambda_1 & & & \\ & \lambda_2 & & \\ & & \ddots & \\ & & & \lambda_N \end{bmatrix} = \mathrm{diag}(\lambda_1, \lambda_2, \ldots, \lambda_N) = \mathbf{D},$$

the eigenvalue–eigenvector equations can be expressed very compactly as

$$\mathbf{A}\mathbf{X} = \mathbf{X}\mathbf{D}.$$

Hence

$$\mathbf{X}^{-1}\mathbf{A}\mathbf{X} = \mathbf{D}. \tag{3.35}$$

(ii) By eqn (3.35),

$$(\mathbf{X}^{-1}\mathbf{A}\mathbf{X})^2 = (\mathbf{X}^{-1}\mathbf{A}\mathbf{X})(\mathbf{X}^{-1}\mathbf{A}\mathbf{X}) = \mathbf{X}^{-1}\mathbf{A}^2\mathbf{X} = \mathbf{D}^2$$
$$= \mathrm{diag}(\lambda_1^2, \lambda_2^2, \ldots, \lambda_N^2).$$

Similarly, $\mathbf{X}^{-1}\mathbf{A}^m\mathbf{X} = \mathbf{D}^m = \mathrm{diag}(\lambda_1^m, \lambda_2^m, \ldots, \lambda_N^m)$.

(iii) By definition of $\exp \mathbf{A}$,

$$\mathbf{X}^{-1}(\exp \mathbf{A})\mathbf{X} = \mathbf{X}^{-1}\left[\mathbf{I}_N + \mathbf{A} + \frac{1}{2!}\mathbf{A}^2 + \ldots\right]\mathbf{X}$$
$$= \mathbf{I}_N + \mathbf{D} + \frac{1}{2!}\mathbf{D}^2 + \ldots$$

i.e.

$$\mathbf{X}^{-1}(\exp \mathbf{A})\mathbf{X} = \exp \mathbf{D}.$$

(iv) Using 2×2 matrices for simplicity, the definition of

$\exp(t\mathbf{D})$, t a scalar, gives

$$\exp(t\mathbf{D}) = \mathbf{I}_2 + t\mathbf{D} + \frac{1}{2!}(t\mathbf{D})^2 + \ldots + \frac{1}{m!}(t\mathbf{D})^m + \ldots$$

$$= \begin{bmatrix} 1 & 0 \\ 0 & 1 \end{bmatrix} + \begin{bmatrix} t\lambda_1 & 0 \\ 0 & t\lambda_2 \end{bmatrix} + \ldots + \frac{1}{m!}\begin{bmatrix} t^m\lambda_1^m & 0 \\ 0 & t^m\lambda_2^m \end{bmatrix} + \ldots$$

$$= \begin{bmatrix} \left(1 + t\lambda_1 + \cdots + \frac{1}{m}t^m\lambda_1^m + \cdots\right) & \\ & \left(1 + t\lambda_2 + \cdots + \frac{1}{m!}t^m\lambda_2^m + \cdots\right) \end{bmatrix},$$

i.e.

$$\exp(t\mathbf{D}) = \begin{bmatrix} \exp(t\lambda_1) & 0 \\ 0 & \exp(t\lambda_2) \end{bmatrix}. \tag{3.36}$$

The eigenvalue–eigenvector solution of $d\mathbf{V}/dt = \mathbf{A}\mathbf{V}$

As shown earlier, the solution of

$$\frac{d\mathbf{V}}{dt} = \mathbf{A}\mathbf{V},$$

where $\mathbf{A}$ is independent of t, is

$$\mathbf{V}(t) = \{\exp(t\mathbf{A})\}\mathbf{V}(0). \tag{3.37}$$

Put

$$\mathbf{V}(t) = \mathbf{X}\mathbf{Y}(t), \tag{3.38}$$

where $\mathbf{X}$ is the modal matrix of $\mathbf{A}$, of order $(N-1)$, say. Then, by eqns (3.37) and (3.38),

$$\mathbf{X}\mathbf{Y}(t) = \{\exp(t\mathbf{A})\}\mathbf{V}(0) = \{\exp(t\mathbf{A})\}\mathbf{X}\mathbf{Y}(0).$$

Hence by (iii) and (iv),

$$\mathbf{Y}(t) = \mathbf{X}^{-1}\{\exp(t\mathbf{A})\}\mathbf{X}\mathbf{Y}(0)$$

$$= \{\exp(t\mathbf{D})\}\mathbf{Y}(0)$$

$$= \begin{bmatrix} \exp(t\lambda_1) & & & \\ & \exp(t\lambda_2) & & \\ & & \ddots & \\ & & & \exp(t\lambda_{N-1}) \end{bmatrix}\mathbf{Y}(0),$$

from which it follows that the sth component $Y_s(t)$ of $\mathbf{Y}(t)$ is

$$Y_s(t) = \{\exp(t\lambda_s)\} Y_s(0).$$

Therefore eqn (3.38) gives the solution

$$\begin{bmatrix} V_1(t) \\ V_2(t) \\ \vdots \\ V_{N-1}(t) \end{bmatrix} = \mathbf{X} \begin{bmatrix} Y_1(0)\exp(t\lambda_1) \\ Y_2(0)\exp(t\lambda_2) \\ \vdots \\ Y_{N-1}(0)\exp(t\lambda_{N-1}) \end{bmatrix}, \tag{3.39}$$

where $\mathbf{XY}(0) = \mathbf{V}(0)$. For known eigenvectors the latter could be solved for the components of $\mathbf{Y}(0)$ by Gaussian elimination.

Solution (3.39) will also hold for problems with derivative boundary-conditions at $x = 0$ and L except that $\mathbf{V}(t) = [V_0(t), V_1(t), \ldots, V_N(t)]^T$, $Nh = L$, and matrix $\mathbf{A}$ will be of order $(N+1)$.

Equation (3.39) can be used to give an approximate solution for sufficiently large positive values of t when the eigenvalues are negative, and I am indebted to D. Drew, Brunel University, for the following example.

Consider Worked Example 2.2, for which $h = 0.1$ and $\lambda_s = -(4/h^2)\sin^2 s\pi/20$, so that

$$\lambda_1 = -9.7887, \lambda_2 = -38.197, \ldots, \lambda_9 = -390.2.$$

Then

$$\begin{bmatrix} V_1(t) \\ V_2(t) \\ \vdots \\ V_9(t) \end{bmatrix} = \mathbf{X} \begin{bmatrix} Y_1(0)\exp(-9.8t) \\ Y_2(0)\exp(-38.2t) \\ \vdots \end{bmatrix} \simeq \mathbf{X} \begin{bmatrix} Y_1(0)\exp(-9.8t) \\ 0 \\ 0 \\ \vdots \\ 0 \end{bmatrix}, \tag{3.40}$$

for $t > 0.1$, to about 3 significant place accuracy.

As $\mathbf{V}(0) = \mathbf{XY}(0)$, $\mathbf{Y}(0) = \mathbf{X}^{-1}\mathbf{V}(0)$.

But matrix $\mathbf{A}$ is symmetric, so $\mathbf{X}^T\mathbf{X} = \mathbf{I}$, assuming each eigenvector has been normalized to 1, i.e. divided by the square root of the sum of the squares of its components. Therefore,

$$\mathbf{Y}(0) = \mathbf{X}^T\mathbf{V}(0). \tag{3.41}$$

If $\mathbf{x}_1^T = [x_{11}, x_{21}, \ldots, x_{91}]$, eqn (3.41) gives

$$Y_1(0) = [x_{11}, x_{21}, \ldots, x_{91}]\mathbf{V}(0)$$
$$= \sum_{i=1}^{9} x_{i1} V_i(0).$$

Hence by (3.40),

$$V_i(t) \simeq x_{i1} Y_1(0)\exp(-9.7887t)$$
$$= x_{i1}[\exp(-9.7887t)] \sum_{i=1}^{9} x_{i1} V_i(0). \qquad (3.42)$$

In Worked Example 2.2, which is symmetric about $i = 5$, $V_i(0) = 0.2i$, $i = 1(1)5$, and the normalized component x_{i1} of $\mathbf{x}_1$ is $x_{i1} = (\sin i\pi/10)/\sqrt{5}$, $i = 1(1)9$. These values give that $\sum_1^9 x_{i1} V_i(0) = 4.0863/\sqrt{5}$. For $t = 0.1$,

$$V_i(0.1) \simeq \frac{1}{5}\left(\sin\frac{i\pi}{10}\right)[\exp(-0.9789)](4.0863)$$
$$= 0.3071 \sin i\pi/10.$$

Table 3.3 shows that these approximate values differ from the Crank–Nicolson solution for $r = 1$ only in the fourth decimal place.

If matrix $\mathbf{A}$ is non-symmetric but satisfies the conditions of Exercise 16, Chapter 2, the same method can be used by transforming $\mathbf{A}$ to the symmetric matrix $\mathbf{DAD}^{-1}$, where $\mathbf{D}$ is the diagonal matrix of Exercise 16.

TABLE 3.3

x	0.1	0.2	0.3	0.4	0.5
Approximate V_i	0.0949	0.1805	0.2484	0.2920	0.3071
$C-N$ solution	0.0948	0.1803	0.2482	0.2918	0.3069
Solution of P.D.E.	0.0934	0.1776	0.2444	0.2873	0.3021

Miscellaneous methods for improving accuracy

(i) *Reduction of the local truncation error* (Douglas equations)

All derivatives can be expressed exactly in terms of infinite series of forward, backward, or central-differences. For example

$$\frac{\partial^2 U}{\partial x^2} = \frac{1}{h^2}(\delta_x^2 U - \tfrac{1}{12}\delta_x^4 U + \tfrac{1}{90}\delta_x^6 U + \ldots) \tag{3.43}$$

where the subscript x denotes differencing in the x-direction and the central-differences are defined by

$$\delta_x U_{i,j} = U_{i+\frac{1}{2},j} - U_{i-\frac{1}{2},j}$$

and

$$\delta_x^2 U_{i,j} = \delta_x(\delta_x U_{i,j}) = U_{i+1,j} - 2U_{i,j} + U_{i-1,j}, \text{ etc.} \tag{3.44}$$

In the approximation methods already considered the right-hand side of (3.43) has been truncated after the first term. If it is truncated after two or more terms the accuracy of the approximation method will always be improved but this normally increases the number of unknowns in an implicit method and complicates the boundary procedure. For equations involving second-order derivatives however it is possible to eliminate the fourth-order central-differences yet leave the number of unknowns unchanged. For example, if the equation

$$\frac{\partial U}{\partial t} = \frac{\partial^2 U}{\partial x^2}$$

is approximated at the point $(i, j+\frac{1}{2})$ by

$$\frac{1}{k}(u_{i,j+1} - u_{i,j}) = \frac{1}{2}\left\{\left(\frac{\partial^2 u}{\partial x^2}\right)_{i,j+1} + \left(\frac{\partial^2 u}{\partial x^2}\right)_{i,j}\right\}$$

$$= \frac{1}{2h^2}(\delta_x^2 - \tfrac{1}{12}\delta_x^4 + \tfrac{1}{90}\delta_x^6 + \ldots)(u_{i,j+1} + u_{i,j}),$$

then the terms involving δ_x^4 can be eliminated by operating on both sides with $(1 + \tfrac{1}{12}\delta_x^2)$. This gives that

$$(1 + \tfrac{1}{12}\delta_x^2)(u_{i,j+1} - u_{i,j}) = \tfrac{1}{2}r(\delta_x^2 u_{i,j+1} + \delta_x^2 u_{i,j}) + O(\delta_x^6)$$

which can be written as

$$\{1 + (\tfrac{1}{12} - \tfrac{1}{2}r)\delta_x^2\}u_{i,j+1} = \{1 + (\tfrac{1}{12} + \tfrac{1}{2}r)\delta_x^2\}u_{i,j}$$

when terms of order δ_x^6 are neglected. By (3.44) it follows that the differential equation at the point $(i, j + \frac{1}{2})$ can be approximated by the implicit algebraic equation

$$(1 - 6r)u_{i-1,j+1} + (10 + 12r)u_{i,j+1} + (1 - 6r)u_{i+1,j+1}$$
$$= (1 + 6r)u_{i-1,j} + (10 - 12r)u_{i,j} + (1 + 6r)u_{i+1,j}, \quad (3.45)$$

where $r = k/h^2$. The resulting tridiagonal system of equations can be solved by the algorithm on p. 24 and requires exactly the same amount of arithmetic as the Crank–Nicolson method. Whereas, however, the local truncation error of the Crank–Nicolson equation is $O(h^2) + O(k^2)$, it is $O(h^4) + O(k^2)$ for the Douglas equation. As proved in Chapter 3, Exercise 6, the equations are stable and consistent for all positive r. The numerical solution of Example 2.1 by eqns (3.45) for $h = 0.1$ and $r = 1$ at $t = 0.1$ is compared with the Crank–Nicolson solution for $r = 1$ in Table 3.4.

An explicit difference equation of h^4 accuracy is developed in Exercise 7.

(ii) Use of three time-level difference equations

The finite-difference approximation of a parabolic equation needs only two time-levels. Three (or more) time-level schemes can be constructed but naturally this is done only to achieve some advantage over two-level schemes, such as a smaller local truncation error, greater stability, or the transformation of a non-linear problem to a linear one as is demonstrated further on in this chapter. For example, the three-level difference equation

$$\frac{3}{2}\frac{(u_{i,j+1} - u_{i,j})}{k} - \frac{1}{2}\frac{(u_{i,j} - u_{i,j-1})}{k} = \frac{(u_{i+1,j+1} - 2u_{i,j+1} + u_{i-1,j+1})}{h^2}$$

TABLE 3.4

$x =$	0.1	0.2	0.3	0.4	0.5
Solution of P.D.E.	0.0934	0.1776	0.2444	0.2873	0.3021
Douglas solution	0.0941	0.1789	0.2463	0.2895	0.3044
C–N solution	0.0948	0.1803	0.2482	0.2918	0.3069

approximating $\partial U/\partial t = \partial^2 U/\partial x^2$ has a truncation error of the same order as the Crank–Nicolson equation, namely $O(k^2) + O(h^2)$, but is the better one to use when the initial data is discontinuous or varies very rapidly with x. The Crank–Nicolson approximation should be used when the initial data and its derivatives are continuous (reference 25). In order to solve the first set of equations for $u_{i,2}$ it is necessary to calculate a solution along the first time-level by some other method, it being assumed that the initial data along $t = 0$ are known. This first time-level solution must be of the same accuracy as that given by the three-levels equation. A three-level variation of the Douglas equation is

$$\frac{1}{12k}\{\tfrac{3}{2}(u_{i+1,j+1}-u_{i+1,j})-\tfrac{1}{2}(u_{i+1,j}-u_{i+1,j-1})\}$$

$$+\frac{5}{6k}\{\tfrac{3}{2}(u_{i,j+1}-u_{i,j})-\tfrac{1}{2}(u_{i,j}-u_{i,j-1})\}$$

$$+\frac{1}{12k}\{\tfrac{3}{2}(u_{i-1,j+1}-u_{i-1,j})-\tfrac{1}{2}(u_{i-1,j}-u_{i-1,j-1})\}$$

$$=\frac{1}{h^2}(u_{i+1,j+1}-2u_{i,j+1}+u_{i-1,j+1}),$$

and like the Douglas equation its truncation error is $O(k^2) + O(h^4)$. A number of such schemes for constant and variable coefficients and for one and two space dimensions are discussed in references 18 and 25.

(iii) *Deferred correction method*

In this method the approximating difference equations are solved as usual. Their solution is then used to calculate a correction term, at each mesh point of the solution domain, which is added to the approximating difference equation at each mesh point. The corrected equations are then re-solved and the process repeated if necessary. The correction terms are numbers obtained by differencing the numerical solution in either the x-direction or the t-direction, or both directions. One method for deriving a correction term for the Crank–Nicolson equations is given in Chapter 3, Exercise 9, but the following is better as it is based on a general result. Define the averaging operator μ by $\mu f_{j+\frac{1}{2}} = \tfrac{1}{2}(f_j + f_{j+1})$ and use the following results which are proved in most introductory books to numerical analysis, namely,

$$k\frac{\partial}{\partial t} \equiv 2\sinh^{-1}(\tfrac{1}{2}\delta_t) \quad \text{and} \quad \mu_t \equiv (1 + \tfrac{1}{4}\delta_t^2)^{\frac{1}{2}}.$$

From these it follows that

$$\tfrac{1}{2}k\left\{\left(\frac{\partial U}{\partial t}\right)_{i,j} + \left(\frac{\partial U}{\partial t}\right)_{i,j+1}\right\} = k\mu_t\left(\frac{\partial U}{\partial t}\right)_{i,j+\frac{1}{2}}$$

$$= \{(1 + \tfrac{1}{4}\delta_t^2)^{\frac{1}{2}} 2\sinh^{-1}(\tfrac{1}{2}\delta_t)\}U_{i,j+\frac{1}{2}}.$$

The expansion of the right-hand side into positive powers of δ_t leads to

$$\tfrac{1}{2}k\left\{\left(\frac{\partial U}{\partial t}\right)_{i,j} + \left(\frac{\partial U}{\partial t}\right)_{i,j+1}\right\} = (\delta_t + \tfrac{1}{12}\delta_t^3 - \tfrac{1}{120}\delta_t^5 + \ldots)U_{i,j+\frac{1}{2}}$$

which can be rearranged as

$$\delta_t U_{i,j+\frac{1}{2}} = \tfrac{1}{2}k\left\{\left(\frac{\partial U}{\partial t}\right)_{i,j} + \left(\frac{\partial U}{\partial t}\right)_{i,j+1}\right\} + C_t U_{i,j+\frac{1}{2}},$$

giving

$$U_{i,j+1} - U_{i,j} = \tfrac{1}{2}k\frac{\partial}{\partial t}(U_{i,j} + U_{i,j+1}) + C_t U_{i,j+\frac{1}{2}}, \qquad (3.46)$$

where

$$C_t \equiv -\tfrac{1}{12}\delta_t^3 + \tfrac{1}{120}\delta_t^5 + \ldots.$$

Equation (3.46) is a general result relating the value of a continuous function at the $(j+1)$th time-level to its value at the jth time-level in terms of first time-derivatives and central-differences in the t-direction. For the equation

$$\frac{\partial U}{\partial t} = \frac{\partial^2 U}{\partial x^2}$$

it follows that

$$\frac{\partial}{\partial t} \equiv \frac{\partial^2}{\partial x^2}.$$

Hence by (3.46)

$$U_{i,j+1} - U_{i,j} = \tfrac{1}{2}k\frac{\partial^2}{\partial x^2}(U_{i,j} + U_{i,j+1}) + C_t U_{i,j+\frac{1}{2}}.$$

Using eqn (3.43) it is seen that

$$U_{i,j+1} - U_{i,j} = \frac{1}{2}\frac{k}{h^2}(\delta_x^2 - \tfrac{1}{12}\delta_x^4 + \tfrac{1}{90}\delta_x^6 + \ldots)(U_{i,j+1} + U_{i,j}) + C_t U_{i,j+\frac{1}{2}}$$

$$= \tfrac{1}{2}r(\delta_x^2 U_{i,j+1} + \delta_x^2 U_{i,j}) + C, \qquad (3.47)$$

where

$$C = \tfrac{1}{2}r\{(-\tfrac{1}{12}\delta_x^4 U_{i,j+1} + \tfrac{1}{90}\delta_x^6 U_{i,j+1} + \ldots)$$
$$+ (-\tfrac{1}{12}\delta_x^4 U_{i,j} + \tfrac{1}{90}\delta_x^6 U_{i,j} + \ldots)\} + (-\tfrac{1}{12}\delta_t^3 U_{i,j+\frac{1}{2}} + \tfrac{1}{120}\delta_t^5 U_{i,j+\frac{1}{2}}).$$

The first approximation to the solution of (3.47) would be found by putting $C = 0$ and solving the Crank–Nicolson equations

$$u_{i,j+1} - u_{i,j} = \tfrac{1}{2}r(\delta_x^2 u_{i,j+1} + \delta_x^2 u_{i,j})$$

over the solution domain $[0 < x < 1] \times [0 < t \leq T]$, say. The correction term at each mesh point would then be calculated from a truncated approximation to C such as

$$C' = -\frac{r}{24}(\delta_x^4 u_{i,j+1} + \delta_x^4 u_{i,j}) - \tfrac{1}{12}\delta_t^3 u_{i,j+\frac{1}{2}}.$$

This method, in effect, includes higher-order difference terms in the approximations to the derivatives but keeps the matrix of coefficients of the approximation equations tridiagonal which allows the algorithm on p. 24 to be used.

(iv) *Richardson's deferred approach to the limit*

For this method two or more solutions approximating the problem must be known for two or more different mesh sizes and the difference between the solution of the partial differential equation and the solution of the approximating equations must be known as a function of the mesh lengths.

Let U represent the solution of the differential equation and $u_{r,s}$ represent the solution of the finite-difference equations for a mesh of size (h_r, k_s). Now assume, for example, that the discretization error

$$U - u(h, k) = Ak + Bh^2 + Ck^2 + Dh^4 + \ldots,$$

as it is for the classical explicit method for finite t.

If two solutions $u_{1,1}$ and $u_{2,2}$ are known then

$$U - u_{1,1} = Ak_1 + Bh_1^2 + Ck_1^2 + Dh_1^4 + \ldots$$

and

$$U - u_{2,2} = Ak_2 + Bh_2^2 + Ck_2^2 + Dh_2^4 + \ldots.$$

Hence B (or A, but not both) can be eliminated to give that

$$U = \frac{1}{h_2^2 - h_1^2}(h_2^2 u_{1,1} - h_1^2 u_{2,2}) + A\frac{k_1 h_2^2 - k_2 h_1^2}{h_2^2 - h_1^2} + \ldots.$$

If the term involving A is negligible then U is an improvement on $u_{1,1}$ and $u_{2,2}$. For the special case $h_2 = 2h_1$, $k_2 = 2k_1$,

$$U = \tfrac{1}{3}(4u_{1,1} - u_{2,2}) + \tfrac{2}{3}kA + \ldots.$$

If three different solutions are known then A and B can be eliminated. For the Crank–Nicolson equations,

$$U - u(h, k) = Ah^2 + Bk^2 + Ch^4 + \ldots.$$

If $h_2 = \lambda h_1$ and $k_2 = \lambda k_1$ it is easily shown that

$$U = u_{2,2} + \frac{\lambda^2}{1 - \lambda^2}(u_{2,2} - u_{1,1}) + O(h^4).$$

Solution of non-linear parabolic equations

There is no difficulty in formally applying finite-difference methods to non-linear parabolic equations. The difficulties are associated with the difference equations themselves. If they are linear they can usually be solved quite easily, although we still have the problem of determining the conditions that must be satisfied for stability and convergence because the coefficients of the unknowns will be functions of the solution at earlier time-levels. If they are non-linear we have also the problem of their solution. Direct methods, in general, are difficult, so they are usually solved iteratively after being linearized in some way. Taylor's expansion provides a standard way of doing this and the method is usually referred to as Newton's method.

Linearization by Newton's method

Let

$$f_i(u_1, u_2, \ldots, u_N) = 0, \quad i = 1(1)N, \tag{3.48}$$

represent N equations in the N dependent variables $u_1, u_2, \ldots, u_N$.

Let V_i be a known approximation to the exact solution value u_i, $i = 1(1)N$.

Put $u_i = V_i + \varepsilon_i$ and substitute into eqn (3.48). Then by Taylor's expansion to first-order terms in ε_i, $i = 1(1)N$,

$$f_i(V_1, V_2, \ldots, V_N) + \left[\frac{\partial f_i}{\partial u_1} \varepsilon_1 + \frac{\partial f_i}{\partial u_2} \varepsilon_2 + \ldots + \frac{\partial f_i}{\partial u_N} \varepsilon_N \right]_{u_i = V_i} = 0,$$

$$i = 1(1)N. \quad (3.49)$$

The subscript notation on the second bracket indicates that the dependent variables $u_1, u_2, \ldots, u_N$ appearing in the coefficients of $\varepsilon_1, \varepsilon_2, \ldots, \varepsilon_N$ are replaced by $V_1, V_2, \ldots, V_N$ respectively after the differentiations. Equation (3.49) represents N linear equations for the N unknowns $\varepsilon_1, \varepsilon_2, \ldots, \varepsilon_N$ because V_1, V_2, V_N are known. When the ε's have been calculated the process is repeated, the starting values of the dependent variables for the next iteration being $(V_i + \varepsilon_i)$, $i = 1(1)N$. This process of successive approximations is continued until the u_i's have been found to the required degree of accuracy, such as $|\varepsilon_i| < 10^{-8}$, $i = 1(1)N$. Some numerical results for a particular problem are given on p. 148.

Example 3.1

The function U satisfies the non-linear equation

$$\frac{\partial U}{\partial t} = \frac{\partial^2 U^2}{\partial x^2}, \quad 0 < x < 1, \quad t > 0,$$

the initial condition $U = 4x(1-x)$, $0 < x < 1$, $t = 0$, and the boundary conditions $U = 0$ at $x = 0$ and 1, $t \geq 0$.

If the equation is approximated at the point $\{ih, (j+\frac{1}{2})k\}$ in the x–t plane by the difference scheme

$$\frac{1}{k} \delta_t u_{i,j+\frac{1}{2}} = \frac{1}{2h^2} (\delta_x^2 u_{i,j+1}^2 + \delta_x^2 u_{i,j}^2),$$

use Newton's method to derive a set of linear equations giving an improved value $(V_i + \varepsilon_i)$ to the approximate value V_i at the mesh points defined by $x_i = \frac{1}{6}i$, $i = 1(1)5$, $t = k = \frac{1}{36}$.

If the V_i are taken equal to the initial values at $x_i = \frac{1}{6}i$,

$i = 1(1)5$, $t = 0$, show that these equations reduce to

$$-19\varepsilon_1 + 8\varepsilon_2 + \tfrac{14}{9} = 0,$$

$$5\varepsilon_1 - 25\varepsilon_2 + 9\varepsilon_3 - \tfrac{22}{9} = 0,$$

and

$$16\varepsilon_2 - 27\varepsilon_3 - \tfrac{34}{9} = 0.$$

The approximation equation is

$$\frac{1}{k}(u_{i,j+1} - u_{i,j}) = \frac{1}{2h^2}\{(u_{i-1,j+1}^2 - 2u_{i,j+1}^2 + u_{i+1,j+1}^2)$$

$$+ (u_{i-1,j}^2 - 2u_{i,j}^2 + u_{i+1,j}^2)\}.$$

Put $p = h^2/k$ and denote $u_{i,j+1}$ by u_i. The equation can then be written as

$$u_{i-1}^2 - 2(u_i^2 + pu_i) + u_{i+1}^2 + \{u_{i-1,j}^2 - 2(u_{i,j}^2 - pu_{i,j}) + u_{i+1,j}^2\}$$

$$= 0 \equiv f_i(u_{i-1}, u_i, u_{i+1}).$$

By eqn (3.49),

$$\left[\frac{\partial f_i}{\partial u_{i-1}}\varepsilon_{i-1} + \frac{\partial f_i}{\partial u_i}\varepsilon_i + \frac{\partial f_i}{\partial u_{i+1}}\varepsilon_{i+1}\right]_{u_i = V_i} + f_i(V_{i-1}, V_i, V_{i+1}) = 0,$$

hence

$$2V_{i-1}\varepsilon_{i-1} - 2(2V_i + p)\varepsilon_i + 2V_{i+1}\varepsilon_{i+1} + [\{V_{i-1}^2 - 2(V_i^2 + pV_i) + V_{i+1}^2\}$$

$$+ \{u_{i-1,j}^2 - 2(u_{i,j}^2 - pu_{i,j}) + u_{i+1,j}^2\}] = 0, \quad (3.50)$$

where V_i is an approximation to $u_{i,j+1}$.
The problem is symmetric with respect to $x = \tfrac{1}{2}$. When the V_i are taken equal to $u_{i,0}$, eqn (3.50) for $j = 0$ reduces to

$$2u_{i-1,0}\varepsilon_{i-1} - 2(2u_{i,0} + p)\varepsilon_i + 2u_{i+1,0}\varepsilon_{i+1}$$

$$+ \{2u_{i-1,0}^2 - 4u_{i,0}^2 + 2u_{i+1,0}^2\} = 0.$$

This gives the equations quoted for $p = 1$ and the initial values indicated in Fig. 3.5 since $\varepsilon_0 = 0$.

Richtmyer's linearization method

Richtmyer, reference 25, considers the equation

$$\frac{\partial U}{\partial t} = \frac{\partial^2 U^m}{\partial x^2}, \quad m \text{ a positive integer} \geqslant 2,$$

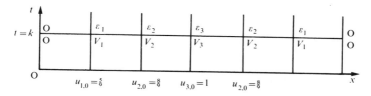

Fig. 3.5

which he approximates by the implicit weighted average difference scheme

$$\frac{1}{k}(u_{i,j+1} - u_{i,j}) = \frac{1}{h^2}[\theta \, \delta_x^2(u_{i,j+1}^m) + (1 - \theta) \, \delta_x^2(u_{i,j}^m)]. \quad (3.51)$$

By Taylor's expansion about the point (i, j),

$$u_{i,j+1}^m = u_{i,j}^m + k \frac{\partial u_{i,j}^m}{\partial t} + \dots$$

$$= u_{i,j}^m + k \frac{\partial u_{i,j}^m}{\partial u_{i,j}} \frac{\partial u_{i,j}}{\partial t} + \dots$$

Hence to terms of order k,

$$u_{i,j+1}^m = u_{i,j}^m + m u_{i,j}^{m-1}(u_{i,j+1} - u_{i,j}),$$

a result which replaces the non-linear unknown $u_{i,j+1}^m$ by an
approximation linear in $u_{i,j+1}$.

Putting $\omega_i = u_{i,j+1} - u_{i,j}$ in eqn (3.51) leads to

$$\frac{1}{k}\omega_i = \frac{1}{h^2}[\theta \, \delta_x^2(u_{i,j}^m + m u_{i,j}^{m-1}\omega_i) + (1 - \theta) \, \delta_x^2 u_{i,j}^m]$$

$$= \frac{1}{h^2}[m\theta \, \delta_x^2 u_{i,j}^{m-1}\omega_i + \delta_x^2 u_{i,j}^m]$$

$$= \frac{1}{h^2}[m\theta(u_{i-1,j}^{m-1}\omega_{i-1} - 2u_{i,j}^{m-1}\omega_i + u_{i+1,j}^{m-1}\omega_{i+1})$$

$$+ (u_{i-1,j}^m - 2u_{i,j}^m + u_{i+1,j}^m)],$$

which gives a set of linear equations for the ω_i. The solution at
the $(j+1)$th time-level is obtained from $u_{i,j+1} = \omega_i + u_{i,j}$. For
known boundary values at $x = 0$ and 1, where $Nh = 1$ and $r =$

k/h^2, the equations in matrix form for $m = 2$ are

$$
\begin{bmatrix}
(1+4r\theta u_{1,j}) & -2r\theta u_{2,j} & & & \\
-2r\theta u_{1,j} & (1+4r\theta u_{2,j}) & -2r\theta u_{3,j} & & \\
& & & \ddots & \\
& & & -2r\theta u_{N-3,j}(1+4r\theta u_{N-2,j})-2r\theta u_{N-1,j} \\
& & & -2r\theta u_{N-2,j}(1+4r\theta u_{N-1,j})
\end{bmatrix}
\begin{bmatrix}
\omega_1 \\ \omega_2 \\ \\ \omega_{N-2} \\ \omega_{N-1}
\end{bmatrix}
$$

$$
=
\begin{bmatrix}
-2ru_{1,j} & ru_{2,j} & & \\
ru_{1,j} & -2ru_{2,j} & ru_{3,j} & \\
& & \ddots & \\
& & & ru_{N-2,j}-2ru_{N-1,j}
\end{bmatrix}
\begin{bmatrix}
u_{1,j} \\ u_{2,j} \\ \\ u_{N-1,j}
\end{bmatrix}
$$

$$
+
\begin{bmatrix}
ru_{0,j}^2 + 2r\theta u_{0,j}(u_{0,j+1}-u_{0,j}) \\
0 \\
0 \\
\vdots \\
0 \\
ru_{N,j}^2 + 2r\theta u_{N,j}(u_{N,j+1}-u_{N,j})
\end{bmatrix}
\qquad (3.52)
$$

These are easily solved by the algorithm on p. 24. Numerical results for a particular problem are given on p. 148.

A *three time-level method*

Lees, reference 16, considered the non-linear equation

$$
b(U)\frac{\partial U}{\partial t} = \frac{\partial}{\partial x}\left\{a(U)\frac{\partial U}{\partial x}\right\}, \quad a(U) > 0, \quad b(U) > 0, \quad (3.53)
$$

and investigated a difference scheme that
(i) achieved linearity in the unknowns $u_{i,j+1}$ by evaluating all coefficients of $u_{i,j+1}$ at a time-level of known solution values,
(ii) preserved stability by averaging $u_{i,j}$ over three time-levels, and
(iii) maintained accuracy by using central-difference approximations.
As

$$
\left(\frac{\partial U}{\partial x}\right)_{i,j} \simeq \frac{1}{h}(U_{i+\frac{1}{2},j} - U_{i-\frac{1}{2},j}) = \frac{1}{h}\delta_x U_{i,j},
$$

an obvious central-difference approximation to (3.53) is

$$b(u_{i,j})\frac{1}{2k}(u_{i,j+1}-u_{i,j-1})=\frac{1}{h}\delta_x\left\{a(u_{i,j})\frac{1}{h}\delta_x u_{i,j}\right\}$$

$$=\frac{1}{h^2}\delta_x\{a(u_{i,j})(u_{i+\frac{1}{2},j}-u_{i-\frac{1}{2},j})\}$$

$$=\frac{1}{h^2}\{a(u_{i+\frac{1}{2},j})(u_{i+1,j}-u_{i,j})-a(u_{i-\frac{1}{2},j})(u_{i,j}-u_{i-1,j})\},$$

but this is certainly unstable for $a=b=1$ (reference Chapter 2, Exercise 26(b)). If, however, $u_{i+1,j}$, $u_{i,j}$, and $u_{i-1,j}$ are replaced by $\frac{1}{3}(u_{i+1,j+1}+u_{i+1,j}+u_{i+1,j-1})$, $\frac{1}{3}(u_{i,j+1}+u_{i,j}+u_{i,j-1})$, and $\frac{1}{3}(u_{i-1,j+1}+u_{i-1,j}+u_{i-1,j-1})$ respectively, and the coefficients $a(u_{i+\frac{1}{2},j})$ and $a(u_{i-\frac{1}{2},j})$ are replaced by $a\{\frac{1}{2}(u_{i+1,j}+u_{i,j})\}$ and $a\{\frac{1}{2}(u_{i,j}+u_{i-1,j})\}$ respectively in order to avoid midpoint values of u, Lees proved that for sufficiently small values of h and k,

$$\max_{i,j}|U_{i,j}-u_{i,j}|\leq A(h^2+k^2),$$

where A is a constant.

For this method the equation considered in the preceding section needs to be written as

$$\frac{\partial U}{\partial t}=\frac{\partial^2 U^m}{\partial x^2}=\frac{\partial}{\partial x}\left(mU^{m-1}\frac{\partial U}{\partial x}\right).$$

A comparison of results for a particular problem

If it is assumed that $U=U(x-vt)$, v constant, is a solution of the equation

$$\frac{\partial U}{\partial t}=\frac{\partial^2 U^2}{\partial x^2},\quad 0<x<1,\quad t>0, \tag{3.54}$$

substitution into (3.54) and integration with respect to $(x-vt)$ gives that

$$\frac{A}{v}\log\left(U-\frac{A}{v}\right)+U=B-\tfrac{1}{2}v(x-vt),$$

where A and B are constants. Choosing $A=1$ and $v=2$, then $B=U(0,0)=1.5$ leads to the particular solution

$$(2U-3)+\log(U-\tfrac{1}{2})=2(2t-x). \tag{3.55}$$

For any given x and t eqn (3.55) can be solved iteratively for U by, for example, the Newton–Raphson method. In a similar way initial values for this particular solution can be found by putting $t = 0$ in (3.55), and $x = 0(0.1)1$, say. Boundary values at $x = 0$ and 1 for known values of t can be found in the same manner. These initial values and boundary values can then be used as the boundary data for approximation methods whose accuracy can be checked against the analytical solution of the differential equation.

The solution of eqns (3.51) for $\theta = \frac{1}{2}$, $m = 2$, $h = 0.1$, $r = k/h^2 = \frac{1}{2}$, and $t = 0.5$, which corresponds to 100 time-steps, by Newton's, Richtmyer's, and the three time-levels methods, are given below. All agree with the analytical solution of the differential equation to five decimal places after rounding, and with each other to $6D$.

x	Solution of the P.D.E.	Newton's, Richtmyer's, and the three time-levels methods
0.1	2.149703	2.149701
0.3	1.997951	1.997948
0.5	1.849962	1.849958
0.7	1.706244	1.706240
0.9	1.567391	1.567389

The stability of three of more time-level difference equations (Fixed mesh lengths)

The following theorem is useful for the matrix method of analysis of the stability of three or more time-level difference equations and is easier to use than one might at first think.

Theorem: If the matrix $\mathbf{A}$ can be written as

$$\mathbf{A} = \begin{bmatrix} \mathbf{A}_{1,1} & \mathbf{A}_{1,2} & \cdots & \mathbf{A}_{1,m} \\ \mathbf{A}_{2,1} & \mathbf{A}_{2,2} & \cdots & \mathbf{A}_{2,m} \\ \vdots & & & \vdots \\ \mathbf{A}_{m,1} & \mathbf{A}_{m,2} & \cdots & \mathbf{A}_{m,m} \end{bmatrix},$$

where each $\mathbf{A}_{i,j}$ is an $n \times n$ matrix, and all the $\mathbf{A}_{i,j}$ have a common set of n linearly independent eigenvectors, then the eigenvalues of $\mathbf{A}$ are given by the eigenvalues of the matrices

$$\begin{bmatrix} \lambda_{1,1}^{(k)} & \lambda_{1,2}^{(k)} & \cdots & \lambda_{1,m}^{(k)} \\ \lambda_{2,1}^{(k)} & \lambda_{2,2}^{(k)} & \cdots & \lambda_{2,m}^{(k)} \\ \vdots & & & \vdots \\ \lambda_{m,1}^{(k)} & \lambda_{m,2}^{(k)} & \cdots & \lambda_{m,m}^{(k)} \end{bmatrix}, \quad k = 1(1)n,$$

where $\lambda_{i,j}^{(k)}$ is the kth eigenvalue of $\mathbf{A}_{i,j}$ corresponding to the kth eigenvector $\mathbf{v}_k$ common to all the $\mathbf{A}_{i,j}$'s.

Proof

Let $\mathbf{v}_k$ be an eigenvector common to all the submatrices $\mathbf{A}_{i,j}$, $i, j = 1(1)m$, and denote the corresponding eigenvalues of $\mathbf{A}_{1,1}, \mathbf{A}_{2,1}, \ldots$ by $\lambda_{1,1}^{(k)}, \lambda_{2,1}^{(k)}, \ldots$ respectively. For simplicity consider $i, j = 1(1)2$ and denote $\mathbf{v}_k$ by $\mathbf{v}$, $\lambda_{i,j}^{(k)}$ by $\lambda_{i,j}$. Then

$$\mathbf{A}_{1,1}\mathbf{v} = \lambda_{1,1}\mathbf{v}, \quad \mathbf{A}_{1,2}\mathbf{v} = \lambda_{1,2}\mathbf{v},$$

$$\mathbf{A}_{2,1}\mathbf{v} = \lambda_{2,1}\mathbf{v}, \quad \mathbf{A}_{2,2}\mathbf{v} = \lambda_{2,2}\mathbf{v}.$$

Multiply these equations respectively by the non-zero constants α_1, α_2, α_1, and α_2 and write them as

$$\begin{bmatrix} \mathbf{A}_{1,1}\mathbf{A}_{1,2} \\ \mathbf{A}_{2,1}\mathbf{A}_{2,2} \end{bmatrix} \begin{bmatrix} \alpha_1\mathbf{v} \\ \alpha_2\mathbf{v} \end{bmatrix} = \begin{bmatrix} (\lambda_{1,1}\alpha_1 + \lambda_{1,2}\alpha_2)\mathbf{v} \\ (\lambda_{2,1}\alpha_1 + \lambda_{2,2}\alpha_2)\mathbf{v} \end{bmatrix}. \tag{3.56}$$

Assume now that

$$\mathbf{A} = \begin{bmatrix} \mathbf{A}_{1,1}\mathbf{A}_{1,2} \\ \mathbf{A}_{2,1}\mathbf{A}_{2,2} \end{bmatrix}$$

has an eigenvalue μ corresponding to the eigenvector

$$\begin{bmatrix} \alpha_1\mathbf{v} \\ \alpha_2\mathbf{v} \end{bmatrix}$$

so that

$$\begin{bmatrix} \mathbf{A}_{1,1}\mathbf{A}_{1,2} \\ \mathbf{A}_{2,1}\mathbf{A}_{2,2} \end{bmatrix} \begin{bmatrix} \alpha_1\mathbf{v} \\ \alpha_2\mathbf{v} \end{bmatrix} = \mu \begin{bmatrix} \alpha_1\mathbf{v} \\ \alpha_2\mathbf{v} \end{bmatrix}. \tag{3.57}$$

By the right-hand sides of eqns (3.56) and (3.57),

$$(\lambda_{1,1} - \mu)\alpha_1 + \lambda_{1,2}\alpha_2 = 0$$

and

$$\lambda_{2,1}\alpha_1 + (\lambda_{2,2} - \mu)\alpha_2 = 0.$$

These two equations will have a non-trivial solution for α_1 and α_2 if and only if

$$\det\begin{bmatrix}(\lambda_{1,1}-\mu) & \lambda_{1,2} \\ \lambda_{2,1} & (\lambda_{2,2}-\mu)\end{bmatrix}=0,$$

i.e. if and only if μ is an eigenvalue of the matrix

$$\begin{bmatrix}\lambda_{1,1} & \lambda_{1,2} \\ \lambda_{2,1} & \lambda_{2,2}\end{bmatrix}.$$

Matrices with common eigenvector systems

Proofs of the following theorems can be found in reference 30.

(i) If the $N\times N$ matrix $\mathbf{A}$ has N distinct eigenvalues λ_s it has N unique linearly independent eigenvectors $\mathbf{v}_s$, $s=1(1)N$. As proved earlier, any polynomial $f(\mathbf{A})$ of $\mathbf{A}$ has the same set of eigenvectors $\mathbf{v}_s$ and a corresponding set of eigenvalues $f(\lambda_s)$.

(ii) All $N\times N$ Hermitian matrices, which includes real symmetric matrices, have N linearly independent eigenvectors.

(iii) If the matrices $\mathbf{A}$ and $\mathbf{B}$ commute and have linear elementary divisors then they have a common system of eigenvectors. In particular, all matrices with distinct eigenvalues, all Hermitian, and therefore all real symmetric matrices, have linear elementary divisors.

(iv) Let $\mathbf{A}$ and $\mathbf{B}$ be matrices with a common system of eigenvectors. Let λ and μ be the eigenvalues of $\mathbf{A}$ and $\mathbf{B}$ respectively corresponding to the common eigenvector $\mathbf{v}$. Then $\mathbf{v}$ is an eigenvector of $\mathbf{AB}$ and $\mathbf{A}^{-1}\mathbf{B}$ and the corresponding eigenvalues are $\lambda\mu$ and $\lambda^{-1}\mu$ respectively. These results are easily proved. By hypothesis, $\mathbf{Bv}=\mu\mathbf{v}$ and $\mathbf{Av}=\lambda\mathbf{v}$. Therefore $\mathbf{ABv}=\mu\mathbf{Av}=\mu\lambda\mathbf{v}$. Also $\mathbf{A}^{-1}\mathbf{Bv}=\mu\mathbf{A}^{-1}\mathbf{v}=(\mu/\lambda)\mathbf{v}$.

Example 3.2

Investigate the stability of the Du Fort and Frankel approximation to the equation

$$\frac{\partial U}{\partial t}=\frac{\partial^2 U}{\partial x^2}, \quad 0<x<1, \quad t>0,$$

for fixed mesh lengths, given that U is known on the boundaries $x = 0$ and 1 for $t > 0$. (See Worked Example 2.7.)

The approximation at the point (ih, jk) is

$$\frac{1}{2k}(u_{i,j+1} - u_{i,j-1}) = \frac{1}{h^2}\{u_{i-1,j} - (u_{i,j-1} + u_{i,j+1}) + u_{i+1,j}\},$$

which may be written as

$$(1 + 2r)u_{i,j+1} = 2r(u_{i-1,j} + u_{i+1,j}) + (1 - 2r)u_{i,j-1},$$

where $r = k/h^2$. For known boundary-values and $Nh = 1$ these equations in matrix form are

$$(1 + 2r)\begin{bmatrix} u_{1,j+1} \\ u_{2,j+1} \\ u_{3,j+1} \\ \vdots \\ u_{N-1,j+1} \end{bmatrix} = 2r\begin{bmatrix} 0 & 1 & & & \\ 1 & 0 & 1 & & \\ & 1 & 0 & 1 & \\ & & & \ddots & \\ & & & 1 & 0 \end{bmatrix}\begin{bmatrix} u_{1,j} \\ u_{2,j} \\ u_{3,j} \\ \vdots \\ u_{N-1,j} \end{bmatrix}$$

$$+ (1 - 2r)\begin{bmatrix} u_{1,j-1} \\ u_{2,j-1} \\ u_{3,j-1} \\ \vdots \\ u_{N-1,j-1} \end{bmatrix} + 2r\begin{bmatrix} u_{0,j} \\ 0 \\ 0 \\ \vdots \\ u_{N,j} \end{bmatrix}$$

giving

$$\mathbf{u}_{j+1} = \frac{2r}{1+2r}\mathbf{A}\mathbf{u}_j + \frac{1-2r}{1+2r}\mathbf{u}_{j-1} + \mathbf{c}_j, \tag{3.58}$$

where $\mathbf{A}$ is as displayed and $\mathbf{c}_j$ is a vector of known values. Put

$$\mathbf{v}_j = \begin{bmatrix} \mathbf{u}_j \\ \mathbf{u}_{j-1} \end{bmatrix}.$$

Then eqn (3.58) and the identity $\mathbf{u}_j = \mathbf{u}_j$ can be written as

$$\begin{bmatrix} \mathbf{u}_{j+1} \\ \hline \mathbf{u}_j \end{bmatrix} = \begin{bmatrix} \dfrac{2r}{1+2r}\mathbf{A} & \dfrac{(1-2r)}{(1+2r)}\mathbf{I}_{N-1} \\ \hline \mathbf{I}_{N-1} & \mathbf{O} \end{bmatrix}\begin{bmatrix} \mathbf{u}_j \\ \hline \mathbf{u}_{j-1} \end{bmatrix} + \begin{bmatrix} \mathbf{c}_j \\ \hline \mathbf{O} \end{bmatrix},$$

where $\mathbf{I}_{N-1}$ is the unit matrix of order $(N-1)$, i.e. as

$$\mathbf{v}_{j+1} = \mathbf{P}\mathbf{v}_j + \mathbf{d}_j,$$

where **P** is the matrix shown and $\mathbf{d}_i$ a column vector of known constants. This technique has reduced a three-level difference equation to a two-level one. The equations will be stable when each eigenvalue of **P** has a modulus ≤ 1. The matrix **A** has $(N-1)$ different eigenvalues so it has $(N-1)$ linearly independent eigenvectors $\mathbf{v}_s$, $s = 1(1)(N-1)$. Although the matrix $\mathbf{I}_{N-1}$ has $(N-1)$ eigenvalues each equal to 1 it has $(N-1)$ linearly independent eigenvectors which may be taken as $\mathbf{v}_s$, $s = 1(1)(N-1)$, because the eigenvalue equation $\mathbf{Bx} = \lambda\mathbf{x}$ is clearly satisfied by $\mathbf{I}_{N-1}\mathbf{v}_s = 1 \cdot \mathbf{v}_s$. Hence the eigenvalues λ of P are the eigenvalues of the matrix

$$\begin{bmatrix} \dfrac{2r\lambda_k}{1+2r} & \dfrac{1-2r}{1+2r} \\ 1 & 0 \end{bmatrix},$$

where λ_k is the kth eigenvalue of **A**. For such a simple case we can work from first principles and find λ by evaluating

$$\det\begin{bmatrix} \left\{\dfrac{2r\lambda_k}{1+2r} - \lambda\right\} & \dfrac{1-2r}{1+2r} \\ 1 & -\lambda \end{bmatrix} = 0,$$

giving

$$\lambda^2 - \frac{2r\lambda_k}{1+2r}\lambda - \frac{1-2r}{1+2r} = 0.$$

By the formula on p. 59

$$\lambda_k = 2\cos(k\pi/N), \quad k = 1(1)(N-1).$$

Hence

$$\lambda = \left\{2r\cos\frac{k\pi}{N} \pm \left(1 - 4r^2\sin^2\frac{k\pi}{N}\right)^{\frac{1}{2}}\right\}\Big/(1+2r).$$

Case (i) $1 > 1 - 4r^2\sin^2\dfrac{k\pi}{N} \geq 0$.

Then

$$|\lambda| < \frac{2r+1}{1+2r} = 1.$$

Case (ii) $1 - 4r^2 \sin^2 \dfrac{k\pi}{N} < 0.$

Then

$$|\lambda|^2 = \frac{1}{(2r+1)^2}\left\{ \left(2r\cos\frac{k\pi}{N}\right)^2 + 4r^2\sin^2\frac{k\pi}{N} - 1 \right\}$$

$$= \frac{4r^2 - 1}{4r^2 + 4r + 1} < 1 \quad \text{since} \quad r > 0.$$

Therefore the equations, for a fixed mesh size, are unconditionally stable for all positive r. It is shown in reference 25 that these equations are also stable for all $r > 0$ as the mesh lengths tend to zero.

Brief introduction to the analytical solution of homogeneous finite-difference equations

Linear equations with constant coefficients

Consider the difference equation

$$u_{j+2} + au_{j+1} + bu_j = 0, \quad j = 0, 1, 2, \ldots, \tag{3.59}$$

where a and b are real constants.

Assume that

$$u_j = Am^j$$

is a solution, where A and m are non-zero constants. Substitution into (3.59) shows that m is a root of the quadratic equation

$$m^2 + am + b = 0. \tag{3.60}$$

Case (i) Roots real and distinct, $m = m_1$ and $m = m_2$, say. One solution is $u_j = Am_1^j$ and another is $u_j = Bm_2^j$ where A and B are arbitrary constants. As eqn (3.59) is linear in u its general solution is

$$u_j = Am_1^j + Bm_2^j.$$

Case (ii) Repeated roots, $m = m_1$ twice, say. Clearly one solution is $u_j = Am_1^j$.

Put $u_j = m_1^j f(j)$. Substitution into (3.59) and the use of $a = -2m_1$, $b = m_1^2$ leads to

$$f(j+2) - 2f(j+1) + f(j) = 0.$$

By inspection it is seen that $f(j) = j$ satisfies this equation. Therefore a second solution of (3.59) is $u_j = Bjm_1^j$. Hence the solution of eqn (3.59) in this case is

$$u_j = (A + Bj)m_1^j.$$

Case (iii) Complex roots.

Because a and b are real the roots of (3.60) will be conjugate complex numbers, $m_1 = re^{i\theta}$ and $m_2 = re^{-i\theta}$, say, where $i = \sqrt{(-1)}$.

Hence $a = -r(e^{i\theta} + e^{-i\theta}) = -2r\cos\theta$ and $b = r^2$. As in Case (i) the solution of (3.59) is

$$u_j = Ar^j e^{ij\theta} + Br^j e^{-ij\theta} = r^j\{(A+B)\cos j\theta + i(A-B)\sin j\theta\}.$$

Since A and B are arbitrary constants and $r = b^{\frac{1}{2}}$, this can be written as

$$u_j = b^{\frac{1}{2}j}(C\cos j\theta + D\sin j\theta),$$

where C and D are arbitrary constants and $\cos\theta = -a/2r = -a/2\sqrt{b}$. Methods for deriving particular integrals for non-homogeneous difference equations are given in *Finite Difference Equations* by H. Levy and F. Lessman (Pitman).

The eigenvalues and vectors of a common tridiagonal matrix

Let

$$\mathbf{A} = \begin{bmatrix} a & b & & & \\ c & a & b & & \\ & c & a & b & \\ & & c & a & b \\ & & & c & a \end{bmatrix}$$

be a square matrix of order N, where a, b, and c may be real or complex numbers.

Let λ represent an eigenvalue of $\mathbf{A}$ and $\mathbf{v}$ the corresponding eigenvector with components $v_1, v_2, \ldots, v_N$. Then the eigenvalue equation $\mathbf{A}\mathbf{v} = \lambda\mathbf{v}$ gives

$$(a - \lambda)v_1 + bv_2 = 0$$
$$cv_1 + (a - \lambda)v_2 + bv_3 = 0$$
$$\cdots$$
$$cv_{j-1} + (a - \lambda)v_j + bv_{j+1} = 0$$

and

$$cv_{N-1} + (a - \lambda)v_N = 0.$$

If we define $v_0 = v_{N+1} = 0$ then these N equations can be combined into the single difference equation

$$cv_{j-1} + (a - \lambda)v_j + bv_{j+1} = 0, \quad j = 1(1)N. \tag{3.61}$$

As shown, previously, the solution of (3.61) is

$$v_j = Bm_1^j + Cm_2^j, \tag{3.62}$$

where B and C are arbitrary constants and m_1, m_2 are the roots of the equation

$$c + (a - \lambda)m + bm^2 = 0. \tag{3.63}$$

(It is proved later that the roots cannot be equal.)
By eqn (3.62) it follows, since $v_0 = v_{N+1} = 0$, that,

$$0 = B + C$$

and

$$0 = Bm_1^{N+1} + Cm_2^{N+1}.$$

Hence

$$\left(\frac{m_1}{m_2}\right)^{N+1} = 1 = e^{i2s\pi}, \quad s = 1(1)N,$$

where $i = \sqrt{(-1)}$. Therefore

$$\frac{m_1}{m_2} = e^{i2s\pi/(N+1)}. \tag{3.64}$$

By eqn (3.63)

$$m_1m_2 = \frac{c}{b}, \tag{3.65}$$

and elimination of m_2 between (3.64) and (3.65) leads to

$$m_1 = \left(\frac{c}{b}\right)^{\frac{1}{2}}e^{is\pi/(N+1)}.$$

Similarly,

$$m_2 = \left(\frac{c}{b}\right)^{\frac{1}{2}}e^{-is\pi/(N+1)}.$$

Again, by eqn (3.63)

$$m_1 + m_2 = (\lambda - a)/b,$$

giving that

$$\lambda = a + b\left(\frac{c}{b}\right)^{\frac{1}{2}}(e^{is\pi/(N+1)} + e^{-is\pi/(N+1)}).$$

Hence the N eigenvalues are given by

$$\lambda_s = a + 2b\left(\frac{c}{b}\right)^{\frac{1}{2}}\cos\frac{s\pi}{N+1}, \quad s = 1(1)N.$$

The jth component of the eigenvector is

$$v_j = Bm_1^j + Cm_2^j = B\left(\frac{c}{b}\right)^{\frac{1}{2}j}(e^{ijs\pi/N+1} - e^{-ijs\pi/N+1})$$

$$= 2iB\left(\frac{c}{b}\right)^{\frac{1}{2}j}\sin\frac{js\pi}{N+1},$$

so the eigenvector $\mathbf{v}_s$ corresponding to λ_s can be taken as

$$\mathbf{v}_s^T = \left\{\left(\frac{c}{b}\right)^{\frac{1}{2}}\sin\frac{s\pi}{N+1}, \frac{c}{b}\sin\frac{2s\pi}{N+1}, \left(\frac{c}{b}\right)^{\frac{3}{2}}\right.$$

$$\left. \times\sin\frac{3s\pi}{N+1}, \ldots, \left(\frac{c}{b}\right)^{N/2}\sin\frac{Ns\pi}{N+1}\right\}.$$

It is easily shown that the roots of eqn (3.63) cannot be equal because if we assume $m_1 = m_2$ the solution of (3.63) is then

$$v_j = (B + Cj)m_1^j$$

and $v_0 = v_{N+1} = 0$ implies that $B = C = 0$, giving $\mathbf{v} = 0$, which is not possible.

An analytical solution of the classical explicit approximation to $\partial U/\partial t = \partial^2 U/\partial x^2$

Consider the equation

$$\frac{\partial U}{\partial t} = \frac{\partial^2 U}{\partial x^2}, \quad 0 < x < 1, \quad t > 0,$$

where $U = 0$ at $x = 0$ and 1, $t > 0$, and U is known when $t = 0$, $0 \leq x \leq 1$.

The classical explicit approximation to the differential equation is

$$u_{i,j+1} = ru_{i-1,j} + (1 - 2r)u_{i,j} + ru_{i+1,j}, \tag{3.66}$$

where
$$x = ih, \quad t = jk, \quad r = k/h^2, \quad \text{and} \quad Nh = 1.$$

Assume that a solution of eqn (3.66) is of the form

$$u_{i,j} = f_i g_j. \tag{3.67}$$

Substitution of (3.67) into (3.66) leads to

$$\frac{g_{j+1}}{g_j} = \frac{rf_{i-1} + (1 - 2r)f_i + rf_{i+1}}{f_i}. \tag{3.68}$$

Since the left-hand side of (3.68) is independent of i and the right-hand side is independent of j it follows that both sides must equal a parameter c which is independent of i and j. This gives two homogeneous difference equations for f_i and g_j, namely,

$$g_{j+1} - cg_j = 0 \tag{3.69}$$

and

$$f_{i+1} + \frac{(1 - 2r - c)}{r} f_i + f_{i-1} = 0. \tag{3.70}$$

The solution of eqn (3.69) is

$$g_j = Ac^j. \tag{3.71}$$

As the solution of the partial differential equation is periodic in x it is reasonable to assume that the solution of (3.70) is periodic in i, so that

$$f_i = B \cos i\theta + D \sin i\theta, \tag{3.72}$$

where

$$\cos \theta = (2r + c - 1)/2r. \tag{3.73}$$

Then, by (3.72), the boundary condition $u_{0,j} = 0$ for all j gives that

$$f_0 = 0 = B.$$

Similarly, the condition $u_{N,j} = 0$ for all j gives that

$$f_N = 0 = D \sin N\theta,$$

showing that

$$N\theta = s\pi, \quad s \text{ an integer.}$$

Therefore

$$f_i = D \sin \frac{is\pi}{N}. \tag{3.74}$$

By eqn (3.73)

$$c = 1 - 2r(1 - \cos \theta) = 1 - 4r \sin^2 \frac{s\pi}{2N}.$$

Hence eqns (3.67), (3.71), and (3.74) give that

$$u_{i,j} = E\left(1 - 4r \sin^2 \frac{s\pi}{2N}\right)^j \sin \frac{s\pi i}{N},$$

where E replaces AD and s is an integer. But eqn (3.66) is linear in $u_{i,j}$ so the sum of different solutions is a solution. It follows therefore that a more general solution that will satisfy fairly general initial conditions is

$$u_{i,j} = \sum_{s=1}^{\infty} E_s\left(1 - 4r \sin^2 \frac{s\pi}{2N}\right)^j \sin \frac{s\pi i}{N}. \qquad (3.75)$$

Case (*i*) If the initial function $u_{i,0}$ is known only at the $(N+1)$ mesh points $(i, 0)$, $i = 0(1)N$, only the first $(N+1)$ values of E_s can be found by solving the $(N+1)$ linear equations

$$u_{i,0} = \sum_{s=1}^{N+1} E_s \sin \frac{s\pi i}{N}, \quad i = 0(1)N.$$

Case (*ii*) If $u_{i,0} = \phi(x)$, say, is a continuous function of x in $0 < x < 1$, it follows from eqn (3.75) that

$$\phi(x) = \sum_{s=1}^{\infty} E_s \sin \frac{s\pi i}{N} = \sum_{s=1}^{\infty} E_s \sin \frac{s\pi i h}{Nh}$$

$$= \sum_{s=1}^{\infty} E_s \sin s\pi x.$$

In this case E_s are the coefficients in the Fourier sine series for $\phi(x)$ and their values will be given by

$$E_s = 2\int_0^1 \phi(x)\sin s\pi x \, dx, \quad s = 1, 2, 3, \ldots.$$

Exercises and solutions

1. (a) Define the (S, T) Padé approximant. Calculate the $(2, 0)$ Padé approximant to $\exp \theta$ and show that its leading error term is $\theta^3/6$.

(b) The equation $\partial U/\partial t = \partial^2 U/\partial x^2$, $0 < x < 1$, $t > 0$, for which U is zero at $x = 0$ and 1, $t > 0$, is approximated at the mesh points $x_i = ih$, $i = 1(1)N - 1$, $Nh = 1$, along time-level $t_j = jk$ by the set of difference equations

$$\mathbf{u}_{j+1} = R_{2,0}(r\mathbf{T}_{N-1})\mathbf{u}_j,$$

where $R_{2,0}(\theta)$ is the $(2, 0)$ Padé approximant to $\exp \theta$, $r = k/h^2$ and $\mathbf{T}_{N-1}$ is the matrix of order $(N - 1)$ defined on p. 55. Show:

(i) That this difference scheme is L_0-stable.

(ii) That its extrapolated form is L_0-stable.

Could discontinuities between boundary and initial values possibly induce finite oscillations in the numerical solution?

Solution

(a) $\exp \theta = \{1/(1 + q_1\theta + q_2\theta^2)\} + c_3\theta^3 + \ldots$. The Maclaurin expansion of $\exp \theta$ leads to $(1 + q_1)\theta + (q_1 + q_2 + \frac{1}{2})\theta^2 + (\frac{1}{2}q_1 + q_2 - c_3)\theta^3 + \ldots \equiv 0$ for all θ. Hence $q_1 = -1$, $q_2 = \frac{1}{2}$, $c_3 = \frac{1}{6}$.

(b) $R_{2,0}(-z) = 1/(1 + z + \frac{1}{2}z^2)$, $z > 0$. Clearly, $|R_{2,0}(-z)| < 1$ for all $z > 0$ and $R_{2,0}(-z)$ tends to zero *monotonically* as $z \to \infty$ and discontinuities cannot induce finite oscillations.

(ii) $\mathbf{u}^{(E)} = \alpha\mathbf{u}^{(2)} - (\alpha - 1)\mathbf{u}^{(1)}$, where $\alpha = 2^{S+T}/(2^{S+T} - 1) = \frac{4}{3}$.

$$
\begin{aligned}
S_{2,0}(-z) &= \tfrac{4}{3}[R_{2,0}(-z)]^2 - \tfrac{1}{3}R_{2,0}(-2z) \\
&= 4/3(1 + z + \tfrac{1}{2}z^2)^2 - 1/3(1 + 2z + 2z^2) \\
&= \frac{1 + 2z + 2z^2 - \tfrac{1}{3}z^3 - \tfrac{1}{12}z^4}{1 + 4z + 8z^2 + 9z^3 + \tfrac{25}{4}z^4 + \tfrac{5}{2}z^5 + \tfrac{1}{2}z^6}.
\end{aligned}
$$

Therefore, $|S_{2,0}(-z)| < 1$ for all $z > 0$ and $S_{2,0}(-z) \to 0$ as $z \to \infty$. The graph of $S_{2,0}(-z)$ is as in Fig. 3.6 and shows that $S_{2,0}$

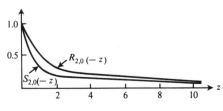

Fig. 3.6

decreases monotonically to zero. Hence the scheme is suitable for problems with discontinuities.

2. Show that the extrapolated form of the classical explicit-difference approximation to $\partial U/\partial t = \partial^2 U/\partial x^2$ is stable for $0 < r \leqslant \frac{1}{4}$, $r = k/h^2$.

Solution

$\mathbf{u}_{j+1} = R_{0,1}(k\mathbf{A})\mathbf{u}_j = (I + k\mathbf{A})\mathbf{u}_j.$ $\mathbf{u}^{(1)}(t_j + 2k) = (\mathbf{I} + 2k\mathbf{A})\mathbf{u}_j$ and $\mathbf{u}^{(2)}(t_j + 2k) = (\mathbf{I} + k\mathbf{A})^2\mathbf{u}_j = (\mathbf{I} + 2k\mathbf{A} + k^2\mathbf{A}^2)\mathbf{u}_j.$ $\mathbf{V}(t_j + 2k) = \{\exp(2k\mathbf{A})\}\mathbf{V}_j = (\mathbf{I} + 2k\mathbf{A} + 2k^2\mathbf{A}^2)\mathbf{V}_j + O(k^3).$ Hence,

$$\mathbf{u}^{(E)}(t_j + 2k) = 2\mathbf{u}^{(2)} - \mathbf{u}^{(1)} = (\mathbf{I} + 2k\mathbf{A} + 2k^2\mathbf{A}^2)\mathbf{u}_j = S_{0,1}(k\mathbf{A})\mathbf{u}_j.$$

Therefore, $S_{0,1}(-z) = 1 - 2z + 2z^2$, where $z = 4r\sin^2 s\pi/2N$, $s = 1(1)N - 1$. For stability, $-1 < 1 - 2z + 2z^2 < 1$. The lower inequality implies that $0 < (z - \frac{1}{2})^2 + \frac{3}{4}$, which is satisfied by all real z. The upper inequality gives $z(z - 1) < 0$. Hence, $0 < z < 1$, i.e. $0 < r < 1/4\sin^2 s\pi/2N > 1/4$, which is satisfied by $0 < r \leqslant 1/4$.

3. The function U satisfies the equation

$$\frac{\partial U}{\partial t} = \frac{\partial^2 U}{\partial x^2}, \quad 0 < x < X, \quad t > 0,$$

and the boundary conditions $U(0, t) = U(X, t) = 0$, $t > 0$. If the space derivative is replaced by

$$\frac{1}{h^2}\{U(x - h, t) - 2U(x, t) + U(x + h, t)\} + O(h^2),$$

where $Nh = X$, show that the partial differential equation, when applied at each of the $(N - 1)$ internal mesh points (ih, t), $i = 1(1)N - 1$, can be approximated by a system of first-order ordinary differential equations $d\mathbf{V}(t)/dt = \mathbf{A}\mathbf{V}(t)$, and define vector $\mathbf{V}(t)$ and matrix $\mathbf{A}$.

Show that the solution of this system satisfies the time-step relation

$$\mathbf{V}(t + k) = \{\exp(k\mathbf{A})\}\mathbf{V}(t).$$

This relation is approximated by

$$\mathbf{u}(t + k) = \{\mathbf{I} - k(1 - \theta)\mathbf{A}\}^{-1}(\mathbf{I} + k\theta\mathbf{A})\mathbf{u}(t) = L_k\mathbf{u}(t),$$

where $\mathbf{I}$ is the unit matrix of order $(N-1)$ and θ a scalar parameter. If the vectors $\mathbf{u}^{(1)}(t+2k)$ and $\mathbf{u}^{(2)}(t+2k)$ are defined by the computational procedures

$$\mathbf{u}^{(1)}(t+2k) = \mathbf{L}_{2k}\mathbf{u}(t) \quad \text{and} \quad \mathbf{u}^{(2)}(t+2k) = \mathbf{L}_k^2\mathbf{u}(t),$$

show that the vector of extrapolated values $\mathbf{u}^{(E)}(t+2k)$, defined by

$$\mathbf{u}^{(E)}(t+2k) = \alpha\mathbf{u}^{(2)} + (1-\alpha)\mathbf{u}^{(1)},$$

where α is a scalar parameter, is second-order accurate in t if $\alpha = 2$ for all θ, or $\theta = \frac{1}{2}$ for all α.

Show that the method is L_0-stable for $\alpha = 2$, $\theta = -1$.

Solution

$$
\begin{aligned}
\mathbf{u}^{(1)}(t+2k) &= \{\mathbf{I} - 2k(1-\theta)\mathbf{A}\}^{-1}(\mathbf{I}+2k\theta\mathbf{A})\mathbf{u}(t) \\
&= \{\mathbf{I} + 2k\mathbf{A} + 4(1-\theta)k^2\mathbf{A}^2 + 8(1-\theta)^2k^3\mathbf{A}^3 + \ldots\}\mathbf{u}(t).
\end{aligned}
$$

$$
\begin{aligned}
\mathbf{u}^{(2)}(t+2k) &= \{\mathbf{I} - k(1-\theta)\mathbf{A}\}^{-2}(\mathbf{I}+k\theta\mathbf{A})^2\mathbf{u}(t) \\
&= \{\mathbf{I} + 2k\mathbf{A} + (3-2\theta)k^2\mathbf{A}^2 + 2(1-\theta)(2-\theta)k^3\mathbf{A}^3 + \ldots\}\mathbf{u}(t).
\end{aligned}
$$

$$
\begin{aligned}
\mathbf{u}^{(E)}(t+2k) = [\mathbf{I} &+ 2k\mathbf{A} + \{(3-2\theta)\alpha + 4(1-\alpha)(1-\theta)k^2\mathbf{A}^2 \\
&+ \{2\alpha(1-\theta)(2-\theta) + 8(1-\alpha)(1-\theta)^2\}k^3\mathbf{A}^3 + \ldots]\mathbf{u}(t).
\end{aligned}
$$

$$\mathbf{V}(t+2k) = \exp(2k\mathbf{A})\mathbf{V}(t) = (\mathbf{I}+2k\mathbf{A}+2k^2\mathbf{A}^2+\tfrac{4}{3}k^3\mathbf{A}^3+\ldots)\mathbf{V}(t).$$

$\mathbf{u}^{(E)}(t+2k)$ and $\mathbf{V}(t+2k)$ will be equal to terms in k^2 if $(3-2\theta)\alpha + 4(1-\alpha)(1-\theta) = 2$, i.e. $(\alpha-2)(2\theta-1) = 0$. Hence the result.

Putting $\mathbf{u}^{(E)}(t+2k) = S(k\mathbf{A})\mathbf{u}(t)$, it follows that

$$
\begin{aligned}
S(k\mathbf{A}) = \alpha\{\mathbf{I} &- (1-\theta)k\mathbf{A}\}^{-2}(\mathbf{I}+\theta k\mathbf{A})^2 \\
&+ (1-\alpha)\{\mathbf{I} - 2(1-\theta)k\mathbf{A}\}^{-1}(\mathbf{I}+2\theta k\mathbf{A}).
\end{aligned}
$$

Hence

$$
\begin{aligned}
S(-z) = \alpha(1 - \theta z)^2/\{1 + (1-\theta)z\}^2 \\
+ (1-\alpha)(1 - 2\theta z)/\{1 + 2(1-\theta)z\}.
\end{aligned}
$$

For $\alpha = 2$, $\theta = -1$,

$$
\begin{aligned}
S(-z) &= (1+6z+6z^2)/(1+8z+20z^2+16z^3) \\
&= 1 - [(2z+14z^2+16z^3)/(1+8z+20z^2+16z^3)]
\end{aligned}
$$

Hence $0 < S(-z) < 1$, $z > 0$, and $S(-z) \to 0$ as $z \to \infty$.

Comments

Gourlay and Morris, reference 10, give numerical results for various values of θ and α and also develop very accurate schemes of third- and fourth-order accuracy in t.

4. The equation $\partial U/\partial t - \partial^2 U/\partial x^2 = 0$, $0 < x < 1$, $t > 0$, for which U is zero at $x = 0$ and 1, $t > 0$, is approximated at the mesh points $x_i = ih$, $i = 1(1)N - 1$, along time-level $t_j = jk$ by the set of difference equations

$$\frac{1}{k}\{\mathbf{u}_{j+1} - R_{1,0}(r\mathbf{T}_{N-1})\mathbf{u}_j\} = 0,$$

where $R_{1,0}(\theta)$ is the $(1, 0)$ Padé approximant to $\exp \theta$, $r = k/h^2$ and matrix $\mathbf{T}_{N-1}$ is as previously defined.

Show that the principal part of the local truncation error at the point (ih, jk) is

$$\left[-\tfrac{1}{12}h^2\frac{\partial^4 U}{\partial x^4} - \tfrac{1}{2}k\frac{\partial^2 U}{\partial t^2}\right]_{i,j}, \quad i = 2(1)N - 2.$$

Solution

$$T_{i,j} = i\text{th row of } \frac{1}{k}\{(\mathbf{I} - r\mathbf{T}_{N-1})\mathbf{U}_{j+1} - \mathbf{U}_j\}$$

$$= \frac{1}{k}\{-rU_{i-1,j+1} + (1+2r)U_{i,j+1} - rU_{i+1,j+1} - U_{i,j}\},$$

where

$$U_{i-1,j+1} = \left[U + \left(-h\frac{\partial}{\partial x} + k\frac{\partial}{\partial t}\right)U + \frac{1}{2}\left(-h\frac{\partial}{\partial x} + k\frac{\partial}{\partial t}\right)^2 U + \ldots\right]_{i,j},$$

$$U_{i,j+1} = \left[U + k\frac{\partial U}{\partial t} + \frac{1}{2}k^2\frac{\partial^2 U}{\partial t^2} + \ldots\right]_{i,j}$$

$$U_{i+1,j+1} = \left[U + \left(h\frac{\partial}{\partial x} + k\frac{\partial}{\partial t}\right)U + \frac{1}{2}\left(h\frac{\partial}{\partial x} + k\frac{\partial}{\partial t}\right)^2 U + \ldots\right]_{i,j}.$$

First- and second-order derivative terms $= \left[\dfrac{\partial U}{\partial t} - \dfrac{\partial^2 U}{\partial x^2} + \tfrac{1}{2}k\dfrac{\partial^2 U}{\partial t^2}\right]_{i,j}.$

Third-order terms $= \left[-k\dfrac{\partial^3 U}{\partial x^2 \partial t} + \tfrac{1}{6}k^2\dfrac{\partial^3 U}{\partial t^3}\right]_{i,j}.$

$$\text{Fourth-order terms} = \left[-\tfrac{1}{12}h^2\frac{\partial^4 U}{\partial x^4}\right]_{i,j}.$$

Hence the result since $\dfrac{\partial}{\partial t} \equiv \dfrac{\partial^2}{\partial x^2}$ so $\dfrac{\partial^3 U}{\partial x^2 \partial t} = \dfrac{\partial^2 U}{\partial t^2}$.

5. As in Exercise 3, show that the solution $\mathbf{V}(t)$ of the system of ordinary differential equations approximating the solution of $\partial U/\partial t = \partial^2 U/\partial x^2$, $0 < x < X$, $t > 0$, $U(0, t) = U(X, t) = 0$, at the mesh points $x_i = ih$, $i = 1(1)N - 1$, satisfies the recurrence relation

$$\mathbf{V}(t + k) = \{\exp(k\mathbf{A})\}\mathbf{V}(t).$$

If the exponential of $k\mathbf{A}$ is approximated by its $(2, 0)$ Padé approximant, show that the vector $\mathbf{u}(t)$ approximating $\mathbf{V}$ is the solution of the difference equations

$$\mathbf{L}_k\mathbf{u}(t_j + k) = \mathbf{u}(t_j),$$

where

$$\mathbf{L}_k = \mathbf{I}_{N-1} - k\mathbf{A} + \tfrac{1}{2}k^2\mathbf{A}^2 = \begin{bmatrix} g & d & a & & & & \\ d & c & d & a & & & \\ a & d & c & d & a & & \\ & \cdot & \cdot & \cdot & \cdot & \cdot & \\ & & a & d & c & d & a \\ & & & a & d & c & d \\ & & & & a & d & g \end{bmatrix},$$

$g = 1 + 2r + \tfrac{5}{2}r^2$, $c = 1 + 2r + 3r^2$, $d = -r - 2r^2$, $a = \tfrac{1}{2}r^2$, $r = k/h^2$, $\mathbf{u}(t_j) = [u_{1,j}, u_{2,j}, \ldots, u_{N-1,j}]^T$ and $t_j = jk$.

Defining $\mathbf{u}^{(1)}(t + 2k) = \mathbf{L}_k^{-2}\mathbf{u}(t)$ and $\mathbf{u}^{(2)}(t + 2k) = \mathbf{L}_{2k}^{-1}\mathbf{u}(t)$, show that both $\mathbf{u}^{(1)}$ and $\mathbf{u}^{(2)}$ are second-order accurate in t, i.e. with a leading error term $O(k^3)$ in t.

Find the value of α if

$$\mathbf{u}^{(E)}(t + 2k) = \alpha\mathbf{u}^{(1)} - (\alpha - 1)\mathbf{u}^{(2)}$$

is third-order accurate in t.

Comments

In Exercise 1, the extrapolated scheme is shown to be L_0-stable and to exclude spurious oscillations. Twizell and Khaliq, reference 28, show that the method is highly accurate, with maximum errors for $U(x, 0) = 1$, $X = 2$ and $t = 1.2$, of 0.74×10^{-4}, $(r = 10)$, 0.41×10^{-3}, $(r = 40)$, and 0.36×10^{-3}, $(r = 160)$.

Solution

$R_{2,0}(k\mathbf{A}) = (\mathbf{I} - k\mathbf{A} + \frac{1}{2}k^2\mathbf{A}^2)^{-1} = \mathbf{L}_k^{-1}$. Matrix $\mathbf{A}^2$ is given in the section on extrapolation, so $\mathbf{L}_k$ is easily written down.

$$\mathbf{u}^{(1)}(t + 2k) = (\mathbf{I} - k\mathbf{A} + \frac{1}{2}k^2\mathbf{A}^2)^{-2}\mathbf{u}(t)$$
$$= (\mathbf{I} + 2k\mathbf{A} + 2k^2\mathbf{A}^2 + k^3\mathbf{A}^3 - \frac{1}{4}k^4\mathbf{A}^4)\mathbf{u}(t) + O(k^5).$$

$$\mathbf{u}^{(2)}(t + 2k) = (\mathbf{I} - 2k\mathbf{A} + 2k^2\mathbf{A}^2)^{-1}\mathbf{u}(t)$$
$$= (\mathbf{I} + 2k\mathbf{A} + 2k^2\mathbf{A}^2 - 4k^4\mathbf{A}^4)\mathbf{u}(t) + O(k^5).$$

$$\mathbf{u}(t + 2k) = (\exp 2k\mathbf{A})\mathbf{u}(t)$$
$$= (\mathbf{I} + 2k\mathbf{A} + 2k^2\mathbf{A}^2 + \frac{4}{3}k^3\mathbf{A}^3 + \frac{2}{3}k^4\mathbf{A}^4)\mathbf{u}(t) + O(k^5).$$

$\alpha = \frac{4}{3}$, so $\mathbf{u}^{(E)}(t + 2k) = \frac{4}{3}\mathbf{u}^{(1)} - \frac{1}{3}\mathbf{u}^{(2)} + O(k^4) + O(h^2)$.

6. The Douglas equations approximating $\partial U/\partial t = \partial^2 U/\partial x^2$, $0 < x < 1$, $t > 0$, are given by

$$(1 - 6r)u_{i-1,j+1} + (10 + 12r)u_{i,j+1} + (1 - 6r)u_{i+1,j+1}$$
$$= (1 + 6r)u_{i-1,j} + (10 - 12r)u_{i,j} + (1 + 6r)u_{i+1,j},$$
$$i = 1(1)N - 1, \quad Nh = 1.$$

Given that the boundary values are known for all j, prove that the equations are unconditionally stable in the Lax–Richtmyer sense for all $r = k/h^2 > 0$. (See Exercise 8(c)).

Solution

The equations in matrix form are

$$\{(10 + 12r)\mathbf{I} + (1 - 6r)\mathbf{A}\}\mathbf{u}_{j+1} = \{(10 - 12r)\mathbf{I} + (1 + 6r)\mathbf{A}\}\mathbf{u}_j + \mathbf{b}_j,$$

where $\mathbf{b}_j$ is known, and matrix $\mathbf{A}$ is

$$\begin{bmatrix} 0 & 1 & & & & \\ 1 & 0 & 1 & & & \\ & 1 & 0 & 1 & & \\ & & 1 & 0 & 1 & \\ & & & & \ddots & \\ & & & & 1 & 0 \end{bmatrix}$$

of order $N - 1$. Since the coefficient matrices of $\mathbf{u}_{j+1}$ and $\mathbf{u}_j$ are symmetric and commute, it follows that

$$\mathbf{B} = \{(10 + 12r)\mathbf{I} + (1 - 6r)\mathbf{A}\}^{-1}\{(10 - 12r)\mathbf{I} + (1 + 6r)\mathbf{A}\}$$

is real and symmetric. Hence, $\|\mathbf{B}\|_2 = \rho(\mathbf{B})$. If λ is an eigenvalue of $\mathbf{A}$, the equations will be stable when

$$-1 \leqslant \frac{(10 - 12r) + (1 + 6r)\lambda}{(10 + 12r) + (1 - 6r)\lambda} \leqslant 1,$$

for all λ, i.e.,

$$-(10 + 12r) - (1 - 6r)\lambda \leqslant (10 - 12r) + (1 + 6r)\lambda$$
$$\leqslant (10 + 12r) + (1 - 6r)\lambda,$$

provided $(10 + 12r) + (1 - 6r)\lambda > 0$. Hence $-10 \leqslant \lambda \leqslant 2$. The eigenvalues of $\mathbf{A}$ are $\lambda_s = 2\cos s\pi/N$. This value satisfies all the necessary inequalities for $r > 0$.

7. Show that $U_{i,j+1} = \{\exp(k(\partial/\partial t))\}U_{i,j}$, where $U_{i,j} = U(x_i, t_j)$, $x_i = ih$, $i = 0, \pm 1, \pm 2, \ldots$, and $t_j = jk$, $j = 0, 1, 2, \ldots$. Using the result

$$D_x \equiv \frac{2}{h}\sinh^{-1}(\tfrac{1}{2}\delta_x) = \frac{1}{h}(\delta_x - \tfrac{1}{24}\delta_x^3 + \tfrac{3}{640}\delta_x^5 + \ldots),$$

where $D_x \equiv (\partial/\partial x)$ and $\delta_x U_{i,j} = U_{i+\frac{1}{2},j} - U_{i-\frac{1}{2},j}$, show that the *exact* difference replacement of the equation $(\partial U/\partial t) = (\partial^2 U/\partial x^2)$ is given by

$$U_{i,j+1} = \{1 + r\delta_x^2 + \tfrac{1}{2}r(r - \tfrac{1}{6})\,\delta_x^4 + \tfrac{1}{6}r(r^2 - \tfrac{1}{2}r + \tfrac{1}{15})\,\delta_x^6 + \ldots\}U_{i,j},$$

where $r = k/h^2$.

If only second-order central-differences are retained show that the function $u_{i,j}$ approximating $U_{i,j}$ is the solution of the classical explicit equation

$$u_{i,j+1} = ru_{i-1,j} + (1 - 2r)u_{i,j} + ru_{i+1,j}.$$

If only second- and fourth-order differences are retained show that $u_{i,j}$ is the solution of the explicit equation

$$u_{i,j+1} = \tfrac{1}{2}(2 - 5r + 6r^2)u_{i,j} + \tfrac{2}{3}r(2 - 3r)(u_{i+1,j} + u_{i-1,j})$$
$$- \tfrac{1}{12}r(1 - 6r)(u_{i+2,j} + u_{i-2,j}).$$

(This equation is stable for $r \leqslant \tfrac{2}{3}$.)

Solution

By Taylor's series,

$$U_{i,j+1} = U(x_i, t_j + k) = U_{i,j} + k \frac{\partial U_{i,j}}{\partial t} + \tfrac{1}{2}k^2 \frac{\partial^2 U_{i,j}}{\partial x^2} + \dots$$

$$= \left(1 + k\frac{\partial}{\partial t} + \frac{1}{2!}k^2\frac{\partial^2}{\partial t^2} + \frac{1}{3!}k^3\frac{\partial^3}{\partial t^3} + \dots\right)U_{i,j} = \left\{\exp\left(k\frac{\partial}{\partial t}\right)\right\}U_{i,j}.$$

For the equation

$$\frac{\partial U}{\partial t} = \frac{\partial^2 U}{\partial x^2}, \quad \frac{\partial}{\partial t} \equiv \frac{\partial^2}{\partial x^2}.$$

Hence

$$U_{i,j+1} = \left\{\exp\left(k\frac{\partial^2}{\partial x^2}\right)\right\}U_{i,j} = \{\exp(kD_x^2)\}U_{i,j}$$

$$= (1 + kD_x^2 + \tfrac{1}{2}k^2 D_x^4 + \tfrac{1}{6}k^3 D_x^6 + \dots)U_{i,j}$$

where

$$D_x \equiv \frac{1}{h}(\delta_x - \tfrac{1}{24}\delta_x^3 + \tfrac{3}{640}\delta_x^5 + \dots)$$

Evaluate D_x^2, D_x^4, and D_x^6 to terms in δ_x^6 and substitute into the expression for $u_{i,j+1}$ etc. The results stated follow from

$$\delta_x^2 u_{i,j} = u_{i-1,j} - 2u_{i,j} + u_{i+1,j}$$

and

$$\delta_x^4 u_{i,j} = \delta_x^2(u_{i-1,j} - 2u_{i,j} + u_{i+1,j})$$

$$= u_{i-2,j} - 4u_{i-1,j} + 6u_{i,j} - 4u_{i+1,j} + u_{i+2,j}.$$

8. (a) The equation

$$\frac{\partial U}{\partial t} = \frac{\partial}{\partial x}\left\{a(x)\frac{\partial U}{\partial x}\right\}, \quad a(x) > 0,$$

is approximated at the point (ih, jk) by the difference equation

$$\frac{1}{k}\Delta_t u_{i,j} = \frac{1}{h}\delta_x\left(a_i \frac{1}{h}\delta_x u_{i,j}\right).$$

Show that this gives the explicit difference equation

$$u_{i,j+1} = ra_{i-\frac{1}{2}}u_{i-1,j} + \{1 - r(a_{i-\frac{1}{2}} + a_{i+\frac{1}{2}})\}u_{i,j} + ra_{i+\frac{1}{2}}u_{i+1,j},$$

where $r = k/h^2$.

(b) Using the result

$$U(x, t+k) = \left\{ \exp\left(k \frac{\partial}{\partial t}\right) \right\} U(x, t),$$

show that the equation

$$\frac{\partial U}{\partial t} = \frac{\partial^2 U}{\partial x^2}$$

may be replaced exactly by the equation

$$\exp(-\tfrac{1}{2}kD_x^2) U_{i,j+1} = \exp(\tfrac{1}{2}kD_x^2) U_{i,j},$$

where

$$D_x^2 U_{i,j} = \frac{\partial^2 U_{i,j}}{\partial x^2} = \frac{1}{h^2}(\delta_x^2 - \tfrac{1}{12}\delta_x^4 + \tfrac{1}{90}\delta_x^6 + \ldots) U_{i,j}.$$

Hence show that to second-order differences the differential equation may be approximated by the (Crank–Nicolson) equation

$$-ru_{i-1,j+1} + (2+2r)u_{i,j+1} - ru_{i+1,j+1} = ru_{i-1,j} + (2-2r)u_{i,j} + ru_{i+1,j}.$$

(c) Eliminate the fourth-order term in

$$D_x^2 \equiv \frac{1}{h^2}(\delta_x^2 - \tfrac{1}{12}\delta_x^4 + \tfrac{1}{90}\delta_x^6 + \ldots).$$

Hence use the first part of (b) to show that $\partial U/\partial t = \partial^2 U/\partial x^2$ may be approximated by the Douglas difference equation

$$(1-6r)u_{i+1,j+1} + (10+12r)u_{i,j+1} + (1-6r)u_{i+1,j+1}$$
$$= (1+6r)u_{i-1,j} + (10-12r)u_{i,j} + (1+6r)u_{i+1,j},$$

where $r = k/h^2$. (As mentioned previously this uses the same six grid points as the Crank–Nicolson equation but gives a more accurate solution.)

Solution

(a) As $\delta_x u_{i,j} = u_{i+\frac{1}{2},j} - u_{i-\frac{1}{2},j}$, the approximation leads to

$$\frac{1}{k}(u_{i,j+1} - u_{i,j}) = \frac{1}{h^2}\delta_x\{a_i(u_{i+\frac{1}{2},j} - u_{i-\frac{1}{2},j})\}$$

$$= \frac{1}{h^2}\{a_{i+\frac{1}{2}}(u_{i+1,j} - u_{i,j}) - a_{i-\frac{1}{2},j}(u_{i,j} - u_{i-1,j})\} \quad \text{etc.}$$

(b) $U_{i,j+1} = \{\exp(k(\partial/\partial t))\}U_{i,j}$. Operate on both sides with $\exp(-\frac{1}{2}k(\partial/\partial t))$. Hence $\{\exp(-\frac{1}{2}k(\partial/\partial t))\}U_{i,j+1} = \{\exp(\frac{1}{2}k(\partial/\partial t))\}U_{i,j}$. By the differential equation $(\partial/\partial t) \equiv D_x^2$, therefore $\{\exp(-\frac{1}{2}kD_x^2)\}U_{i,j+1} = \{\exp(\frac{1}{2}kD_x^2)\}U_{i,j}$. Substituting for D_x^2 in terms of δ_x^2 and expanding to terms in δ_x^2 shows that

$$(1 - \tfrac{1}{2}r\delta_x^2)u_{i,j+1} = (1 + \tfrac{1}{2}r\delta_x^2)u_{i,j}$$

approximates the partial differential equation to this order of accuracy. Hence the result.

(c) Operate on both sides of

$$D_x^2 \equiv \frac{1}{h^2}(\delta_x^2 - \tfrac{1}{12}\delta_x^4 + \ldots)$$

with $(1 + \tfrac{1}{12}\delta_x^2)$ to give that

$$(1 + \tfrac{1}{12}\delta_x^2)D_x^2 \equiv \frac{1}{h^2}\delta_x^2 + O(\delta_x^6).$$

By part (b), $(1 - \frac{1}{2}kD_x^2 + \ldots)U_{i,j+1} = (1 + \frac{1}{2}kD_x^2 + \ldots)U_{i,j}$. Operate on both sides with $(1 + \tfrac{1}{12}\delta_x^2)$ and use the previous identity. Neglecting terms of order δ^6 and above yields the approximation equation

$$(1 + \tfrac{1}{12}\delta_x^2 - \tfrac{1}{2}r\delta_x^2)u_{i,j+1} = (1 + \tfrac{1}{12}\delta_x^2 + \tfrac{1}{2}r\delta_x^2)u_{i,j}.$$

Hence the result.

9. The Crank–Nicolson method approximates the equation

$$\left(\frac{\partial U}{\partial t}\right)_{i,j-\frac{1}{2}} = \left(\frac{\partial^2 U}{\partial x^2}\right)_{i,j-\frac{1}{2}} \quad \text{by} \quad \left(\frac{\partial U}{\partial t}\right)_{i,j-\frac{1}{2}} = \frac{1}{2}\left\{\left(\frac{\partial^2 U}{\partial x^2}\right)_{i,j} + \left(\frac{\partial^2 U}{\partial x^2}\right)_{i,j-1}\right\}.$$

Assuming the following central-difference formulae for the derivatives f' and f'' for a mesh length h, namely,

$$hf'_{\frac{1}{2}} = (\delta - \tfrac{1}{24}\delta^3 + \tfrac{3}{640}\delta^5 - \ldots)f_{\frac{1}{2}},$$

and

$$h^2 f''_0 = (\delta^2 - \tfrac{1}{12}\delta^4 + \tfrac{1}{90}\delta^6 - \ldots)f_0,$$

show that the Crank–Nicolson equation leads to

$$(u_{i,j} - u_{i,j-1})/k$$
$$= (u_{i+1,j} - 2u_{i,j} + u_{i-1,j} + u_{i+1,j-1} - 2u_{i,j-1} + u_{i-1,j-1})/2h^2 + C,$$

where the correction term C is given by

$$C = \frac{1}{k}(\tfrac{1}{24}\delta^3 u_{i,j-\frac{1}{2}} - \tfrac{3}{640}\delta^5 u_{i,j-\frac{1}{2}} + \ldots)_t + \frac{1}{2h^2}\{(-\tfrac{1}{12}\delta^4 u_{i,j} + \tfrac{1}{90}\delta^6 u_{i,j} - \ldots)_x$$
$$+ (-\tfrac{1}{12}\delta^4 u_{i,j-1} + \tfrac{1}{90}\delta^6 u_{i,j-1} - \ldots)_x\}.$$

Comment

The correction term can be used to improve the accuracy of the finite-difference solution obtained initially by neglecting it. The function values of this first approximation are differenced in the t direction to give $\delta^3 u_{i,j-\frac{1}{2}}$ and in the x-direction to give

$$\delta^4 u_{i,j}, \quad \delta^4 u_{i,j-1}, \quad \delta^6 u_{i,j} \quad \text{and} \quad \delta^6 u_{i,j-1}.$$

The correction term for each equation is then calculated from these differences, which of course are numbers, and the corrected finite-difference equations re-solved for the $u_{i,j}$.

10. The function U satisfies the non-linear equation

$$\frac{\partial U}{\partial t} = \frac{\partial^2 U^2}{\partial x^2}, \quad 0 < x < 1, \quad t > 0,$$

the initial condition $U = 1$, $0 \le x \le 1$, $t = 0$, and the boundary conditions

$$\frac{\partial U}{\partial x} = 0 \text{ at } x = 0, \ t > 0, \quad \frac{\partial U}{\partial x} = -U \text{ at } x = 1, \ t > 0.$$

If the equation is approximated by the difference equation

$$\frac{1}{k}\delta_t u_{i,j+\frac{1}{2}} = \frac{1}{2h^2}(\delta_x^2 u_{i,j+1}^2 + \delta_x^2 u_{i,j}^2),$$

and the derivative boundary conditions are approximated by the usual central-difference formulae, show that the corresponding non-linear approximation equations are

$$u_0^2 + pu_0 - u_1^2 + \{u_{0,j}^2 - pu_{0,j} - u_{1,j}^2\} = 0,$$
$$u_{i-1}^2 - 2(u_i^2 + pu_i) + u_{i+1}^2 + \{u_{i-1,j}^2 - 2(u_{i,j}^2 - pu_{i,j}) + u_{i+1,j}^2\} = 0,$$
$$i = 1(1)(N-1),$$

and

$$u_{N-1}^2 - 2hu_{N-1}u_N - (1-2h^2)u_N^2 - pu_N$$
$$+ \{u_{N-1,j}^2 - 2hu_{N-1,j}u_{N,j} - (1-2h^2)u_{N,j}^2 + pu_{N,j}\} = 0,$$

where $p = h^2/k$, $Nh = 1$ and u_i denotes the unknown $u_{i,j+1}$, $i = 0(1)N$.

Briefly describe Newton's method for deriving a set of linear equations giving an improved value $(V_i + \varepsilon_i)$ to the approximate solution values V_i, $i = 1(1)N$, of the N non-linear equations

$$f_i(u_1, u_2, \ldots, u_N) = 0, \quad i = 1(1)N.$$

Apply this method to the approximation equations previously obtained and write out the linear equations giving the first iteration values at mesh points along the time-level $t = k$, taking V_i, $i = 0(1)N$, equal to the initial value 1.

Solution

$$\frac{1}{k}(u_{i,j+1} - u_{i,j})$$

$$= \frac{1}{2h^2}(u_{i-1,j+1}^2 - 2u_{i,j+1}^2 + u_{i+1,j+1}^2 + u_{i-1,j}^2 - 2u_{i,j}^2 + u_{i+1,j}^2).$$

Put $p = h^2/k$ and denote $u_{i,j+1}$ by u_i. Then

$$u_{i-1}^2 - 2(u_i^2 + pu_i) + u_{i+1}^2 + \{u_{i-1,j}^2 - 2(u_{i,j}^2 - pu_{i,j}) + u_{i+1,j}^2\} = 0.$$

By the B.C.,

$$\left(\frac{\partial U}{\partial x}\right)_{0,j} \simeq \frac{u_{1,j} - u_{-1,j}}{2h} = 0 \Rightarrow u_{-1,j} = u_{1,j}.$$

Put $i = 0$ in the approximation equation and eliminate $u_{-1,j}$ to give

$$u_0^2 + pu_0 - u_1^2 + (u_{0,j}^2 - pu_{0,j} - u_{1,j}^2) = 0.$$

By the B.C. at

$$i = N, \quad \frac{1}{2h}(u_{N+1,j+1} - u_{N-1,j+1}) = -u_{N,j} \quad \text{etc.}$$

By Newton's method,

$$\left(\frac{\partial f_i}{\partial u_1}\varepsilon_1 + \ldots + \frac{\partial f_i}{\partial u_N}\varepsilon_N\right)_{u_i=V_i} + f_i(V_1, V_2, \ldots, V_N) = 0.$$

Hence

$$(2V_0+p)\varepsilon_0-2V_1\varepsilon_1+\{V_0^2+pV_0-V_1^2+(u_{0,j}^2-pu_{0,j}-u_{1,j}^2)\}=0,$$

$$2V_{i-1}\varepsilon_{i-1}-2(2V_i+p)\varepsilon_i+2V_{i+1}\varepsilon_{i+1}$$
$$+\{V_{i-1}^2-2(V_i^2+pV_i)+V_{i-1}^2$$
$$+(u_{i-1,j}^2-2[u_{i,j}^2-pu_{i,j}]+u_{i+1,j}^2)\}=0,$$

and

$$(2V_{N-1}-2hV_N)\varepsilon_{N-1}+\{-2hV_{N-1}-2(1-2h^2)V_N-p\}\varepsilon_N$$
$$+\{V_{N-1}^2-2hV_{N-1}V_N-(1-2h^2)V_N^2-pV_N$$
$$+u_{N-1,j}^2-2hu_{N-1,j}u_{N,j}-(1-2h^2)u_{N,j}^2+pu_{N,j}\}=0.$$

Put $V_i=1$, $i=0(1)N$, $j=0$, and $u_{i,0}=1$, $i=0(1)N$ etc.

11. The equation

$$\frac{\partial U}{\partial t}=\frac{\partial^2 U}{\partial x^2}, \quad 0<x<1, \quad t>0,$$

is approximated at the point (ih, jk) by the difference equation

$$\frac{3}{2}\left(\frac{u_{i,j+1}-u_{i,j}}{k}\right)-\frac{1}{2}\left(\frac{u_{i,j}-u_{i,j-1}}{k}\right)=\frac{1}{h^2}\delta_x^2 u_{i,j+1},$$

where $x=ih$, $t=jk$, and $Nh=1$. Investigate the stability of this system of equations by the matrix method, for known boundary values and a fixed mesh size.

Solution

Equations are $-2ru_{i-1,j+1}+(3+4r)u_{i,j+1}-2ru_{i+1,j+1}=4u_{i,j}-u_{i,j-1}$, $i=1(1)(N-1)$. For known boundary values they can be written as

$$\mathbf{u}_{j+1}=4\mathbf{A}^{-1}\mathbf{u}_j-\mathbf{A}^{-1}\mathbf{u}_{j-1}+\mathbf{A}^{-1}\mathbf{c}_{j+1},$$

where $\mathbf{c}_{j+1}$ is a vector of known boundary values and

$$\mathbf{A}=\begin{bmatrix}(3+4r) & -2r & & \\ -2r & (3+4r) & -2r & \\ & & & \\ & & -2r & (3+4r)\end{bmatrix}$$

is of order $(N-1)$. This equation and $\mathbf{u}_j = \mathbf{u}_j$ can be written as

$$\begin{bmatrix} \mathbf{u}_{j+1} \\ \mathbf{u}_j \end{bmatrix} = \begin{bmatrix} 4\mathbf{A}^{-1} & -\mathbf{A}^{-1} \\ \mathbf{I} & \mathbf{O} \end{bmatrix} \begin{bmatrix} \mathbf{u}_j \\ \mathbf{u}_{j-1} \end{bmatrix} + \begin{bmatrix} \mathbf{A}^{-1}\mathbf{c} \\ \mathbf{O} \end{bmatrix},$$

i.e. $\mathbf{v}_{j+1} = \mathbf{P}\mathbf{v}_j + \mathbf{c}'$. $\mathbf{A}$ has distinct eigenvalues and all the sub-matrices of $\mathbf{P}$ commute with each other so the eigenvalues λ of $\mathbf{P}$ are the eigenvalues of

$$\begin{bmatrix} \dfrac{4}{\mu_k} & -\dfrac{1}{\mu_k} \\ 1 & 0 \end{bmatrix}$$

where μ_k is the kth eigenvalue of $\mathbf{A}$. As $\det(\mathbf{P}-\lambda\mathbf{I})=0= \lambda^2 - (4/\mu_k)\lambda + (1/\mu_k)$, $\lambda = \{2 \pm \sqrt{(4-\mu_k)}\}/\mu_k$ where

$$\mu_k = 3 + 8r \sin^2(k\pi/2N), \quad k = 1(1)(N-1)$$

by p. 59. Hence

$$\lambda = \left\{ 2 \pm \left(1 - 8r \sin^2 \frac{k\pi}{2N}\right)^{\frac{1}{2}} \right\} \bigg/ \left(3 + 8r \sin^2 \frac{k\pi}{2N}\right).$$

When the roots are real, $|\lambda| < (2+1)/(3+\delta)$, $\delta > 0$. Hence $|\lambda| < 1$. When the roots are complex, $|\lambda| = \mu_k^{-\frac{1}{2}} < 1$. Therefore the equations are unconditionally stable.

12. Show that the components of the eigenvector $\mathbf{v}$, correspond-ing to the eigenvalue λ of the matrix $\mathbf{A}$ of order N defined by

$$\mathbf{A} = \begin{bmatrix} -2 & 2 & & & & \\ 1 & -2 & 1 & & & \\ & 1 & -2 & 1 & & \\ & & & \cdot & & \\ & & & \cdot & & \\ & & & 1 & -2 & 1 \\ & & & & 1 & -2 \end{bmatrix}$$

are given by the solution of the difference equations

$$v_{j-1} - (2+\lambda)v_j + v_{j+1} = 0, \quad j = 2(1)N,$$

satisfying the conditions

$$-(2+\lambda)v_1 + 2v_2 = 0 = v_{N+1}.$$

Hence prove that the jth component of $\mathbf{v}$ can be expressed as

$$v_j = B \cos j\theta + C \sin j\theta,$$

where B and C are constants. Deduce that the *r*th eigenvalue of
A is $-4\sin^2((2r-1)\pi/4N)$.

Solution

Expand $\mathbf{Av} = \lambda\mathbf{v}$. The solution of $v_{j-1} - (2+\lambda)v_j + v_{j+1} = 0$ is $v_j = Dm_1^j + Em_2^j$ where m_1, m_2 are the roots of $m^2 - (2+\lambda)m + 1 = 0$. Therefore $m_1 m_2 = 1$, $m_1 + m_2 = 2 + \lambda$, λ real by part (a). Put $m_1 = re^{i\theta}$. Then $r = 1$, giving $\lambda = 2(-1 + \cos\theta)$, $v_j = $ B $\cos j\theta + $ C $\sin j\theta$. As $v_{N+1} = 0$, B/C $= -\tan(N+1)\theta$. Substitute for v_1, v_2, and λ in terms of θ into $-(2+\lambda)v_1 + 2v_2 = 0$ to get C $\sin\theta \cos N\theta = 0$. Hence $\theta = (2r-1)\pi/2N$, *r* an integer.

13. Verify that $U = e^{-\pi^2 t}\sin\pi x$ is a solution of the equation

$$\frac{\partial U}{\partial t} = \frac{\partial^2 U}{\partial x^2}, \quad 0 < x < 1, \quad t > 0,$$

which satisfies the boundary conditions $U = 0$ at $x = 0$ and 1, $t > 0$, and the initial condition $U = \sin\pi x$ when $t = 0$, $0 \leq x \leq 1$. The differential equation is approximated at the point (ih, jk) by the explicit equation

$$\frac{1}{k}\Delta_t u_{i,j} = \frac{1}{h^2}\delta_x^2 u_{i,j}.$$

Show that the analytical solution of the difference equation satisfying the same boundary and initial values is

$$u_{i,j} = \left(1 - 4r\sin^2\frac{\pi h}{2}\right)^j \sin\pi x_i,$$

where $r = k/h^2$.

Given that $0 < r \leq \frac{1}{2}$, deduce that $u_{i,j}$ converges to $U_{i,j}$ as *h* tends to zero, for finite values of *t*.

Solution

As in the text. Because of the initial function, only the first term of the series solution is needed.

$$|u_{i,j} - U_{i,j}| = \left|\left(1 - 4r\sin^2\frac{\pi h}{2}\right)^j - e^{-\pi^2 kj}\right||\sin\pi x| = |a^j - b^j||\sin\pi x|,$$

where

$$|a| = \left|1 - 4r \sin^2 \frac{\pi h}{2}\right| \leqslant 1$$

since $r \leqslant \frac{1}{2}$ and $|b| = |e^{-\pi^2 k}| < 1$. Hence

$$|u_{i,j} - U_{i,j}| = |a - b| \, |a^{j-1} + a^{j-2}b + \ldots + b^{j-1}| \, |\sin \pi x| < j \, |a - b|,$$

since there are j terms in the second series, $= j \, |1 - 2r + 2r \cos \pi h - e^{-\pi^2 k}|$. Replace $\cos \pi h$ and $e^{-\pi^2 k}$ by their Maclaurin expansions to get that

$$|u_{i,j} - U_{i,j}| < t\pi^4 h^2 A,$$

where A is bounded because it cannot exceed $|a - b|$. Hence the result.

4 Hyperbolic equations

First-order quasi-linear equations and characteristics

Consider the equation

$$a\frac{\partial U}{\partial x} + b\frac{\partial U}{\partial y} = c,$$

where a, b, and c are, in general, functions of x, y, and U but not of $\partial U/\partial x$ and $\partial U/\partial y$, i.e. the first-order derivatives occur only to the first degree although the equation need not be linear in U. Such an equation is said to be quasi-linear. It is customary to put $\partial U/\partial x = p$ and $\partial U/\partial y = q$ and to write the equation as

$$ap + bq = c. \tag{4.1}$$

The following analysis shows that at each point of the solution domain of such an equation there is a direction along which the integration of eqn (4.1) transforms to the integration of an ordinary differential equation. In other words, *in this direction* the expression to be integrated will be independent of partial derivatives in other directions, such as p and q.

Assume we know the solution values U of eqn (4.1) at every point on a curve C in the x–y plane, where C does *not* coincide with the curve Γ on which initial values of U are specified. The question to be asked at this stage is this. Can we determine values for p and q on C from the values of U on C so that they satisfy eqn (4.1)?

If we can, then in directions tangential to C from points on C we shall automatically satisfy the differential relationship

$$dU = \frac{\partial U}{\partial x}dx + \frac{\partial U}{\partial y}dy = p\,dx + q\,dy, \tag{4.2}$$

where dy/dx is the slope of the tangent to C at $P(x, y)$ on C.

Equations (4.1) and (4.2) are two equations for p and q.

Elimination of p between them gives that

$$dU = \frac{c - bq}{a} dx + q \, dy,$$

which can be written as

$$q(a \, dy - b \, dx) + (c \, dx - a \, dU) = 0. \tag{4.3}$$

This equation is explicitly independent of p because the coefficients a, b, and c are functions of x, y, and U only. It can also be made independent of q by choosing the curve C so that its slope dy/dx satisfies the equation

$$a \, dy - b \, dx = 0. \tag{4.4}$$

By eqns (4.4) and (4.3) it then follows that

$$c \, dx - a \, dU = 0. \tag{4.5}$$

Equation (4.4) is a differential equation for the curve C and (4.5) is a differential equation for the solution values of U along C. The curve C is called a characteristic curve or characteristic. These equations are easy to remember because they can be written as

$$\frac{dx}{a} = \frac{dy}{b} = \frac{dU}{c}.$$

This also shows that U may be found from either the equation $dU = (c/a) \, dx$ or the equation $dU = (c/b) \, dy$.

Example 4.1

Consider the equation

$$y \frac{\partial U}{\partial x} + \frac{\partial U}{\partial y} = 2, \tag{4.6}$$

where U is known along the initial segment Γ defined by $y = 0$, $0 \leqslant x \leqslant 1$.

The differential equation of the family of characteristic curves is

$$\frac{dx}{y} = \frac{dy}{1}.$$

Hence the equation of this family is $x = \frac{1}{2}y^2 + A$, where the parameter A is a constant for each characteristic. For the characteristic through $R(x_R, 0)$, $A = x_R$, so the equation of this particular characteristic is $y^2 = 2(x - x_R)$.

The solution along a characteristic curve is given by

$$\frac{dy}{1} = \frac{dU}{2},$$

which integrates to $U = 2y + B$, where B is constant along a particular characteristic. If $U = U_R$ at $R(x_R, 0)$ then $B = U_R$ and the solution along the characteristic $y^2 = 2(x - x_R)$ is $U = 2y + U_R$.

Since initial values for U are known only on the line segment OF, Fig. 4.1, where $0 \leqslant x_R \leqslant 1$, it follows that the solution is defined only in the region bounded by, and including, the terminal characteristics $y^2 = 2x$ and $y^2 = 2(x - 1)$. In this region the solution is clearly unique. Outside this region the solution is undefined.

If the initial curve Γ coincides with a characteristic, say, for example, the characteristic $y^2 = 2x$ through $O(0, 0)$, then the solution along this characteristic is $U = 2y + U_0$, where U_0 is the specified initial value of U at O. In other words, the initial values

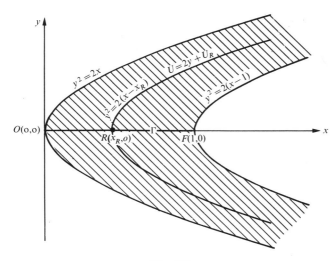

Fig. 4.1

for U on the initial curve $y^2 = 2x$ cannot now be arbitrarily prescribed as is obviously possible when Γ does not coincide with a characteristic curve. In this case it is also easily shown that the solution is not unique at points off $y^2 = 2x$. Consider, for example,
$$U = 2y + U_0 + A(y^2 - 2x),$$

where A is an arbitrary constant. This is clearly the solution along $y^2 = 2x$, whatever the value of A. It is also a solution of eqn (4.6) when $y^2 \neq 2x$ as can easily be verified by direct differentiation. Since A is an arbitrary constant there is an infinite number of different solutions at points off the characteristic $y^2 = 2x$.

A method for numerical integration along a characteristic

Let U be specified on the initial curve Γ which must not be a characteristic curve.

Let $R(x_R, y_R)$ be a point on Γ and $P(x_P, y_P)$ be a point on the characteristic curve C through R such that $x_P - x_R$ is small, Fig. 4.2. The differential equation for the characteristic is
$$a \, dy = b \, dx,$$

which gives either dy or dx when the other quantities are known.

The differential equation for the solution along a characteristic is either
$$a \, dU = c \, dx \quad \text{or} \quad b \, dU = c \, dy,$$

which gives dU for known dx or dy and known a, b, and c.

Denote a first approximation to U by $u^{(1)}$, a second approximation by $u^{(2)}$, etc.

First approximations

Assume that x_P is known. Then by the equations (4.4), (4.5)
$$a_R\{y_P^{(1)} - y_R\} = b_R(x_P - x_R)$$

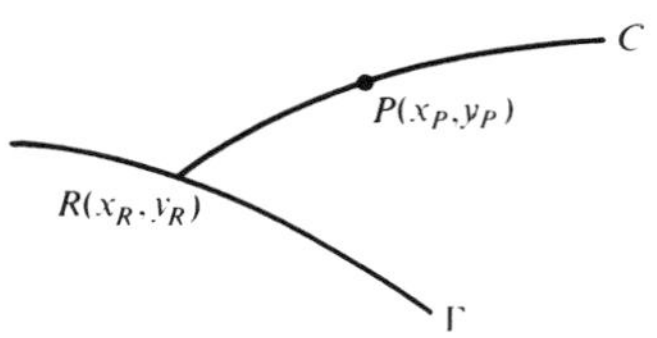

Fig. 4.2

gives a first approximation $y_P^{(1)}$ to y_P and

$$a_R\{u_P^{(1)} - u_R\} = c_R(x_P - x_R) \text{ gives } u_P^{(1)}.$$

Second and subsequent approximations

Replace the coefficients a, b, and c by known mean values over the arc RP. Then

$$\tfrac{1}{2}(a_R + a_P^{(1)})(y_P^{(2)} - y_R) = \tfrac{1}{2}(b_R + b_P^{(1)})(x_P - x_R) \text{ gives } y_P^{(2)},$$

and

$$\tfrac{1}{2}(a_R + a_P^{(1)})(u_P^{(2)} - u_R) = \tfrac{1}{2}(c_R + c_P^{(1)})(x_P - x_R) \text{ gives } u_P^{(2)}.$$

This second procedure can be repeated iteratively until successive iterates agree to a specified number of decimal places.

Example 4.2

The function U satisfies the equation

$$\sqrt{x}\frac{\partial U}{\partial x} + U\frac{\partial U}{\partial y} = -U^2$$

and the condition $U = 1$ on $y = 0$, $0 < x < \infty$.

Show that the Cartesian equation of the characteristic through the point $R(x_R, 0)$, $x_R > 0$, is $y = \log(2\sqrt{x} + 1 - 2\sqrt{x_R})$. Use a finite-difference method to calculate a first approximation to the solution and to the value of y at the point $P(1.1, y)$, $y > 0$, on the characteristic through the point $R(1, 0)$.

Calculate a second approximation to these values by an iterative method. Compare the results with those given by the analytical formulae for y and U. We have that

$$\frac{\mathrm{d}x}{\sqrt{x}} = \frac{\mathrm{d}y}{U} = \frac{\mathrm{d}U}{-U^2}.$$

Hence $y = -\log AU$. As $U = 1$ at $(x_R, 0)$, $A = 1$, so

$$y = \log\frac{1}{U}. \tag{4.7}$$

Similarly, $2\sqrt{x} = 1/U + B$. As $U = 1$ at $(x_R, 0)$, $B = 2\sqrt{x_R} - 1$. Therefore,

$$\frac{1}{U} = 2\sqrt{x} + 1 - 2\sqrt{x_R}. \tag{4.8}$$

Elimination of the parameter U between eqns (4.7) and (4.8) shows that the Cartesian equation of the characteristic through $(x_R, 0)$ is

$$y = \log(2\sqrt{x} + 1 - 2\sqrt{x_R}). \tag{4.9}$$

The solution along the characteristic is given either by

$$U = e^{-y} \quad \text{or} \quad U = \frac{1}{(2\sqrt{x} + 1 - 2\sqrt{x_R})}. \tag{4.10}$$

First approximations at (1.1, y). (Fig. 4.2)

We have that $\sqrt{x}\, dy = U\, dx$ and $\sqrt{x}\, dU = -U^2\, dx$.
Hence

$$\sqrt{x_R}(y_P^{(1)} - 0) = U_R\, dx \text{ giving } y_P^{(1)} = \frac{1}{\sqrt{1}}(0.1) = 0.1,$$

and

$$\sqrt{x_R}(u_P^{(1)} - 1) = -U_R^2\, dx \text{ giving } u_P^{(1)} = 1 - 0.1 = 0.9.$$

Second approximations

Using average values for the coefficients,

$$\tfrac{1}{2}(u_R + u_P^{(1)})\, dx = \tfrac{1}{2}(\sqrt{x_R} + \sqrt{x_P})(y_P^{(2)} - y_R),$$

giving

$$(1 + 0.9)(0.1) = (1 + 1.0488)(y_P^{(2)} - 0),$$

from which $y_P^{(2)} = 0.0927$. Also

$$\tfrac{1}{2}(\sqrt{x_R} + \sqrt{x_P})(u_P^{(2)} - u_R) = -\tfrac{1}{2}\{u_R^2 + (u_P^{(1)})^2\}\, dx,$$

giving

$$(1 + 1.0488)(u_P^{(2)} - 1) = -(1 + 0.81)(0.1),$$

from which $U_P^{(2)} = 0.9117$.

Note: The differential equations could have been written as

$$dy = \frac{U}{\sqrt{x}}\, dx \quad \text{and} \quad dU = -\frac{U^2}{\sqrt{x}}\, dx$$

and the second approximations obtained from

$$y_P^{(2)} = \frac{1}{2}\left\{\left(\frac{u}{\sqrt{x}}\right)_R + \left(\frac{u}{\sqrt{x}}\right)_P\right\} dx$$

and

$$u_P^{(2)} - u_R = \frac{1}{2}\left\{-\left(\frac{u^2}{\sqrt{x}}\right)_R - \left(\frac{u^2}{\sqrt{x}}\right)_P\right\} dx.$$

These approximations yield $y_P^{(2)} = 0.0929$ and $u_P^{(2)} = 0.9114$.

Analytical values

By eqn (4.9),

$$y_P = \log_e\{2(1.0488) + 1 - 2\} = 0.0934.$$

By eqn (4.10),

$$U_P = \frac{1}{1.0976} = 0.9111.$$

Finite-difference methods on a rectangular mesh for first-order equations

Lax–Wendroff explicit method

In the theory of fluid flow the equations of motion, of continuity, and of energy can be combined into one conservation equation of the form

$$\frac{\partial \mathbf{U}}{\partial t} + \frac{\partial \mathbf{F}(\mathbf{U})}{\partial x} = 0, \tag{4.11}$$

where $\mathbf{U}$ and $\mathbf{F}$ are each column vectors with three components. The Lax–Wendroff method, as illustrated below for a single dependent variable, can be used to approximate eqn (4.11) by an explicit difference equation of second-order accuracy.

Consider

$$\frac{\partial U}{\partial t} + a \frac{\partial U}{\partial x} = 0,$$

a a positive constant. By Taylor's expansion,

$$U_{i,j+1} = U(x_i, t_j + k) = U_{i,j} + k\left(\frac{\partial U}{\partial t}\right)_{i,j} + \tfrac{1}{2}k^2\left(\frac{\partial^2 U}{\partial t^2}\right)_{i,j} + \dots,$$

where $x_i = ih$ and $t_j = jk$, $i = 0, \pm 1, \pm 2, \dots, j = 0, 1, 2, \dots.$
The differential equation can now be used to eliminate the

t-derivatives because it gives that

$$\frac{\partial}{\partial t} \equiv -a\frac{\partial}{\partial x},$$

so

$$U_{i,j+1} = U_{i,j} - ka\left(\frac{\partial U}{\partial x}\right)_{i,j} + \tfrac{1}{2}k^2 a^2\left(\frac{\partial^2 U}{\partial x^2}\right)_{i,j} + \ldots$$

Finally, the replacement of the *x*-derivatives by central-difference approximations gives, to terms in k^2, the explicit difference equation

$$u_{i,j+1} = u_{i,j} - \frac{ka}{2h}(u_{i+1,j} - u_{i-1,j}) + \frac{k^2 a^2}{2h^2}(u_{i-1,j} - 2u_{i,j} + u_{i+1,j})$$

$$= \tfrac{1}{2}ap(1 + ap)u_{i-1,j} + (1 - a^2 p^2)u_{i,j} - \tfrac{1}{2}ap(1 - ap)u_{i+1,j},$$
$$(4.12)$$

where $p = k/h$. This may be used for both initial-value and initial-value boundary-value problems. It is often used to obtain numerical solutions to differential equations in fluid-flow problems when the dependent variables change rapidly with time. In such problems k must be kept small so the advantages of implicit methods are lost, namely stable equations and accurate results for fairly large values of k. As shown in Exercise 3, eqn (4.12) is stable for $0 < ap \leq 1$ and its local truncation error is

$$\tfrac{1}{6}k^2\frac{\partial^3 U}{\partial t^3} + \tfrac{1}{6}ah^2\frac{\partial^3 U}{\partial x^3} + \ldots$$

Example 4.3

The function U satisfies the equation

$$\frac{\partial U}{\partial t} + \frac{\partial U}{\partial x} = 0, \quad 0 < x < \infty, \quad t > 0,$$

the boundary condition

$$U(0, t) = 2t, \quad t > 0,$$

and the initial conditions

$$U(x, 0) = x(x - 2), \quad 0 \leq x \leq 2,$$
$$U(x, 0) = 2(x - 2), \quad 2 \leq x.$$

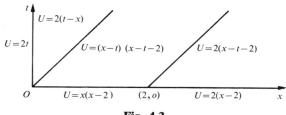

Fig. 4.3

Calculate (i) the analytical solution and (ii) a numerical solution using the explicit Lax–Wendroff equation.

(i) $dt = dx = dU/0$. Hence U is constant along the straight line characteristics $t = x + \text{constant}$. Therefore, the equation of the characteristic from $R(x_R, 0)$ is $t = x - x_R$ and if $U(x, 0)$ is $\phi(x)$, say, then the solution along this characteristic is $U(x, t) = \phi(x_R) = \phi(x - t)$. Similarly, if $U(0, t) = \psi(t)$, the solution along the characteristic $t - t_S = x$ from $S(0, t_s)$ is $U(x, t) = \psi(t - x)$. Hence the conditions of the problem give the solution shown in Fig. 4.3.

(ii) Equation (4.12) for $h = \tfrac{1}{4}$ and $k = \tfrac{1}{8}$ is

$$u_{i,j+1} = \tfrac{3}{8}u_{i-1,j} + \tfrac{3}{4}u_{i,j} - \tfrac{1}{8}u_{i+1,j}.$$

Table 4.1 displays the analytical solution of the differential equation below each mesh point and the corresponding Lax–Wendroff solution to $3D$ above each mesh point for $x = 0(1)7$ and $t = 0(\tfrac{1}{2})3$. The boundary and initial conditions make $U(x, t)$ continuous at $(0, 0)$ and $(2, 0)$, $\partial U/\partial x$ continuous at $(2, 0)$, and ensures that $\partial U/\partial x$ and $\partial U/\partial t$ satisfy the differential equation at $(0, 0)$.

The Lax–Wendroff method for a set of simultaneous equations

The Lax–Wendroff approximation is easily extended to the set of simultaneous equations

$$\frac{\partial \mathbf{U}}{\partial t} + \mathbf{A}\frac{\partial \mathbf{U}}{\partial x} = 0, \tag{4.13}$$

where

$$\mathbf{U} = [U_1, U_2, \ldots, U_N]^T, \quad \frac{\partial \mathbf{U}}{\partial t} = \left[\frac{\partial U_1}{\partial t}, \frac{\partial U_2}{\partial t}, \ldots, \frac{\partial U_N}{\partial t}\right]^T,$$

etc., and $\mathbf{A}$ is an $N \times N$ matrix with real constant elements.

TABLE 4.1

t		x 1	2	3	4	5	6	7
		6·000	3·996	2·008	−0·060			
(0, 4)	8·0	6·0	4·0	2·0	0	−1·0	0	2·0
		5·001	3·005	0·946	−0·763	−0·695		
(0, 3·5)	7·0	5·0	3·0	1·0	−0·75	−0·75	1·0	3·0
		3·999	2·010	−0·050	−1·011	0·054	2·000	4·000
(0, 3)	6·0	4·0	2·0	0	−1·0	0	2·0	4·0
		3·000	0·965	−0·754	−0·713	1·004	3·000	5·000
(0, 2·5)	5·0	3·0	1·0	−0·75	−0·75	1·0	3·0	5·0
		2·007	−0·039	−1·009	0·041	2·000	4·000	6·000
(0, 2)	4·0	2·0	0	−1·0	0	2·0	4·0	6·0
		0·985	−0·750	−0·732	1·001	3·000	5·000	7·000
(0, 1·5)	3·0	1·0	−0·75	−0·75	1·0	3·0	5·0	7·0
		−0·022	−1·002	0·024	2·000	4·000	6·000	8·000
(0, 1)	2·0	0	−1·0	0	2·0	4·0	6·0	8·0
		−0·750	−0·750	1·000	3·000	5·000	7·000	9·000
(0, 0·5)	1·0	−0·75	−0·75	1·0	3·0	5·0	7·0	9·0
		−1·0	0	2·0	4·0	6·0	8·0	10·0
(0, 0)		(1, 0)	(2, 0)	(3, 0)	(4, 0)	(5, 0)	(6, 0)	(7, 0)

By Taylor's expansion,

$$\mathbf{U}_{i,j+1} = \mathbf{U}_{i,j} + k\,\frac{\partial \mathbf{U}_{i,j}}{\partial t} + \tfrac{1}{2}k^2 \frac{\partial}{\partial t}\!\left(\frac{\partial \mathbf{U}}{\partial t}\right)_{i,j} + \ldots .$$

By eqn (4.13),

$$\frac{\partial}{\partial t} \equiv -\mathbf{A}\,\frac{\partial}{\partial x} .$$

Hence,

$$\mathbf{U}_{i,j+1} = \mathbf{U}_{i,j} - k\mathbf{A}\left(\frac{\partial \mathbf{U}}{\partial x}\right)_{i,j} + \tfrac{1}{2}k^2\mathbf{A}^2\left(\frac{\partial^2 \mathbf{U}}{\partial x^2}\right)_{i,j} + \dots .$$

The standard central-difference approximations for $\partial \mathbf{U}/\partial x$ and $\partial^2 \mathbf{U}/\partial x^2$ then give the approximating difference equations

$$\mathbf{u}_{i,j+1} = \mathbf{u}_{i,j} - \tfrac{1}{2}p\mathbf{A}(\mathbf{u}_{i+1,j} - \mathbf{u}_{i-1,j}) + \tfrac{1}{2}p^2\mathbf{A}^2(\mathbf{u}_{i-1,j} - 2\mathbf{u}_{i,j} + \mathbf{u}_{i+1,j}).$$

(See Exercise 4.)

The formal development of an approximation to (4.11) is as follows. By Taylor's series,

$$\mathbf{U}_{i,j+1} = \mathbf{U}_{i,j} + k\frac{\partial \mathbf{U}}{\partial t} + \tfrac{1}{2}k^2\frac{\partial}{\partial t}\left(\frac{\partial \mathbf{U}}{\partial t}\right)_{i,j} + \dots .$$

In virtue of eqn (4.11) it follows that

$$\mathbf{U}_{i,j+1} = \mathbf{U}_{i,j} - k\left(\frac{\partial \mathbf{F}}{\partial x}\right)_{i,j} - \tfrac{1}{2}k^2\frac{\partial}{\partial t}\left(\frac{\partial \mathbf{F}}{\partial x}\right)_{i,j} + \dots . \qquad (4.14)$$

But

$$\frac{\partial}{\partial t}\frac{\partial \mathbf{F}}{\partial x} = \frac{\partial}{\partial x}\frac{\partial \mathbf{F}}{\partial t} = \frac{\partial}{\partial x}\left\{\frac{\partial \mathbf{F}}{\partial \mathbf{U}}\frac{\partial \mathbf{U}}{\partial t}\right\} = -\frac{\partial}{\partial x}\left\{\frac{\partial \mathbf{F}}{\partial \mathbf{U}}\frac{\partial \mathbf{F}}{\partial x}\right\},$$

where

$$\frac{\partial \mathbf{F}}{\partial \mathbf{U}} = A(\mathbf{U}),$$

the Jacobian matrix of $\mathbf{F}$ with respect to $\mathbf{U}$, is defined by

$$A_{m,n} = \frac{\partial F_m}{\partial U_n}.$$

For example, if

$$\mathbf{U} = \begin{bmatrix} U_1 \\ U_2 \end{bmatrix} \quad \text{and} \quad \mathbf{F}(\mathbf{U}) = \begin{bmatrix} F_1(U) \\ F_2(U) \end{bmatrix}$$

$$\text{then} \quad A(\mathbf{U}) = \begin{bmatrix} \partial F_1/\partial U_1 & \partial F_1/\partial U_2 \\ \partial F_2/\partial U_1 & \partial F_2/\partial U_2 \end{bmatrix}.$$

Therefore eqn (4.14) can be written as

$$\mathbf{U}_{i,j+1} = \mathbf{U}_{i,j} - k\left(\frac{\partial \mathbf{F}}{\partial x}\right)_{i,j} + \tfrac{1}{2}k^2\frac{\partial}{\partial x}\left\{A(\mathbf{U})\frac{\partial \mathbf{F}}{\partial x}\right\}_{i,j} + \dots ,$$

and using the central-difference approximation for the last term as shown in Chapter 3, Exercise 8(a), we obtain the Lax–Wendroff approximation

$$\mathbf{u}_{i,j+1} = \mathbf{u}_{i,j} - \tfrac{1}{2}p(\mathbf{F}_{i+1,j} - \mathbf{F}_{i-1,j})$$
$$+ \tfrac{1}{2}p^2\{A_{i+\frac{1}{2},j}(\mathbf{F}_{i+1,j} - \mathbf{F}_{i,j}) - A_{i-\frac{1}{2},j}(\mathbf{F}_{i,j} - \mathbf{F}_{i-1,j})\}.$$

To avoid midpoint evaluations it is usual to approximate $A_{i+\frac{1}{2},j}$ by $\frac{1}{2}(A_{i,j} + A_{i+1,j})$ and $A_{i-\frac{1}{2},j}$ by $\frac{1}{2}(A_{i-1,j} + A_{i,j})$.

Further details concerning stability, well posedness, and the solution of eqn (4.11) are given in references 25 and 18.

The Courant–Friedrichs–Lewy (C.F.L.) condition for first-order equations

Assume that a first-order hyperbolic differential equation has been approximated by a difference equation of the form

$$u_{i,j+1} = au_{i-1,j} + bu_{i,j} + cu_{i+1,j}.$$

Then u_P, Fig. 4.4, depends on the values of u at the mesh points A, B, and C. Assume now that the characteristic curve through P of the hyperbolic equation meets the line AC at D and consider AC as an initial line segment. If the initial values along AC are altered then the solution value at P of the finite-difference equation will change, but these alterations will not affect the solution value at P of the differential equation which depends on the initial value at D. In this case u_P cannot converge to U_P as $h \to 0$, $k \to 0$. For convergence D must lie between A and C. (The C.F.L. condition.) Consider, for example, the Lax–Wendroff approximation (4.12). The slope dt/dx of the characteristic of the corresponding differential equation is given by $dt/1 = dx/a$. For convergence of the difference equation, slope of $PD \geqslant$ slope of

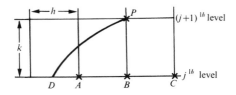

Fig. 4.4

PA, i.e. $1/a \geqslant k/h$, giving $ap \leqslant 1$, which coincides with the condition for stability, namely, $0 < ap \leqslant 1$, since $a > 0$, $p > 0$.

Wendroff's implicit approximation

An implicit approximation of second-order accuracy to the equation

$$a\frac{\partial U}{\partial x} + b\frac{\partial U}{\partial y} = c$$

at the point *P*, Fig. 4.5, is given by approximating $(\partial U/\partial x)_P$ and $(\partial U/\partial y)_P$ by

$$\frac{1}{2}\left\{\left(\frac{\partial U}{\partial x}\right)_H + \left(\frac{\partial U}{\partial x}\right)_F\right\} \quad \text{and} \quad \frac{1}{2}\left\{\left(\frac{\partial U}{\partial y}\right)_G + \left(\frac{\partial U}{\partial y}\right)_E\right\}$$

respectively, then approximating these derivatives by central-difference formulae to give the Wendroff equation

$$\frac{a}{2}\left\{\frac{u_B - u_A}{h} + \frac{u_C - u_D}{h}\right\} + \frac{b}{2}\left\{\frac{u_D - u_A}{k} + \frac{u_C - u_B}{k}\right\} = c,$$

which can be written as

$$(b - ap)u_D + (b + ap)u_C = (b + ap)u_A + (b - ap)u_B + 2kc.$$

This is unconditionally stable. (See Exercise 5.) It cannot be used for pure initial-value problems, i.e. conditions on $t = 0$ only, because it would give an infinite number of simultaneous equations. If, however, initial values are known on *Ox*, $x \geqslant 0$, and boundary values on *Ot*, $t \geqslant 0$, the equation can be used explicitly by writing it as

$$u_C = u_A + \frac{b - ap}{b + ap}(u_B - u_D) + \frac{2kc}{b + ap}, \qquad (4.15)$$

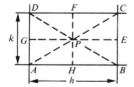

Fig. 4.5

and will give the solution in the quarter plane $x > 0$, $t > 0$. (See p. 191 for a numerical example.) It will also give approximation values in the strip $0 < x \leqslant 1$, $t > 0$, when initial values are known on Ox, $0 \leqslant x \leqslant 1$, and boundary values are known on Ot, $t > 0$.

Propagation of discontinuities in first-order equations

Consider the equation

$$\frac{\partial U}{\partial x} + \frac{\partial U}{\partial y} = 1, \quad y \geqslant 0, \quad -\infty < x < \infty,$$

where U is known at points $P(x_P, 0)$ on the x-axis. The characteristic direction is given by $dx = dy$ and along the characteristics, $dU = dy$. Hence the characteristic through P is $y = x - x_P$ and the solution along it is $U = U_P + y$.

Discontinuous initial values

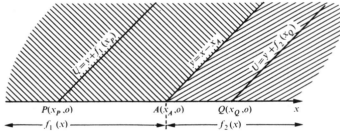

Fig. 4.6

Let
$$U(x, 0) = f_1(x), \quad -\infty < x < x_A,$$
and
$$U(x, 0) = f_2(x), \quad x_A < x < \infty.$$

To the left of the characteristic $y = x - x_A$ the solution $U_{(L)}$ is $U_{(L)} = f_1(x_P) + y$ along $y = x - x_P$. To the right of the straight line $y = x - x_A$ the solution $U_{(R)}$ is $U_{(R)} = f_2(x_Q) + y$ along $y = x - x_Q$.

Hence, for the same value of y in both solutions,

$$U_{(L)} - U_{(R)} = f_1(x_P) - f_2(x_Q).$$

Clearly, as x_P and x_Q both tend to x_A, $U_{(L)} - U_{(R)}$ is discontinuous along $y = x - x_A$ where

$$\lim_{x_P \to x_A} f_1(x_P) \neq \lim_{x_Q \to x_A} f_2(x_Q).$$

This shows that when the initial values are discontinuous at a particular point A, say, then the solution is discontinuous along the characteristic curve C from A. Moreover, the effect of this initial discontinuity does not diminish as we move away from A along C. With parabolic and elliptic differential equations the effect of an initial discontinuity is quite different as it tends to be localized and to diminish fairly rapidly with distance from the point of discontinuity.

Discontinuous initial derivatives

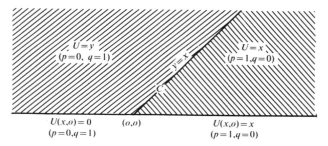

Fig. 4.7

Let

$$U(x, 0) = 0, \quad -\infty < x \leqslant 0,$$

and

$$U(x, 0) = x, \quad 0 < x < \infty.$$

This makes the initial derivative $p(x, 0) = \partial U(x, 0)/\partial x$ discontinuous at $(0, 0)$. The initial derivative $q(x, 0) = [\partial U(x, y)/\partial y]_{(x,0)}$ is also discontinuous at $(0, 0)$ because $p + q = 1$ from the differential equation. $U(x, 0)$ itself is clearly continuous at $(0, 0)$. As before, $U - U_P = y = x - x_P$ along the characteristic $y = x - x_P$ from $P(x_P, 0)$. Therefore the solution $U_{(L)}$ to the left of $y = x$ is

$$U_{(L)} = y, \quad -\infty < x \leqslant 0, \quad y \geqslant 0,$$

and the solution $U_{(R)}$ to the right of $y = x$ is

$$U_{(R)} = x, \quad 0 < x < \infty, \quad y \geqslant 0$$

It is seen from this that $U_{(L)} = U_{(R)}$ along $y = x$ but that $p_{(L)} = \partial U_{(L)}/\partial x = 0$ and $p_{(R)} = \partial U_{(R)}/\partial x = 1$, i.e. the solution is continuous along the characteristic C from the point of discontinuity but the initial discontinuities in the partial derivatives are propagated undiminished across the solution domain along this characteristic. In many cases discontinuous partial derivatives arise from sudden changes in the direction of the boundary-initial curve as illustrated in the next section.

Discontinuities and finite difference methods

As shown above, a discontinuity in the initial data of a first-order equation is propagated across the solution domain along the characteristic from the point of discontinuity. In such a case one would expect the 'method of characteristics' on p. 178 to give a more accurate numerical solution than finite-difference methods because the solution corresponding to a particular initial value is developed along a characteristic that does not intersect the characteristic from a point of discontinuity. But the programming of the method of characteristics, especially for problems involving a set of simultaneous first-order equations, is much more difficult than the programming of difference methods. For this reason a great deal of research has been devoted to the formulation of finite-difference schemes that simulate the propagation of discontinuities along characteristics in the sense that rapid changes are confined to narrow regions. (See references 15 and 25.)

The 'blurring' of discontinuities that occurs with difference methods is illustrated in the following example. Consider the equation

$$\frac{\partial U}{\partial x} + \frac{\partial U}{\partial y} = 1, \quad 0 < x < \infty, \quad y > 0, \qquad (4.16)$$

where $U(0, y) = 0, \quad 0 < y < \infty,$
$\qquad U(x, 0) = 0, \quad 0 \leqslant x \leqslant 3,$
$\qquad U(x, 0) = x - 3, \quad 3 \leqslant x < \infty.$

In general, if $U = U_R$ at the initial point $R(x_R, y_R)$ then the characteristic through R and the solution along it are given by

$U - U_R = x - x_R = y - y_R$. Therefore:
 (i) $U = x$ along the characteristic from $(0, y_R)$, $y_R > 0$,
 (ii) $U = y$ along the characteristic from $(x_R, 0)$ $0 \leqslant x_R \leqslant 3$, and
 (iii) $U = x - 3$ along the characteristic from $(x_R, 0)$, $x_R \geqslant 3$.

The numerical values for this analytical solution are shown in Table 4.2 for $x = 0(1)8$ and $y = 0(1)5$. Although the boundary values and initial values are continuous at $(0, 0)$ and $(3, 0)$ it is seen that $p = \partial U/\partial x$ and $q = \partial U/\partial y$ are discontinuous at these points. The discontinuities at $(0, 0)$ arise from the sudden change in the direction of the boundary curve yOx at O. (N.B. As $p + q = 1$ from the differential equation, $\delta p = -\delta q$.) The table also illustrates numerically how the discontinuities in p and q in the solution domain occur on the characteristics from $(0, 0)$ and $(3, 0)$, i.e. the discontinuities are propagated cleanly along these characteristics.

Let us now approximate eqn (4.16) by Wendroff's equation (4.15). By Fig. 4.5,

$$u_C = u_A + \frac{h - k}{h + k}(u_B - u_D) + \frac{2hk}{h + k}.$$

Case (i) Take $h = k = 1$. Then

$$u_C = u_A + 1, \quad \text{i.e.} \quad u_{i+1,j+1} = u_{i,j} + 1,$$

and the application of this equation to the boundary and initial values of Table 4.2 shows that the numerical solution coincides

TABLE 4.2

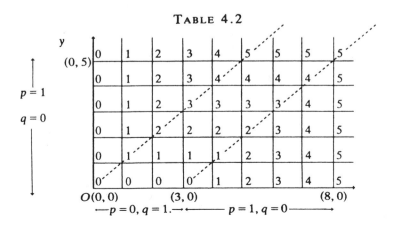

with the differential equation solution. This happens because $u_{i+1,j+1} = u_{i,j} + 1$ propagates the finite-difference solution forward along the characteristics from the initial points and boundary points and because the local truncation error at the point $(i + \frac{1}{2}, j + \frac{1}{2})$ is zero in this case.

Case (*ii*) Take $h = 1$, $k = \frac{1}{2}$. Then

$$u_C = u_A + \tfrac{1}{3}(u_B - u_D) + \tfrac{2}{3}.$$

The numerical solution to $2D$ is displayed in Table 4.3. Every second row corresponds to a row of Table 4.2. The values of $q_{i,j} = (u_{i,j+1} - u_{i,j})/k$ are shown in Table 4.4. The analytical values for $q_{i,j}$ for the three different solution domains are indicated at the top of Table 4.4. The values for u clearly improve in accuracy as one moves away from the characteristics through $(0, 0)$ and $(3, 0)$ and Table 4.4 shows that the discontinuity in q along these characteristics is 'diffused' over an area each side of the characteristics.

TABLE 4.3

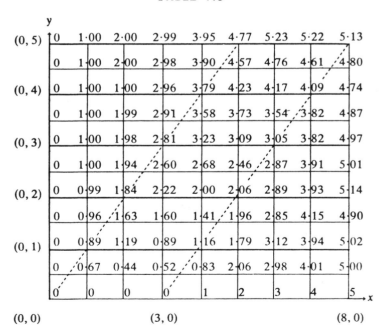

y								
(0, 5)	0	1·00	2·00	2·99	3·95	4·77	5·23	5·22	5·13
	0	1·00	2·00	2·98	3·90	4·57	4·76	4·61	4·80
(0, 4)	0	1·00	1·00	2·96	3·79	4·23	4·17	4·09	4·74
	0	1·00	1·99	2·91	3·58	3·73	3·54	3·82	4·87
(0, 3)	0	1·00	1·98	2·81	3·23	3·09	3·05	3·82	4·97
	0	1·00	1·94	2·60	2·68	2·46	2·87	3·91	5·01
(0, 2)	0	0·99	1·84	2·22	2·00	2·06	2·89	3·93	5·14
	0	0·96	1·63	1·60	1·41	1·96	2·85	4·15	4·90
(0, 1)	0	0·89	1·19	0·89	1·16	1·79	3·12	3·94	5·02
	0	0·67	0·44	0·52	0·83	2·06	2·98	4·01	5·00
	0	0	0	0	1	2	3	4	5 → x

(0, 0)   (3, 0)   (8, 0)

TABLE 4.4

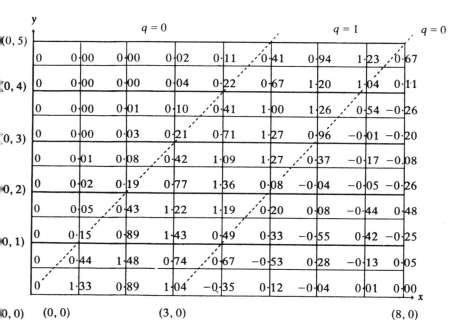

Reduction of a first-order equation to a system of ordinary differential equations

This section can be omitted if the reader wishes to study immediately the numerical solution of second-order hyperbolic equations.

Consider the first-order hyperbolic equation

$$\frac{\partial U}{\partial t} + a\frac{\partial U}{\partial x} = 0, \quad x>0, \quad t>0, \tag{4.17}$$

where a is a real positive constant and U satisfies the initial condition

$$U(x, 0) = g(x), \quad x \geqslant 0,$$

and the boundary condition

$$U(0, t) = b(t), \quad t > 0.$$

Assume that the solution is needed in that strip R of the first quadrant $x > 0$, $t > 0$, defined by $R = [0 < x < X] \times [t > 0]$. (Read as the set of all points (x, t) such that $0 < x < X$ and $t > 0$.) Subdivide the interval $0 \leqslant x \leqslant X$ into N parts each of width h, so that $Nh = X$, and discretize t in steps of length k. A typical mesh point (x_i, t_j) is then defined by

$$(x_i, t_j) = (ih, jk), \quad 0 \leqslant i \leqslant N, \quad j = 0, 1, 2, \ldots .$$

If the x-derivative at (x, t) is replaced by the backward-difference formula

$$\frac{\partial U}{\partial x} = \frac{1}{h} \{U(x, t) - U(x - h, t)\} + O(h), \tag{4.18}$$

and x is considered as a constant, eqn (4.17) can be written as the ordinary differential equation

$$\frac{dU(t)}{dt} = -\frac{a}{h} \{U(x, t) - U(x - h, t)\} + O(h).$$

On writing down this ordinary equation at the N mesh points $x_i = ih$, $i = 1(1)N$, along time-level t, it follows that the values $V_i(t)$ approximating $U_i(t)$ are the exact solution values of the N equations

$$\frac{dV_1}{dt} = -\frac{a}{h} \{V_1 - b(t)\}$$

$$\frac{dV_2}{dt} = -\frac{a}{h} \{V_2 - V_1\}$$

$$\vdots$$

$$\frac{dV_N}{dt} = -\frac{a}{h} \{V_N - V_{N-1}\}.$$

These can be expressed as

$$\frac{dV(t)}{dt} = -a\mathbf{C}V(t) + a\mathbf{b}(t), \tag{4.19}$$

where $\mathbf{V}(t) = [V_1(t), V_2(t), \ldots, V_N(t)]^T$, $\mathbf{C}$ is the $N \times N$ lower bidiagonal matrix

$$\mathbf{C} = \frac{1}{h} \begin{bmatrix} 1 & & & & \\ -1 & 1 & & & \\ & -1 & 1 & & \\ & & \cdot & \cdot & \cdot \\ & & & \cdot & \cdot \\ & & & -1 & 1 \end{bmatrix}$$

and $\mathbf{b}(t) = [b(t)/h, 0, 0, \ldots, 0]^T$ is known from the boundary condition. When $\mathbf{b}(t)$ is a constant vector the solution of (4.19) is, by eqn (3.5),

$$\mathbf{V}(t) = \mathbf{C}^{-1}\mathbf{b}(t) + \{\exp(-at\mathbf{C})\}\{\mathbf{g} - \mathbf{C}^{-1}\mathbf{b}(t)\}, \qquad (4.20)$$

and satisfies the recurrence relationship

$$\mathbf{V}(t+k) = \mathbf{C}^{-1}\mathbf{b}(t) + \{\exp(-ak\mathbf{C})\}\{\mathbf{V}(t) - \mathbf{C}^{-1}\mathbf{b}(t)\}. \qquad (4.21)$$

The approximating finite-difference equations

Replacement of the exponential in eqn (4.21) by an (S, T) Padé approximant, $S \geq T$, leads to an unconditionally stable implicit difference scheme that can always be put into an explicit form because the initial conditions and boundary conditions are known along the two axes of the first quadrant.

For example, the $(1, 0)$ Padé approximant approximates (4.21) by the difference equations

$$\mathbf{u}_{j+1} = \mathbf{C}^{-1}\mathbf{b}_j + (\mathbf{I} + ak\mathbf{C})^{-1}\{\mathbf{u}_j - \mathbf{C}^{-1}\mathbf{b}_j\},$$

giving the implicit scheme

$$(\mathbf{I} + ak\mathbf{C})\mathbf{u}_{j+1} = \mathbf{u}_j + ak\mathbf{b}_j.$$

In detail, the equations are

$$(1 + ap)u_{1,j+1} = u_{1,j} + apb_j$$

and

$$-apu_{i-1,j+1} + (1 + ap)u_{i,j+1} = u_{i,j}, \quad i = 2(1)N,$$

where $p = k/h$. A small modification enables us to write them more compactly. The boundary value has been treated as a constant over $t_j \leq t \leq t_{j+1}$ and could be given its value at t_j or t_{j+1}, or even averaged. For implicit schemes it is usually assigned its value at t_{j+1}. All N equations of the $(1, 0)$ approximant can then be written as

$$-apu_{i-1,j+1} + (1+ap)u_{i,j+1} = u_{i,j}, \quad i = 1(1)N. \qquad (4.22)$$

They can obviously be expressed explicitly as

$$u_{i,j+1} = (apu_{i-1,j+1} + u_{i,j})/(1+ap), \qquad (4.23)$$

and are shown to be unconditionally stable in Exercise 7. The principal part of the local truncation error of (4.22) at (ih, jk) is

$$\left[-\tfrac{1}{2}ah \frac{\partial^2 U}{\partial x^2} - \tfrac{1}{2}k \frac{\partial^2 U}{\partial t^2} \right]_{i,j}.$$

(See Exercise 6.)

A comment on the non-stiffness of the equations

Although matrix $\mathbf{C}$ has N eigenvalues λ_s each equal to $1/h$, it has N linearly independent eigenvectors $\mathbf{v}_s$, so the vector of initial values $\mathbf{g}$, with its N components $g_1, g_2, \ldots, g_N$, can be expressed as $\mathbf{g} = \sum\limits_{s=1}^{N} c_s \mathbf{v}_s$, c_s constants. Hence the solution (4.20) can be written as

$$\mathbf{V}(t) = \mathbf{C}^{-1}\mathbf{b}(t) + \left\{ \sum_{s=1}^{N} c_s[\exp(-at\lambda_s)]\mathbf{v}_s \right\} - \{\exp(-at\mathbf{C})\}\mathbf{C}^{-1}\mathbf{b}(t).$$

This proves that the system of ordinary differential equations is non-stiff because the components $\{\exp(-at\lambda_s)\}\mathbf{v}_s$ of $\mathbf{V}(t)$ do not decay at different rates, as occurs with parabolic equations. The components are also non-oscillatory because a, t and $\lambda_s = 1/h$ are real.

The Crank–Nicolson scheme

It is easily shown that the $(1, 1)$ Padé approximant to $\exp(-at\mathbf{C})$ in eqn (4.20) leads to the explicit equations,

$$u_{1,j+1} = [(1 - \tfrac{1}{2}ap)u_{1,j} + \tfrac{1}{2}ap(u_{0,j+1} + u_{0,j})]/(1 + \tfrac{1}{2}ap),$$

and

$$u_{i,j+1} = [\tfrac{1}{2}apu_{i-1,j+1} + (1 - \tfrac{1}{2}ap)u_{i,j} + \tfrac{1}{2}apu_{i-1,j}]/(1 + \tfrac{1}{2}ap),$$

$$i = 2(1)N. \quad (4.24)$$

These are A_0-stable and the principal part of the local truncation error at (ih, jk) is

$$\left[-\tfrac{1}{2}ah \frac{\partial^2 U}{\partial x^2} - \tfrac{1}{12}k^2 \frac{\partial^3 U}{\partial t^3} \right]_{i,j}.$$

The k-components of the local truncation errors of eqns (4.23) and (4.24) can be improved by at least one power of k by extrapolating in t, as described in Chapter 3, but this may not be worth the computational effort involved if the h-component is not improved to at least $O(h^2)$, as described in the next section.

An improved approximation to $\partial U/\partial x$

By Maclaurin's expansion,

$$U(x-h, t) = U(x, t) - h\frac{\partial U(x, t)}{\partial x} + \tfrac{1}{2}h^2 \frac{\partial^2 U(x, t)}{\partial x^2} + O(h^3),$$

$$U(x-2h, t) = U(x, t) - 2h\frac{\partial U(x, t)}{\partial x} + 2h^2 \frac{\partial^2 U(x, t)}{\partial x^2} + O(h^3).$$

Elimination of $h^2 \partial^2 U/\partial x^2$ leads to

$$\frac{\partial U(x, t)}{\partial x} = \frac{1}{2h}\{3U(x, t) - 4U(x-h, t) + U(x-2h, t)\} + O(h^2).$$

$$(4.25)$$

As this involves three points along any one time-row it can be used only for $i = 2, 3, \ldots, N$. At $i = 1$ we shall use the previous backward difference replacement (4.18). The equation

$$\frac{\partial U}{\partial t} + a\frac{\partial U}{\partial x} = 0, \quad 0 < x < X, \quad t > 0,$$

is then approximated at the mesh points ih, $i = 1(1)N$, along time-level t, by the system of ordinary equations

$$\frac{dV_1}{dt} = -\frac{2a}{2h}\{V_1 - b(t)\}$$

$$\frac{dV_2}{dt} = -\frac{a}{2h}\{3V_2 - 4V_1 + b(t)\}$$

$$\frac{dV_3}{dt} = -\frac{a}{2h}\{3V_3 - 4V_2 + V_1\}$$

$$\vdots$$

$$\frac{dV_N}{dt} = -\frac{a}{2h}\{3V_N - 4V_{N-1} + V_{N-2}\}.$$

These can be written in matrix form as

$$\frac{d\mathbf{V}(t)}{dt} = -\tfrac{1}{2}a\,\mathbf{D}\mathbf{V}(t) + \tfrac{1}{2}a\mathbf{b}(t), \tag{4.26}$$

where the matrix $\mathbf{D}$ of order N is

$$\mathbf{D} = \frac{1}{h}\begin{bmatrix} 2 & & & & & \\ -4 & 3 & & & & \\ 1 & -4 & 3 & & & \\ & 1 & -4 & 3 & & \\ & & & \ddots & \ddots & \ddots \\ & & & 1 & -4 & 3 \end{bmatrix} \quad \text{and} \quad \mathbf{b}(t) = \frac{1}{h}\begin{bmatrix} 2b(t) \\ -b(t) \\ 0 \\ 0 \\ \vdots \\ 0 \end{bmatrix}.$$

The solution of eqn (4.26) satisfying the initial condition $\mathbf{V}(0) = \mathbf{g}$ is, by eqn (3.5), for $\mathbf{b}(t)$ constant

$$\mathbf{V}(t) = \mathbf{D}^{-1}\mathbf{b}(t) + \{\exp(-\tfrac{1}{2}at\mathbf{D})\}\{\mathbf{g} - \mathbf{D}^{-1}\mathbf{b}(t)\}. \tag{4.27}$$

Gustafsson's theorems, reference 11, justify this mixture of approximations to $\partial U/\partial x$ provided the Padé approximant chosen to approximate the exponential is such that when applied to the set of ordinary differential equations given by the lower-order approximation (4.18), the resulting difference equations are un-conditionally stable. The difference equations given by (4.26) will also be unconditionally stable and converge at the same rate as the more accurate approximation.

Equations (4.26) are non-stiff because matrix $\mathbf{D}$ has one eigenvalue $2/h$ and $(N-1)$ eigenvalues $3/h$, so the range of eigenvalues is small.

The solution (4.27) satisfies the recurrence relationship

$$\mathbf{V}(t+k) = \mathbf{D}^{-1}\mathbf{b}(t) + \{\exp(-\tfrac{1}{2}ak\mathbf{D})\}\{\mathbf{V}(t) - \mathbf{D}^{-1}\mathbf{b}(t)\}, \quad (4.28)$$

on treating $\mathbf{b}(t)$ as constant in $t_j \leqslant t \leqslant t_{j+1}$.

The (1, 0) Padé approximant

Approximating the exponential in (4.28) by its $(1, 0)$ Padé approximant $(\mathbf{I}+\tfrac{1}{2}ak\mathbf{D})^{-1}$ and replacing $\mathbf{b}(t)$ by $\mathbf{b}(t+k)$ gives the L_0-stable system

$$(\mathbf{I}+\tfrac{1}{2}ak\mathbf{D})\mathbf{u}_{j+1} - \tfrac{1}{2}ak\mathbf{b}_{j+1} = \mathbf{u}_j.$$

In detail, the equations are

$$(1+ap)u_{1,j+1} - apb_{j+1} = u_{1,j}$$
$$-2apu_{1,j+1} + (1+\tfrac{3}{2}ap)u_{2,j+1} + \tfrac{1}{2}apb_{j+1} = u_{2,j}$$
$$\tfrac{1}{2}apu_{i-2,j+1} - 2apu_{i-1,j+1} + (1+\tfrac{3}{2}ap)u_{i,j+1} = u_{i,j}, \quad i = 3(1)N.$$

Although they look implicit they can be solved explicitly for $u_{i,j+1}$, $i = 1(1)N$.

The principal part of the local truncation error is

$$\left[-\tfrac{1}{2}ah\frac{\partial^2 U}{\partial x^2} - \tfrac{1}{2}k\frac{\partial^2 U}{\partial t^2} \right] \quad \text{at } (h, jk)$$

and

$$\left[-\tfrac{1}{3}ah^2\frac{\partial^3 U}{\partial x^3} - \tfrac{1}{2}k\frac{\partial^2 U}{\partial t^2} \right] \quad \text{at } (ih, jk), \quad i = 2(1)N.$$

This scheme is worth extrapolating in t because the local truncation error is $O(h^2)$ in x and the equations L_0-stable. As in Chapter 3 we calculate $(\mathbf{I}-2k\mathbf{A})\mathbf{u}^{(1)} = \mathbf{u}(t)$, $(\mathbf{I}-k\mathbf{A})^2\mathbf{u}^{(2)} = \mathbf{u}(t)$, where $\mathbf{A} = -\tfrac{1}{2}a\mathbf{D}$, then $\mathbf{u}^{(E)}(t+2k) = 2\mathbf{u}^{(2)} - \mathbf{u}^{(1)}$.

All the equations can be solved explicitly and the principal part of the local truncation error is

$$\left[-\tfrac{1}{2}ah\frac{\partial^2 U}{\partial x^2} + \tfrac{4}{3}k^2\frac{\partial^3 U}{\partial t^2} \right] \quad \text{at } (1, j)$$

and

$$\left[-\tfrac{1}{3}ah^2\frac{\partial^3 U}{\partial x^3} + \tfrac{4}{3}k^2\frac{\partial^3 U}{\partial t^3} \right] \quad \text{at } (i, j), \quad i = 2(1)N.$$

Twizell and Khaliq, references 35 and 13 did numerical experiments with the $(1, 1)$, $(2, 0)$, $(2, 1)$, and $(2, 2)$ Padé approximants and observed that the $(1, 1)$ and $(2, 0)$ approximants with the low-order (4.18) replacement of the space derivative, and the $(2, 1)$ and $(2, 2)$ approximants with the higher-order (4.25) replacement of the space derivative, gave better results than any other method in the literature existing to 1983.

A word of caution on the central-difference approximation to $\partial U/\partial x$

If $\partial U/\partial x$ is approximated at (ih, jk) by

$$[U(x + h, t) - U(x - h, t)]/2h,$$

then the system of differential equations approximating

$$\frac{\partial U}{\partial t} + a\frac{\partial U}{\partial x} = 0, \quad a > 0, \quad 0 < x < X, \quad t > 0, \tag{4.29}$$

$U(x, 0) = g(x)$, $U(0, t) = b(t)$, at the mesh points ih, $i = 1(1)N$, along time-level t is

$$\frac{d\mathbf{V}(t)}{dt} = -\tfrac{1}{2}a\mathbf{B}\mathbf{V}(t) + \tfrac{1}{2}a\mathbf{b}(t), \tag{4.29a}$$

where $\mathbf{V}(t) = [V_1(t), V_2(t), \ldots, V_N(t)]^T$,

$$\mathbf{B} = \frac{1}{h}\begin{bmatrix} 0 & 1 & & & & \\ -1 & 0 & 1 & & & \\ & -1 & 0 & 1 & & \\ & & & \cdot & & \\ & & & & \cdot & \\ & & & -1 & 0 & 1 \\ & & & & -1 & 0 \end{bmatrix} \quad \text{of order } N,$$

and $h\mathbf{b}(t) = [b(t), 0, 0, \ldots, -V_{N+1}(t)]^T$. V_{N+1} is the solution at $\{(N+1)h, t\}$ and would be known only for periodic boundary conditions. That part of the solution corresponding to the term

$$-\tfrac{1}{2}a\mathbf{B}\mathbf{V}(t) \quad \text{is} \quad \{\exp(-\tfrac{1}{2}at\mathbf{B})\}\mathbf{g}. \tag{4.30}$$

The eigenvalues λ_s of **B** are $\lambda_s = 2(\sqrt{-1})\cos s\pi/N + 1$, $s = 1(1)N$, and are all different. Hence the N eigenvectors $\mathbf{v}_s$ of **B** are linearly independent so the initial vector $\mathbf{g} = [g_1, g_2, \ldots, g_N]^T$ can be written as $\mathbf{g} = \sum_1^N c_s \mathbf{v}_s$. Equation (4.30) then gives that the complementary function of the differential equation is

$$\sum_1^N c_s \{\exp(-\tfrac{1}{2}at\lambda_s)\}\mathbf{v}_s.$$

But λ_s is complex and $\exp(\sqrt{-1}\,\theta) = \cos\theta + \sqrt{-1}\sin\theta$, showing that the solution $\mathbf{V}(t)$ contains oscillatory terms. These are due entirely to the central-difference approximation to $\partial U/\partial x$ and would not occur with backward-difference approximations.

The numerical solution shown in Fig. 4.8 was calculated by Khaliq, reference 13, for eqn (4.29a), at $t = 1$ with $a = 1$, $g(x) = \sin 4\pi x$, $0 \le x \le 1$, $b(t) = -\sin 4\pi t$, $h = 1/80$ and $p = k/h = 4$, using the (1, 0) Padé approximant. The solution of the partial differential equation is $U(x, t) = \sin 4\pi(x - t)$. The diagram clearly shows

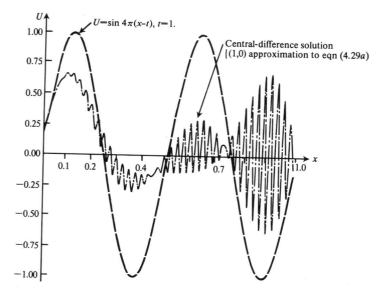

Fig. 4.8

that the effect of the oscillatory terms in the finite-difference solution increases with increasing x.

Second-order quasi-linear hyperbolic equations

Consider the second-order partial differential equation

$$a \frac{\partial^2 U}{\partial x^2} + b \frac{\partial^2 U}{\partial x \, \partial y} + c \frac{\partial^2 U}{\partial y^2} + e = 0, \qquad (4.31)$$

where a, b, c, and e may be functions of x, y, U, $\partial U/\partial x$, and $\partial U/\partial y$ but *not* of $\partial^2 U/\partial x^2$, $\partial^2 U/\partial x \, \partial y$, and $\partial^2 U/\partial y^2$, i.e. the second-order derivatives occur only to the first degree. Such an equation is said to be quasi-linear.

It will be shown that at each point of the solution domain there are two directions along which the integration of the partial differential equation transforms to the integration of an equation involving total differentials only. In other words, in these directions the equation to be integrated is not complicated by the presence of partial derivatives in other directions. Furthermore, it will be seen that this leads to a natural classification of partial differential equations.

Denote the partial derivatives by

$$\frac{\partial U}{\partial x} = p, \quad \frac{\partial U}{\partial y} = q, \quad \frac{\partial^2 U}{\partial x^2} = r, \quad \frac{\partial^2 U}{\partial x \, \partial y} = s \quad \text{and} \quad \frac{\partial^2 U}{\partial y^2} = t.$$

Let C be a curve in the x–y plane on which it is assumed that we know the solution values U of eqn (4.31), together with values for p and q related to U through the equation

$$dU = \frac{\partial U}{\partial x} dx + \frac{\partial U}{\partial y} dy = p \, dx + q \, dy,$$

where dy/dx is the slope of the tangent to C. (Curve C is *not* an initial curve on which initial values for U, p, and q are specified. The reason for this will be apparent later.) The question to ask now is this. Is it possible to find values for r, s, and t on C that will satisfy the partial differential equation, namely,

$$ar + bs + ct + e = 0. \qquad (4.32)$$

As r, s, and t on C must satisfy the equations

$$dp = \frac{\partial p}{\partial x} dx + \frac{\partial p}{\partial y} dy = r\, dx + s\, dy \qquad (4.33)$$

and

$$dq = \frac{\partial q}{\partial x} dx + \frac{\partial q}{\partial y} dy = s\, dx + t\, dy, \qquad (4.34)$$

it is seen that eqns (4.32), (4.33), and (4.34) are three equations for r, s, and t. Elimination of r and t from eqn (4.32) by means of eqns (4.33) and (4.34) leads to

$$\frac{a}{dx}(dp - s\, dy) + bs + \frac{c}{dy}(dq - s\, dx) + e = 0,$$

i.e.

$$s\left\{ a\left(\frac{dy}{dx}\right)^2 - b\left(\frac{dy}{dx}\right) + c \right\} - \left\{ a\frac{dp}{dx}\frac{dy}{dx} + c\frac{dq}{dx} + e\frac{dy}{dx} \right\} = 0, \quad (4.35)$$

where dy/dx is the slope of the tangent to C.

By hypothesis, eqn (4.32) is quasi-linear so a, b, c, and e are independent of r, s, and t. Hence eqn (4.35) is independent of r and t. It can also be made independent of s by choosing the curve C so that the slope of the tangent at each point on C is a root of the equation

$$a\left(\frac{dy}{dx}\right)^2 - b\left(\frac{dy}{dx}\right) + c = 0. \qquad (4.36)$$

By eqns (4.36) and (4.35) it follows that along directions tangential to C from points on C,

$$a\frac{dp}{dx}\frac{dy}{dx} + c\frac{dq}{dx} + e\frac{dy}{dx} = 0. \qquad (4.37)$$

Hence we have shown that at each point of the solution domain there are two directions, given by the roots of eqn (4.36), along which there are relationships between the total differentials dp and dq, given by eqn (4.37), that are independent of partial derivatives in other directions. As will be seen later this relationship can be used to solve the original differential equation numerically by a series of step-by-step integrations.

The directions given by the roots of eqn (4.36) are called the *characteristic directions* and the partial differential equation is said to be *hyperbolic, parabolic,* or *elliptic* according to whether these roots are real and distinct, equal, or complex, respectively, i.e. according to whether $b^2 - 4ac \gtreqless 0$. The best-known examples in these classes are the hyperbolic 'wave-equation' $\partial^2 U/\partial t^2 = \partial^2 U/\partial x^2$, the parabolic 'heat-conduction' or 'diffusion' equation $\partial U/\partial t = \partial^2 U/\partial x^2$, and the elliptic 'Laplace equation' $\partial^2 U/\partial x^2 + \partial^2 U/\partial y^2 = 0$.

Assume eqn (4.31) is hyperbolic and that the roots of eqn (4.36) are $dy/dx = f$ and $dy/dx = g$. Then the curve through the point $P(x, y)$ whose slope at every point is f is said to be an f characteristic. Clearly there are two different characteristic curves through every point of the solution domain of a second-order hyperbolic equation.

It should be noted that the classification of a partial differential equation, and consequently its method of solution, may depend on the region in which the solution is to be found. For example, the characteristic directions of the equation

$$y \frac{\partial^2 U}{\partial x^2} + x \frac{\partial^2 U}{\partial x \, \partial y} + y \frac{\partial^2 U}{\partial y^2} = F(x, y, U, p, q)$$

are given by the roots m_1, m_2 of the quadratic

$$ym^2 - xm + y = 0, \quad m = dy/dx,$$

which are real, equal or complex according to whether $x^2 \gtreqless 4y^2$. Thus, the equation is hyperbolic when $|x| > 2|y|$, parabolic along $|x| = 2|y|$, and elliptic for $|x| < 2|y|$.

Solution of hyperbolic equations by the method of characteristics

Summarizing the previous work, the slopes of the characteristic directions associated with the equation

$$a \frac{\partial^2 U}{\partial x^2} + b \frac{\partial^2 U}{\partial x \, \partial y} + c \frac{\partial^2 U}{\partial y^2} + e = 0 \tag{4.38}$$

are given by the roots of the quadratic equation

$$a \left(\frac{dy}{dx}\right)^2 - b \left(\frac{dy}{dx}\right) + c = 0, \tag{4.39}$$

and along these characteristic directions the differentials dp and dq are related by the equation

$$a\frac{dy}{dx}\frac{dp}{dx} + c\frac{dq}{dx} + e\frac{dy}{dx} = 0,$$

which can be written as

$$a\frac{dy}{dx}\,dp + c\,dq + e\,dy = 0. \tag{4.40}$$

Assuming that eqn (4.38) is hyperbolic the roots of eqn (4.39) will be real and distinct. Let them be

$$\frac{dy}{dx} = f \quad \text{and} \quad \frac{dy}{dx} = g. \tag{4.41}$$

Let Γ be a *non-characteristic curve* along which initial values for U, p, and q are known. Let P and Q be points on Γ that are close together and let the f characteristic through P intersect the g characteristic through Q at the point $R(x_R, y_R)$, Fig. 4.9.

As a first approximation we may regard the arcs PR and QR as straight lines of slopes f_P and g_Q respectively. Then eqns (4.41) can be approximated by

$$y_R - y_P = f_P(x_R - x_P) \tag{4.42}$$

and

$$y_R - y_Q = g_Q(x_R - x_Q), \tag{4.43}$$

giving two equations for the two unknowns x_R, y_R.

By eqn (4.40) the differential relationships along the charac-

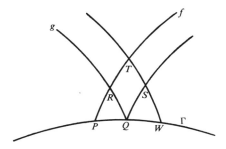

Fig. 4.9

teristics are

$$af\, dp + c\, dq + e\, dy = 0 \tag{4.44}$$

and

$$ag\, dp + c\, dq + e\, dy = 0. \tag{4.45}$$

The first one can be approximated along PR by the equation

$$a_P f_P (p_R - p_P) + c_P (q_R - q_P) + e_P (y_R - y_P) = 0, \tag{4.46}$$

and the second along QR by the equation

$$a_Q g_Q (p_R - p_Q) + c_Q (q_R - q_Q) + e_Q (y_R - y_Q) = 0. \tag{4.47}$$

These are two equations for the two unknowns p_R, q_R, as soon as x_R, y_R have been calculated from (4.42) and (4.43). The value of U at R can then be obtained from

$$dU = \frac{\partial U}{\partial x}\, dx + \frac{\partial U}{\partial y}\, dy = p\, dx + q\, dy,$$

by replacing the values of p and q along PR by their average values and approximating the last equation by

$$u_R - u_P = \tfrac{1}{2}(p_P + p_R)(x_R - x_P) + \tfrac{1}{2}(q_P + q_R)(y_R - y_P). \tag{4.48}$$

This first approximation for u_R can now be improved by replacing the pivotal values of the various coefficients by average values. Equations (4.42) and (4.43) for improved values of x_R and y_R then become

$$y_R - y_P = \tfrac{1}{2}(f_P + f_R)(x_R - x_P) \tag{4.49}$$

and

$$y_R - y_Q = \tfrac{1}{2}(g_Q + g_R)(x_R - x_Q), \tag{4.50}$$

and eqns (4.46), (4.47) for improved values of p_R, q_R become

$$\tfrac{1}{2}(a_P + a_R)\tfrac{1}{2}(f_P + f_R)(p_R - p_P) + \tfrac{1}{2}(c_P + c_R)(q_R - q_P)$$
$$+ \tfrac{1}{2}(e_P + e_R)(y_R - y_P) = 0 \tag{4.51}$$

and

$$\tfrac{1}{2}(a_Q + a_R)\tfrac{1}{2}(g_Q + g_R)(p_R - p_Q) + \tfrac{1}{2}(c_Q + c_R)(q_R - q_Q)$$
$$+ \tfrac{1}{2}(e_Q + e_R)(y_R - y_Q) = 0. \tag{4.52}$$

An improved value for u_R can then be found from eqn (4.48). Repetition of this last cycle of operations will eventually yield u_R

to the accuracy warranted by these finite-difference approximations. Provided Q is close to P the number of iterations will
usually be small.

In this way we can calculate solution values at the grid points R
and S, Fig. 4.9, and thence proceed to the grid-point T, and so
on.

Example 4.4

Use the method of characteristics to derive a solution of the
quasi-linear equation

$$\frac{\partial^2 U}{\partial x^2} - U^2 \frac{\partial^2 U}{\partial y^2} = 0,$$

at the first characteristic grid point between $x = 0.2$ and 0.3,
$y > 0$, where U satisfies the conditions

$$U = 0.2 + 5x^2 \quad \text{and} \quad \frac{\partial U}{\partial y} = 3x,$$

along the initial line $y = 0$, for $0 \leqslant x \leqslant 1$.

Since U is given as a continuous function of x along Ox the
initial value of $p = \partial U/\partial x$ is $10x$. The slopes of the characteristics
are the roots of the equation $m^2 - U^2 = 0$. Hence

$$f = U = -g.$$

In this example the characteristics depend on the solution so
the network of characteristics can be built-up only as the solution
unfolds.

Initially,
$$U = 0.2 + 5x^2 = f = -g,$$
$$p = 10x \quad \text{and} \quad q = 3x.$$

Also $a = 1$, $b = e = 0$, $c = -U^2$, therefore, Fig. 4.10

$$f_P = 0.4, \quad g_Q = -0.65, \quad p_P = 2.0, \quad p_Q = 3.0, \quad U_P = 0.4,$$
$$U_Q = 0.65, \quad q_P = 0.6, \quad q_Q = 0.9, \quad c_P = -0.16, \quad c_Q = -0.4225.$$

By eqns (4.42) and (4.43),

$$y_R = 0.4(x_R - 0.2)$$

and

$$y_R = -0.65(x_R - 0.3),$$

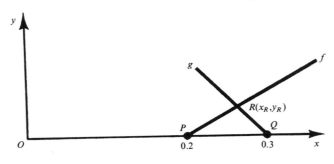

Fig. 4.10

giving, as a first approximation,

$$x_R = 0.26190, \quad y_R = 0.024762,$$

to five significant figures.

The differential relationships along the characteristics are, by eqns (4.46) and (4.47),

$$0.4(p_R - 2.0) - 0.16(q_R - 0.6) = 0,$$

and

$$-0.65(p_R - 3.0) - 0.4225(q_R - 0.9) = 0.$$

Their solution is

$$p_R = 2.45524; \quad q_R = 1.73810.$$

By eqn (4.48),

$$u_R = 0.4 + \tfrac{1}{2}(2.0 + 2.45524)(0.0619) + \tfrac{1}{2}(1.73810 + 0.6)(0.024762)$$
$$= 0.56684.$$

For the second approximation,

$$f_R = -g_R = u_R = 0.56684; \quad c_R = -u_R^2 = -0.32131.$$

By eqns (4.49) and (4.50), more accurate values for x_R, y_R are given by

$$y_R = \tfrac{1}{2}(0.4 + 0.56684)(x_R - 0.2)$$

and

$$y_R = -\tfrac{1}{2}(0.65 + 0.56684)(x_R - 0.3),$$

from which,

$$x_R = 0.25572, \quad y_R = 0.026938.$$

By eqns (4.51) and (4.52),

$$\tfrac{1}{2}(0.4 + 0.56684)(p_R - 2.0) - \tfrac{1}{2}(0.16 + 0.32131)(q_R - 0.6) = 0,$$
$$-\tfrac{1}{2}(0.65 + 0.56684)(p_R - 3.0) - \tfrac{1}{2}(0.4225 + 0.32131)(q_R - 0.9) = 0.$$

These equations give the improved values

$$p_R = 2.53117; \quad q_R = 1.66700.$$

Hence the second approximation to u_R, by eqn (4.48) is

$$u_R = 0.4 + \tfrac{1}{2}\{(2 + 2.53117)(0.05572) + (0.6 + 1.6670)(0.026938)\}$$
$$= 0.55677.$$

It is left to the reader to show that the next iteration gives

$$x_R = 0.25578, \quad y_R = 0.02668,$$
$$p_R = 2.52876, \quad q_R = 1.67637,$$

and

$$u_R = 0.55667.$$

Since, to four decimal places,

$$u_R^{(1)} = 0.5668, \quad u_R^{(2)} = 0.5568 \quad \text{and} \quad u_R^{(3)} = 0.5567,$$

it is obvious that the solution of the finite-difference equations for u_R is 0.5567, to this degree of accuracy. A fourth iteration does, in fact, give $u_R = 0.55666$ to five decimal places.

Additional comments on characteristics

A characteristic as an initial curve

When the curve on which initial values are given is itself a characteristic the equation can have no solution unless the initial conditions satisfy the necessary differential relationship for this characteristic. If they do, the solution will be unique along the initial curve but nowhere else, as is illustrated in the example below. It is also impossible in this case to use the method of characteristics to extend the solution from points on the initial curve to points off it, because the locus of all points such as R in Fig. 4.9 is the initial curve itself.

Consider the equation

$$\frac{\partial^2 U}{\partial x^2} - \frac{\partial^2 U}{\partial x \, \partial y} - 6 \frac{\partial^2 U}{\partial y^2} = 0.$$

The characteristic directions are given by

$$\left(\frac{dy}{dx}\right)^2 + \left(\frac{dy}{dx}\right) - 6 = 0 = \left(\frac{dy}{dx} + 3\right)\left(\frac{dy}{dx} - 2\right),$$

so the characteristics are the straight lines $y + 3x = $ constant, and $y - 2x = $ constant. Let the initial curve be the characteristic $y - 2x = 0$. The differential relationship along this line by eqn (4.40) is $2 \, dp - 6 \, dq = 0$, i.e. $p - 3q = $ constant, and is obviously satisfied by the initial conditions $U = 2$, $p = -2$, $q = 1$. It is easily verified that one solution satisfying these conditions is

$$U = 2 + (y - 2x) + A(y - 2x)^2,$$

where A is an arbitrary constant. This is unique along $y - 2x = 0$ but nowhere else in the x–y plane.

Propagation of discontinuities

It can be proved that the solutions of elliptic and parabolic equations are analytic even when the boundary or initial conditions are discontinuous. Hyperbolic equations however are different in that discontinuities in initial conditions are propagated as discontinuities into the solution domain along the characteristics.

Let Γ, Fig. 4.11, be a non-characteristic curve along which initial values for U, p, and q are known. Let P and Q be two distinct points on Γ and let the f characteristic through P meet the g characteristic through Q at R. Then the solution at R can be calculated in terms of the initial conditions at P and Q. Assuming no two characteristics of the same family intersect it follows that the solution at every point such as S inside the curvilinear triangle PQR is determined by the initial conditions between the points P and Q. Similarly, the solution at each point U inside the curvilinear strip $PRVT$ is determined by the initial conditions at a point along the arc TP (propagating along an f characteristic) and the initial conditions at a point along the arc PQ (propagated along a g characteristic). When the initial conditions along TP are analytically different from the initial condi-

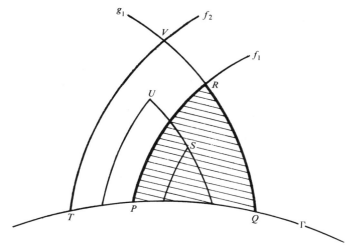

Fig. 4.11

tions along PQ then the solution inside the strip $PRVT$ will be analytically different from the solution inside the curvilinear triangle PQR. As T tends to P the strip tends to the characteristic PR proving that the discontinuity in the initial conditions at P is propagated along a characteristic.

The argument above shows that a characteristic can separate two different solutions. *An extremely important feature of second-order equations is that these two solutions together with their first-order derivatives can be continuous across the dividing characteristic, but their second- and higher-order derivatives discontinuous across the same characteristic.* This is easily seen by recalling that the characteristic directions and the differential relationships in these directions were originally defined from eqn (4.35), namely,

$$s\left\{a\left(\frac{dy}{dx}\right)^2 - b\left(\frac{dy}{dx}\right) + c\right\} - \left\{a\frac{dp}{dx}\frac{dy}{dx} + c\frac{dq}{dx} + e\frac{dy}{dx}\right\} = 0,$$

by making the expressions inside both pairs of braces zero. Then $s = \partial^2 U/\partial x\, \partial y$ is indeterminate and can be given an arbitrary value along a characteristic. Once, however, a value has been assigned to s the values of r and t are not indeterminate but can be calculated uniquely from any pair of eqns (4.32), (4.33), and

(4.34). In general any one of r, s, and t can be chosen arbitrarily and the other two calculated uniquely. (Algebraically this implies that eqns (4.32), (4.33), and (4.34) are not linearly independent but that any one can be written as a linear combination of the other two.) Now assume that continuous values of U, p, and q are prescribed along a characteristic C. Also prescribe a continuous value for s, say, along C. Continuous values for r and t can then be calculated from the equations above, and all the higher-order derivatives found by differentiating the differential equation. Taylor's expansion then gives a solution at points off C. Assume this solution is confined to one side of the characteristic. Repetition of this argument with the same values of U, p, and q but a different value for s leads to the possible existence of a second solution on the other side of the characteristic. It will differ from the first solution but be related to it through the common continuous values of U, p, and q along C. The indeterminacy associated with one of the second-order derivatives when it is not explicitly specified also explains why the solution of a hyperbolic equation is not unique when the initial values for U, p, and q are given on a characteristic. As an illustration consider the wave-equation

$$\frac{\partial^2 U}{\partial x^2} - \frac{\partial^2 U}{\partial y^2} = 0,$$

for which the characteristic directions and differential relationships are given by $dy/dx = dp/dq = \pm 1$. Hence $p - q$ is constant along $y - x = $ constant, and $p + q$ is constant along $y + x = $ constant. A possible solution along the characteristic $y - x = 0$ is therefore $p = 1$, $q = -1$, and $U = 3$. It is easily verified

$$U_1 = 2 + \sin(x - y) + \cos(x - y)$$

and

$$U_2 = 3 + (x - y)^2 + (x - y)$$

both satisfy the differential equation and give the stipulated solution along $y = x$, but that $\partial^2 U_1/\partial x^2 = -1$ and $\partial^2 U_2/\partial x^2 = 2$ on this characteristic.

In terms of the physics of steady supersonic flow of compressible fluids, for which the associated equations are hyperbolic, the characteristics are the Mach lines, along which small disturbances in velocity, pressure and density are propagated, or which separate different flow patterns possessing a common continuous velocity across the geometrical boundary of separation. (Discontinuous changes of state can also be propagated along shock lines. These differ from Mach lines both in position and the associated physics. Whereas the flow across a Mach line satisfies the equations of motion despite possible discontinuities in the normal derivatives of velocity, pressure, density, and entropy, the flow across a shock line does not. Along a shock line the differential equation is replaced by relationships between finite jumps in velocity, pressure, density, and entropy, the Rankine–Hugoniot equations, and these serve as boundary conditions for the differential equations used to calculate the continuous flow on each side of the shock. (See reference 17.)

Rectangular nets and finite-difference methods for second-order hyperbolic equations

In general, the method of characteristics provides the most accurate process for solving hyperbolic equations. It is probably the most convenient method as well when the initial data are discontinuous, because the propagation of the discontinuities into the solution domain along the characteristics is difficult to deal with on any grid other than a grid of characteristics. Problems involving no discontinuities however, can be solved satisfactorily by convergent and stable finite-difference methods using rectangular grids, and the organization of the computations for evaluation on a digital computer is certainly easier than for the method of characteristics.

Explicit methods and the Courant–Friedrichs–Lewy (C.F.L.) condition

Consider the wave-equation

$$\frac{\partial^2 U}{\partial x^2} = \frac{\partial^2 U}{\partial t^2}, \quad t > 0, \tag{4.53}$$

where initially $U(x, 0) = f(x)$ and $\partial U(x, 0)/\partial t = g(x)$.

By eqn (4.36) the slopes dt/dx of the characteristic curves are given by $(dt/dx)^2 = 1$, so the characteristics through $P(x_P, t_P)$, Fig. 4.12 are the straight lines $t - t_P = \pm(x - x_P)$ meeting the x-axis at

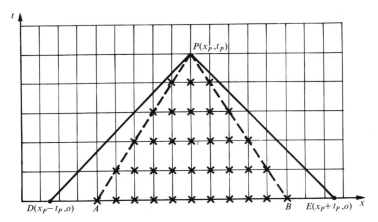

Fig. 4.12

$D(x_P - t_P, 0)$ and $E(x_P + t_P, 0)$. The solution to eqn (4.53) satisfying the given initial conditions is, (see Exercise 12),

$$U(x, t) = \frac{1}{2}\left\{f(x+t) + f(x-t) + \int_{x-t}^{x+t} g(\zeta)\,\mathrm{d}\zeta\right\}.$$

Hence the solution at $P(x_P, t_P)$ is

$$U(x_P, t_P) = \frac{1}{2}\left\{f(x_P+t_P) + f(x_P-t_P) + \int_{x_P-t_P}^{x_P+t_P} g(\zeta)\,\mathrm{d}\zeta\right\},$$

which shows that it depends upon the values of $f(x)$ at D and E and upon the value of $g(x)$ at every point of the closed interval DE, i.e. it depends on the initial data along the *interval of dependence DE*. The area PDE is called the *domain of dependence* of the point P. A central-difference approximation to eqn (4.53) at the mesh points $(x_i, t_j) = (ih, jk)$ of a rectangular mesh covering the solution domain is

$$\frac{u_{i+1,j} - 2u_{i,j} + u_{i-1,j}}{h^2} = \frac{u_{i,j+1} - 2u_{i,j} + u_{i,j-1}}{k^2},$$

i.e.

$$u_{i,j+1} = r^2 u_{i-1,j} + 2(1-r^2)u_{i,j} + r^2 u_{i+1,j} - u_{i,j-1}, \qquad (4.54)$$

where $r = k/h$. This is an explicit formula giving approximation values at mesh points along $t = 2k, 3k, \ldots$, as soon as the mesh values along $t = k$ have been determined. Putting $j = 0$ in eqn (4.54) yields

$$u_{i,1} = r^2 u_{i-1,0} + 2(1 - r^2)u_{i,0} + r^2 u_{i+1,0} - u_{i,-1}$$
$$= r^2 f_{i-1} + 2(1 - r^2)f_i + r^2 f_{i+1} - u_{i,-1},$$

and a central difference approximation to the initial derivative condition gives that

$$\frac{1}{2k}(u_{i,1} - u_{i,-1}) = g_{i,0}.$$

Eliminating $u_{i,-1}$ between these two equations shows that mesh values along $t = k$ can be calculated from the equation

$$u_{i,1} = \tfrac{1}{2}\{r^2 f_{i-1} + 2(1 - r^2)f_i + r^2 f_{i+1} + 2kg_{i,0}\}. \tag{4.55}$$

Equations (4.54) and (4.55) show that u_P at the mesh point P depends on the values of $u_{i,j}$ at the mesh points marked with crosses in Fig. 4.12. This set of mesh points is called the numerical domain of dependence of the point P, and the lines PA, PB are often termed the numerical characteristics. Assume now that the initial conditions along DA and BE are changed. These changes will alter the analytical solution of the partial differential equation at P but *not* the numerical solution given by eqns (4.54) and (4.55). In this case, u_P cannot possibly converge to U_P for all such arbitrary changes, as $h \to 0$ and $k \to 0$, with r remaining constant. When, however, the numerical characteristics PA, PB lie outside the domain of dependence PDE, Courant, Friedrichs, and Lewy, reference 3, have shown that the effect of the initial data along DA and BE upon the solution at P of the finite-difference equations tends to zero as h and k both tend to zero, r remaining constant, P remaining fixed. This C.F.L. condition for convergence is usually expressed by saying that the numerical domain of dependence of the difference equation must include the domain of dependence of the differential equation. Using this condition it is clear that the difference equation (4.54) is convergent for $0 < r \leq 1$. (A proof for $r = 1$ is given in Exercise 18(b) and a proof for $r \leq 1$ is given in reference 7. A numerical example is given in Exercise 13 and stability is considered in Worked Example 2.12.)

Implicit difference methods

Implicit methods cannot be used without simplifying assumptions to solve pure initial value problems because they give an infinite number of simultaneous equations. They can, however, be used effectively for initial-boundary value problems when, for example, initial conditions on $0 \le x \le 1$ for $t = 0$ and boundary conditions on $x = 0$ and 1, $t > 0$, are given. Convergence is usually dealt with via Lax's equivalence theorem.

One satisfactory scheme approximating the wave equation (4.53) at the point (ih, jk) is

$$\frac{1}{k^2} \delta_t^2 u_{i,j} = \frac{1}{h^2} (\tfrac{1}{4} \delta_x^2 u_{i,j+1} + \tfrac{1}{2} \delta_x^2 u_{i,j} + \tfrac{1}{4} \delta_x^2 u_{i,j-1}), \qquad (4.56)$$

where $\delta_t^2 u_{i,j} = u_{i,j+1} - 2u_{i,j} + u_{i,j-1}$ etc. This gives a tridiagonal system of equations that can be solved by the algorithm in Chapter 2. As shown in Exercise 17 it is unconditionally stable for all $r = k/h > 0$. Its truncation error is

$$h^2 \left\{ -\tfrac{1}{12}(4r^2 + 1) \frac{\partial^4 U}{\partial x^4} - \tfrac{1}{720} h^2 (13r^4 + 15r^2 + 1) \frac{\partial^6 U}{\partial x^6} + \ldots \right\}_{i,j},$$

which tends to zero for finite r as $h \to 0$. Hence the scheme is convergent. A more general scheme is given by Mitchell in reference 18. He also uses the following method for deriving implicit approximations to eqn (4.53) that are accurate to fourth-order differences. Expand $U_{i,j+1}$ and $U_{i,j-1}$ about the point (i, j) by Taylor's series to obtain

$$U_{i,j+1} + U_{i,j-1} = 2U_{i,j} + k^2 \left(\frac{\partial^2 U}{\partial t^2} \right)_{i,j} + \tfrac{1}{12} k^4 \left(\frac{\partial^4 U}{\partial t^4} \right)_{i,j} + \ldots.$$

If U is a solution of

$$\frac{\partial^2 U}{\partial t^2} = \frac{\partial^2 U}{\partial x^2}$$

then

$$\frac{\partial^2}{\partial t^2} \equiv \frac{\partial^2}{\partial x^2}$$

and therefore

$$\left(\frac{\partial^4 U}{\partial t^4} \right)_{i,j} = \left(\frac{\partial^4 U}{\partial x^4} \right)_{i,j}, \quad \left(\frac{\partial^6 U}{\partial t^6} \right)_{i,j} = \left(\frac{\partial^6 U}{\partial x^6} \right)_{i,j}, \quad \text{etc.}$$

Hence

$$U_{i,j+1} - 2U_{i,j} + U_{i,j-1} = k^2 \left(\frac{\partial^2 U}{\partial x^2}\right)_{i,j} + \tfrac{1}{12}k^4 \left(\frac{\partial^4 U}{\partial x^4}\right)_{i,j} + \dots \quad (4.57)$$

As

$$\frac{\partial^2 U}{\partial x^2} = \frac{1}{h^2}(\delta_x^2 - \tfrac{1}{12}\delta_x^4 + \tfrac{1}{90}\delta_x^6 + \dots)U,$$

it follows that to fourth-order differences,

$$\frac{\partial^2 U}{\partial x^2} = \frac{1}{h^2}(\delta_x^2 - \tfrac{1}{12}\delta_x^4)U$$

and

$$\frac{\partial^4 U}{\partial x^2} = \frac{\partial^2}{\partial x^2}\left(\frac{\partial^2 U}{\partial x^2}\right) = \frac{1}{h^4}\delta_x^4 U.$$

Substitution of these approximations into (4.57) shows that to this order of accuracy

$$U_{i,j+1} - 2U_{i,j} + U_{i,j-1} = r^2\{1 + \tfrac{1}{12}(r^2 - 1)\delta_x^2\}\delta_x^2 U_{i,j}, \quad (4.58)$$

where $r = k/h$. If we now operate on both sides of this equation with $\{1 + \tfrac{1}{12}(r^2 - 1)\delta_x^2\}^{-\frac{1}{2}}$ and expand each operator up to terms in δ_x^4 by the binomial expansion, we obtain the following implicit difference approximation to the wave equation,

$$u_{i,j+1} - 2u_{i,j} + u_{i,j-1} = \tfrac{1}{24}(r^2 - 1)(\delta_x^2 u_{i,j+1} + \delta_x^2 u_{i,j-1})$$
$$+ \tfrac{1}{12}(11r^2 + 1)\delta_x^2 u_{i,j} + \tfrac{1}{192}(r^2 - 1)(9r^2 - 1)\delta_x^4 u_{i,j}$$
$$- \tfrac{1}{384}(r^2 - 1)^2(\delta_x^4 u_{i,j+1} + \delta_x^4 u_{i,j-1}).$$

Similarly, if both sides of eqn (4.58) are operated on by $\{1 + \tfrac{1}{12}(r^2 - 1)\delta_x^2\}^{-1}$ the corresponding difference equation is

$$u_{i,j+1} - 2u_{i,j} + u_{i,j-1} = \tfrac{1}{12}(r^2 - 1)(\delta_x^2 u_{i,j+1} + \delta_x^2 u_{i,j-1})$$
$$+ \tfrac{1}{6}(5r^2 + 1)\delta_x^2 u_{i,j} + \tfrac{1}{144}(r^2 - 1)^2\{2\delta_x^4 u_{i,j} - (\delta_x^4 u_{i,j+1} + \delta_x^4 u_{i,j-1})\}.$$

These high-order difference approximations would be difficult to implement in practice because of the difficulties associated with the boundary conditions.

Simultaneous first-order equations and their stability

All partial differential equations of second- and higher-order in the independent variable t, say, can be reduced to a system of simultaneous equations of first-order in t which may then be approximated by stable and convergent difference schemes, such as the Lax–Wendroff scheme. A range of problems for which this is a convenient method of solution is given in reference 25. For example, the displacement Y of a point a distance x from one end of a thin beam vibrating transversely satisfies the equation

$$\frac{\partial^2 Y}{\partial t^2} = -a^2 \frac{\partial^4 Y}{\partial x^4}, \quad a \text{ constant.}$$

Let $(\partial Y/\partial t) = v$ and $a(\partial^2 Y/\partial x^2) = w$. Then the original equation can be replaced by the pair of equations

$$\frac{\partial v}{\partial t} = -a^2 \frac{\partial^2 w}{\partial x^2} \quad \text{and} \quad \frac{\partial w}{\partial t} = a \frac{\partial^2 v}{\partial x^2},$$

each of first-order in t.

The wave equation can be dealt with in a similar manner by putting $(\partial U/\partial x) = p$ and $(\partial U/\partial t) = q$. Then eqn (4.53) and the relationship $\partial^2 U/\partial x \, \partial t = \partial^2 U/\partial t \, \partial x$ lead to

$$\frac{\partial q}{\partial t} = \frac{\partial p}{\partial x} \quad \text{and} \quad \frac{\partial p}{\partial t} = \frac{\partial q}{\partial x}. \tag{4.59}$$

One explicit difference approximation to (4.59) which is stable for $k/h \le 1$ is

$$\frac{1}{k}\{q_{i,j+1} - \tfrac{1}{2}(q_{i+1,j} + q_{i-1,j})\} = \frac{1}{2h}(p_{i+1,j} - p_{i-1,j})$$

and

$$\frac{1}{k}\{p_{i,j+1} - \tfrac{1}{2}(p_{i+1,j} + p_{i-1,j})\} = \frac{1}{2h}(q_{i+1,j} - q_{i-1,j}).$$

A numerical example, including the calculation of $u_{i,j}$, is given in Exercise 16 and the stability of the equations is considered in Exercise 15(a). Another approximation to eqns (4.59) is

$$\frac{1}{k}(q_{i,j+1} - q_{i,j}) = \frac{1}{h}(p_{i+\frac{1}{2},j} - p_{i-\frac{1}{2},j})$$

and

$$\frac{1}{k}(p_{i-\frac{1}{2},j+1} - p_{i-\frac{1}{2},j}) = \frac{1}{h}(q_{i,j+1} - q_{i-1,j+1})$$

and is shown to be stable for $k/h \leqslant 1$ in Exercise 15. This scheme, sometimes called the Courant, Friedrichs, and Lewy scheme, is equivalent to the explicit difference equation (4.54) if $p_{i-\frac{1}{2},j}$ is replaced by $(u_{i,j} - u_{i-1,j})/h$ and $q_{i,j}$ by $(u_{i,j} - u_{i,j-1})/k$.

An implicit scheme that is unconditionally stable and equivalent to (4.56) is

$$\frac{1}{k}(q_{i,j+1} - q_{i,j}) = \frac{1}{2h}(p_{i+\frac{1}{2},j} - p_{i-\frac{1}{2},j} + p_{i+\frac{1}{2},j+1} - p_{i-\frac{1}{2},j+1})$$

and

$$\frac{1}{k}(p_{i-\frac{1}{2},j+1} - p_{i-\frac{1}{2},j}) = \frac{1}{2h}(q_{i,j+1} - q_{i-1,j+1} + q_{i,j} - q_{i-1,j}).$$

To illustrate the von Neumann method for investigating stability and to demonstrate that apparently reasonable schemes can be useless, consider the following explicit scheme which approximates the x-derivatives of eqn (4.59) by central-differences and the t-derivatives by forward differences, namely,

$$\frac{1}{k}(q_{i,j+1} - q_{i,j}) = \frac{1}{2h}(p_{i+1,j} - p_{i-1,j})$$

and

$$\frac{1}{k}(p_{i,j+1} - p_{i,j}) = \frac{1}{2h}(q_{i+1,j} - q_{i-1,j}). \tag{4.60}$$

Let the initial perturbations in p and q along $t = 0$ be $Ae^{\beta x\sqrt{(-1)}}$ and $Be^{\beta x\sqrt{(-1)}}$ respectively, where A and B are different constants and $x = ih$. Then we can assume that the perturbations in the calculated values of $p_{i,j}$ and $q_{i,j}$ will be $Ae^{\beta x\sqrt{(-1)}}\xi^j$ and $Be^{\beta x\sqrt{(-1)}}\xi^j$ respectively. Substitution of these perturbations into eqns (4.60) leads to

$$2B(\xi - 1) = rA(e^{\beta h\sqrt{(-1)}} - e^{-\beta h\sqrt{(-1)}})$$

and

$$2A(\xi - 1) = rB(e^{\beta h\sqrt{(-1)}} - e^{-\beta h\sqrt{(-1)}}),$$

where $r = k/h$. Elimination of A/B gives that

$$(\xi - 1)^2 = -r^2 \sin^2 \beta h.$$

Hence

$$\xi = 1 \pm (\sqrt{-1})\sin \beta h$$

and

$$|\xi| = (1 + r^2 \sin^2 \beta h)^{\frac{1}{2}} = 1 + \tfrac{1}{2} r^2 \sin^2 \beta h + \ldots = 1 + O(r^2).$$

On p. 68 it was shown that the perturbations at the finite time-level $t = jk$ would be unbounded as the mesh lengths tend to zero if $|\xi| > 1 + O(k)$, which is so in this case.

Exercises and solutions

1. The function U satisfies the equation

$$x^2 U \frac{\partial U}{\partial x} + e^{-y} \frac{\partial U}{\partial y} = -U^2$$

and the condition $U = 1$ on $y = 0$, $0 < x < \infty$.

Calculate the Cartesian equation of the characteristic through the point $R(x_R, 0)$, $x_R > 0$, and the solution along this characteristic.

Use a finite-difference method to calculate first approximations to the solution and to the value of y at the point $P(1.1, y)$, $y > 0$, on the characteristic through the point $R(1, 0)$.

Calculate second approximations to these values to $4D$ by an iterative method. Compare your final results with those given by the analytical solution.

Solution

$dx/x^2 U = e^y \, dy = -U^{-2} \, dU$. Hence $1/x = A + \log U$. As $U = 1$ at $(x_R, 0)$, $A = 1/x_R$ so $\log U = x^{-1} - x_R^{-1}$. Similarly $e^y = B + U^{-1}$. As $U = 1$ at $(x_R, 0)$, $B = 0$ so $U = e^{-y}$. Eliminating U, $y = x_R^{-1} - x^{-1}$.

First approximations. $dy = dx/x^2 U e^y$, therefore $y_P^{(1)} - 0 = (0.1)/1 = 0.1$. $dU = -U \, dx/x^2$, therefore $u_P^{(1)} - 1 = -(0.1)$ giving $u_P^{(1)} = 0.9$.

Second approximations

$$y_P^{(2)} = \frac{dx}{2}\left\{ \left(\frac{1}{x^2 u e^y}\right)_R + \left(\frac{1}{x^2 u e^y}\right)_P \right\} = 0.05(1 + 0.831) = 0.0915.$$

N.B. If we use $x^2 U \, dy = e^{-y} \, dx$ and approximate by

$$\tfrac{1}{2}\{(x^2 u)_R + (x^2 u)_P\}(y_P^{(2)} - 0) = \tfrac{1}{2}\{e^{-y_R} + e^{-y_P}\}(0.1),$$

then $y_P^{(2)} = 0.0912$.

$$u_P^{(2)} - 1 = -\frac{1}{2}\left\{\left(\frac{u}{x^2}\right)_R + \left(\frac{u}{x^2}\right)_P\right\} dx = -\frac{1}{2}(1.0 + 0.744)(0.1)$$

giving $u_P^{(2)} = 0.9128$.

Analytical values: $y_P = x_R^{-1} - x_P^{-1} = 0.0909$ and $U_P = e^{-y_P} = \exp(-0.0909)$ giving $U_P = 0.9131$.

2. The function U satisfies the equation

$$\frac{\partial U}{\partial x} + \frac{x}{\sqrt{U}}\frac{\partial U}{\partial y} = 2x$$

and the conditions $U = 0$ on $x = 0$, $y \geqslant 0$ and $U = 0$ on $y = 0$, $x > 0$.

Calculate the analytical solutions at the points $(2, 5)$ and $(5, 4)$. Sketch the characteristics through these two points. If the initial condition along $y = 0$ is replaced by $U = x$, calculate approximations to the solution and to the value of y at the point $P(4.05, y)$ on the characteristic through the point $R(4, 0)$. Compare with the analytical values.

Solution

$dx = \sqrt{U}\, dy/x = dU/2x$. Hence $U = x^2 + A$, $y = B + \sqrt{U}$. As $U = 0$ at $R(x_R, 0)$, $A = -x_R^2$ and $B = 0$. Eliminating U between these equations gives that the solution along the characteristic $x^2 - y^2 = x_R^2$ from $(x_R, 0)$ is $U = y^2$. Similarly, as $U = 0$ at $S(0, y_s)$, $A = 0$ and $B = y_s$ and elimination of U shows that the solution along the characteristic $y - y_S = x$ from $S(0, y_s)$ is $U = x^2$. Therefore $U(5, 4) = 4^2 \overset{\star}{=} 16$ and $U(2, 5) = 2^2 = 4$.

Approximations: $dU = 2x\, dx$. Therefore $u_P^{(1)} - 4 = 2 \cdot \frac{1}{2}(x_R + x_P)$ $dx = (8.05)(0.05) = 0.4025$. $\sqrt{U}\, dy = x\, dx$ may be approximated by $\frac{1}{2}(\sqrt{u_R} + \sqrt{u_P})(y_P^{(1)} - 0) = \frac{1}{2}(x_P + x_R)\, dx$ giving $(2 + \sqrt{4.4025})$ $y_P^{(1)} = (4 + 4.05)(0.05)$, from which $y_P^{(1)} = 0.0982$.

Analytical values: As $U_R = x_R$ at $(x_R, 0)$, $A = x_R - x_R^2$, $B = -\sqrt{x_R}$. Hence $y = \sqrt{U} - \sqrt{x_R}$ and $U = x^2 + x_R - x_R^2$. Therefore $U_P = 4.4025$ and $y_P = 0.0982$.

3. (a) Use the von Neumann method to prove that the Lax–Wendroff difference eqns (4.12) are stable for $0 < ap \leqslant 1$.

(b) Show that the principal part of the local truncation error of the Lax–Wendroff equation (4.12) is

$$\left[\tfrac{1}{6}k^2 \frac{\partial^3 U}{\partial t^3} + \tfrac{1}{6}ah^2 \frac{\partial^3 U}{\partial x^3} \right]_{i,j}.$$

(c) Prove that the solution of the differential equation

$$\frac{\partial U}{\partial t} + a \frac{\partial U}{\partial x} = 0,$$

a constant, is the solution of the approximating Lax–Wendroff equation (4.12) when $k/h = 1/a$. Comment on this result in relation to the characteristics of the differential equation.

Solution

(a) The substitution of $E_{i,j} = e^{\sqrt{(-1)}\beta x}\xi^i$ into (4.12) leads to $\xi = (1 - 2a^2p^2 \sin^2 \tfrac{1}{2}\beta h) - 2(\sqrt{-1})ap \sin \tfrac{1}{2}\beta h \cos \tfrac{1}{2}\beta h$. Hence $|\xi|^2 = 1 - 4a^2p^2(1 - a^2p^2)\sin^4\beta h$. Errors will not increase exponentially with j if $|\xi|^2 \leqslant 1$, i.e. $0 \leqslant 4a^2p^2(1 - a^2p^2)$, giving $0 < ap \leqslant 1$.

(b) $T_{i,j} = \{U_{i,j+1} - \tfrac{1}{2}ap(1 + ap)U_{i-1,j} - (1 - a^2p^2)U_{i,j}$
$\qquad\qquad\qquad\qquad + \tfrac{1}{2}ap(1 - ap)U_{i+1,j}\}/k.$

Expand each term by Taylor's series about the point (i, j), etc.

(c) When $k/h = 1/a$, i.e. $ap = 1$,

$$T_{i,j} = (U_{i,j+1} - U_{i-1,j})/k = k\left(\frac{\partial U}{\partial t} + a \frac{\partial U}{\partial x}\right)_{i,j} + \tfrac{1}{2}k^2\left(\frac{\partial^2 U}{\partial t^2} - a^2 \frac{\partial^2 U}{\partial x^2}\right)_{i,j}$$

$$+ \tfrac{1}{6}k^3\left(\frac{\partial^3 U}{\partial t^3} + a^3 \frac{\partial^3 U}{\partial x^3}\right)_{i,j} + \dots.$$

Each bracketed term is zero because $(\partial/\partial t) \equiv -a(\partial/\partial x)$ by the differential equation. Hence the result. By the differential equation, $dt/1 = dx/a = dU/0$. Hence U is constant along the straight-line characteristics of slope $1/a$. The difference equation gives $u_{i,j+1} = u_{i-1,j}$ along the line of slope $1/a$.

4. Develop the Lax–Wendroff equations for the scalars $v_{i,j+1}$,

$w_{i,j+1}$ approximating $V_{i,j+1}$, $W_{i,j+1}$ respectively, where $V(x, t)$ and $W(x, t)$ satisfy the equations

$$\frac{\partial V}{\partial t} + 2\frac{\partial V}{\partial x} + \frac{\partial W}{\partial x} = 0$$

and

$$\frac{\partial W}{\partial t} + 4\frac{\partial V}{\partial x} - \frac{\partial W}{\partial x} = 0.$$

Derive the second-order partial differential equation satisfied by either V or W.

Solution

The equations can be written as

$$\frac{\partial}{\partial t}\begin{bmatrix} V \\ W \end{bmatrix} + \begin{bmatrix} 2 & 1 \\ 4 & -1 \end{bmatrix}\begin{bmatrix} V \\ W \end{bmatrix} = \begin{bmatrix} 0 \\ 0 \end{bmatrix},$$

i.e. as

$$\partial \mathbf{U}/\partial t + \mathbf{A}\,\partial \mathbf{U}/\partial x = \mathbf{O}.$$

The development of the Lax–Wendroff difference equations for $\mathbf{u}$ approximating $\mathbf{U}$ is in the text and leads to

$$\mathbf{u}_{i,j+1} = \mathbf{u}_{i,j} - \tfrac{1}{2}p\mathbf{A}(\mathbf{u}_{i+1,j} - \mathbf{u}_{i-1,j}) + \tfrac{1}{2}p^2\mathbf{A}^2(\mathbf{u}_{i-1,j} - 2\mathbf{u}_{i,j} + \mathbf{u}_{i+1,j}).$$

Hence

$$\begin{bmatrix} v \\ w \end{bmatrix}_{i,j+1} = \begin{bmatrix} v \\ w \end{bmatrix}_{i,j} - \tfrac{1}{2}p\begin{bmatrix} 2 & 1 \\ 4 & -1 \end{bmatrix}\begin{bmatrix} v_{i+1,j} - v_{i-1,j} \\ w_{i+1,j} - w_{i-1,j} \end{bmatrix}$$

$$+ \tfrac{1}{2}p^2\begin{bmatrix} 8 & 1 \\ 4 & 5 \end{bmatrix}\begin{bmatrix} v_{i-1,j} - 2v_{i,j} + v_{i+1,j} \\ w_{i-1,j} - 2w_{i,j} + w_{i+1,j} \end{bmatrix},$$

giving that

$$v_{i,j+1} = p(1+4p)v_{i-1,j} + (1-p^2)v_{i,j} + p(4p-1)v_{i+1,j}$$
$$+ \tfrac{1}{2}p(1+p)w_{i-1,j} - p^2 w_{i,j} + \tfrac{1}{2}p(p-1)w_{i+1,j},$$

$$w_{i,j+1} = p(\tfrac{1}{2}+2p)v_{i-1,j} - 4p^2 v_{i,j} + p(2p-2)v_{i+1,j}$$
$$+ (\tfrac{5}{2}p^2 - \tfrac{1}{2}p)w_{i-1,j} + (1-5p^2)w_{i,j} + (\tfrac{1}{2}p + \tfrac{5}{2}p^2)w_{i+1,j}.$$

Eliminate $\partial V/\partial x$ between the P.D.E.'s, then use $\partial^2 V/\partial x\,\partial t = \partial^2 V/\partial t\,\partial x$ to eliminate derivatives of V. This gives $6\,\partial^2 W/\partial x^2 - \partial^2 W/\partial x\,\partial t - \partial^2 W/\partial t^2 = 0$. (Same equation for V.)

5. The equation

$$a \frac{\partial U}{\partial x} + b \frac{\partial U}{\partial t} = c$$

is approximated at the point $(i+\frac{1}{2}, j+\frac{1}{2})$ by the Wendroff implicit scheme

$$(b+ap)u_{i+1,j+1} + (b-ap)u_{i,j+1} - (b-ap)u_{i+1,j}$$
$$- (b+ap)u_{i,j} - 2kc = 0,$$

where $p = k/h$.

Prove that: (a) The scheme is unconditionally stable, and

(b) the principal part of the local truncation error *at the point*

$(i+\frac{1}{2}, j+\frac{1}{2})$ is $\frac{1}{12}h^2\left(3b \dfrac{\partial^3 U}{\partial x^2 \partial t} + a \dfrac{\partial^3 U}{\partial x^3}\right)_{i+\frac{1}{2},j+\frac{1}{2}}$

$$+ \frac{1}{12}k^2\left(b \frac{\partial^3 U}{\partial t^3} + 3a \frac{\partial^3 U}{\partial x \partial t^2}\right)_{i+\frac{1}{2},j+\frac{1}{2}}$$

Solution

(a) The error function $E_{i,j} = e^{\sqrt{(-1)}\beta x}\xi^j$ is a solution of the equation

$$(b+ap)E_{i+1,j+1} + (b-ap)E_{i,j+1} - (b-ap)E_{i+1,j} - (b+ap)E_{i,j} = 0$$

if

$$\xi = (b \cos \beta h - (\sqrt{-1})ap \sin \beta h)/(b \cos \beta h + (\sqrt{-1})ap \sin \beta h).$$

Hence $|\xi| = 1$ for all real values of a, b, and p.

(b) $T_{i,j} = \{(b+ap)U_{i+1,j+1} + (b-ap)U_{i,j+1} - (b-ap)U_{i+1,j}$
$$- (b+ap)U_{i,j} - 2kc\}/k.$$

Expand each term about the point $(i+\frac{1}{2}, j+\frac{1}{2})$ by Taylor's series.

6. The function U satisfies the equation $\partial U/\partial t + a\, \partial U/\partial x = 0$, $0 < x < X$, $t > 0$, the boundary condition $U(0, t) = b$ and the initial condition $U(x, 0) = g(x)$. Given that a and b are constants, $a > 0$, and that $\partial U/\partial x$ is approximated by $\{U(x, t) - U(x - h, t)\}/h$, show that the problem can be approximated at the mesh points $x_i = ih$,

$i = 1(1)N$, along time-level t, by the system of ordinary equations

$$\frac{d\mathbf{V}}{dt} = -a\mathbf{C}\mathbf{V}(t) + a\mathbf{b},$$

defining matrix $\mathbf{C}$ and vectors $\mathbf{V}$ and $\mathbf{b}$.
   Given that the solution is

$$\mathbf{V}(t) = \mathbf{C}^{-1}\mathbf{b} + \{\exp(-at\mathbf{C})\}(\mathbf{g} - \mathbf{C}^{-1}\mathbf{b})$$

and that the exponential is approximated by its $(1, 0)$ Padé approximant, show that $\mathbf{V}$ is approximated along time-level $(t_j + k)$ by the difference equations

$$-apu_{i-1,j+1} + (1 + ap)u_{i,j+1} - u_{i,j} = 0, \quad i = 1(1)N,$$

where $p = k/h$. Show that the principal part of the local truncation error at (ih, jk) is

$$\left[-\tfrac{1}{2}ah\frac{\partial^2 U}{\partial x^2} - \tfrac{1}{2}k\frac{\partial^2 U}{\partial t^2} \right]_{i,j}.$$

Solution

First part is in the text.

$$T_{i,j} = \frac{1}{k}\{-apU_{i-1,j+1} + (1 + ap)U_{i,j+1} - U_{i,j}\}.$$

Expand all terms about (ih, jk) by Taylor's series and remember that $a\,\partial/\partial x \equiv \partial/\partial t$ so that $a\,\partial^2 U/\partial x\,\partial t = \partial^2 U/\partial t^2$.

7. Show that the backward-difference $(1, 0)$ Padé approximation equations of Exercise 6, namely,

$$\mathbf{u}_{j+1} = \mathbf{C}^{-1}\mathbf{b} + (\mathbf{I} + ak\mathbf{C})^{-1}\{\mathbf{u}_j - \mathbf{C}^{-1}\mathbf{b}\},$$

are (i) L_0-stable
   (ii) Unconditionally stable by von Neumann's method.

Solution

The perturbation $\mathbf{e} = \mathbf{u} - \mathbf{u}^*$ satisfies

$$\mathbf{e}_{j+1} = (\mathbf{I} + ak\mathbf{C})^{-1}\mathbf{e}_j = \ldots = (\mathbf{I} + ak\mathbf{C})^{-j-1}\mathbf{e}_0$$

$$= \sum_{1}^{N} c_s(1 + ak\lambda_s)^{-j-1}\mathbf{v}_s,$$

where

$$\mathbf{Cv}_s = \lambda_s \mathbf{v}_s, \quad \mathbf{e}_0 = \sum_1^N c_s \mathbf{v}_s \quad \text{and} \quad \lambda_s = 1/h.$$

Hence,

$$\mathbf{e}_{j+1} = \sum c_s \mathbf{v}_s/(1 + ap)^{j+1}, \quad a > 0, \quad p = k/h > 0.$$

Clearly, $\mathbf{e}_{j+1} \rightarrow \mathbf{0}$ as $j \rightarrow \infty$ for all p, proving unconditional stability. Also $1/(1 + ap) \rightarrow 0$ as $ap \rightarrow \infty$, proving L_0-stability.

(ii) The difference equations are

$$-apu_{i-1,j+1} + (1 + ap)u_{i,j+1} = u_{i,j}.$$

Putting

$$u_{i,j} = \{\exp(\sqrt{-1}\,\beta ih)\}\xi^j$$

leads to

$$\xi = 1/(d + \sqrt{-1}\,ap\sin\beta h)$$

where

$$d = 1 + 2ap\sin^2\beta h/2.$$

Hence

$$|\xi|^2 = 1/(d^2 + a^2 p^2 \sin^2\beta h) < 1 \quad \text{for all} \quad p > 0.$$

8. The function U satisfies the equation

$$\frac{\partial^2 U}{\partial x^2} - 4x^2 \frac{\partial^2 U}{\partial y^2} = 0$$

and the initial conditions

$$U = x^2, \quad \frac{\partial U}{\partial y} = 0, \quad \text{on} \quad y = 0, \quad -\infty < x < \infty.$$

Show from first principles, with the usual notation, that $dp - 2x\,dq = 0$ along the characteristic of slope $2x$, and that $dp + 2x\,dq = 0$ along the characteristic of slope $(-2x)$.

The characteristic with positive slope through the point $A(0.3, 0)$ intersects the characteristic with negative slope through the point $B(0.4, 0)$ at R. Calculate to $3D$ an approximation to the solution at R.

Solution

First part is bookwork. The characteristic with slope $2x$ through A is $y = x^2 - 0.09$. The other characteristic through B is $y = 0.16 - x^2$. Hence $x_R^2 = 1/8$, so $x_R = 0.3536$, $y_R = 0.035$. The equations, $p_R - 0.6 - (x_A + x_R)q_R = 0 = p_R - 0.8 + (x_B + x_R)q_R$ give $p_R = 0.6929$, $q_R = 0.1421$. Then

$$dU \simeq u_R - u_A = \tfrac{1}{2}(p_A + p_R)(x_R - x_A) + \tfrac{1}{2}(q_A + q_R)(y_R - y_A)$$

gives $u_R = 0.127$.

9. The function U is a solution of the equation

$$\frac{\partial^2 U}{\partial x^2} + \left(\frac{\partial U}{\partial x} - U\right)\frac{\partial^2 U}{\partial x\,\partial y} - U\frac{\partial U}{\partial x}\frac{\partial^2 U}{\partial y^2} + x = 0$$

and satisfies the initial conditions $U = 1 + x^2$,

$$\frac{\partial U}{\partial y} = 1, \quad \text{on} \quad y = 0, \quad -\infty < x < \infty.$$

Show from first principles, for this particular example, that $p\,dp - Up\,dq + x\,dy = 0$ along the characteristic of slope p, and that $-U\,dp - Up\,dq + x\,dy = 0$ along the characteristic of slope $(-U)$, where

$$p = \frac{\partial U}{\partial x} \quad \text{and} \quad q = \frac{\partial U}{\partial y}.$$

The characteristic with positive slope through the point $A(0.5, 0)$ intersects the characteristic with negative slope through the point $B(0.6, 0)$ at the point $R(x_R, y_R)$. Calculate first approximation values for the co-ordinates of R.

Write down the equations giving first approximation values for p and q at R, but do not solve them. Explain how to calculate a first approximation value for U at R.

Given that the first approximation values for U, p, and q at R are $u_R^{(1)} = 1.3711$, $p_R^{(1)} = 1.1009$, and $q_R^{(1)} = 1.1038$, calculate second approximation values for the co-ordinates of R.

Solution

First part is bookwork. First approximation equations to (x_R, y_R) are $y_R^{(1)} = x_R^{(1)} - 0.5 = -1.36(x_R^{(1)} - 0.6)$. Hence $x_R^{(1)} = 0.5576$. $y_R^{(1)} =$

0.0576. First approximations to (p_R, q_R) are given by

$$p_A(p_R^{(1)} - p_A) - p_A u_A(q_R - q_A) + x_A(y_R^{(1)} - y_A) = 0,$$
$$-u_B(p_R^{(1)} - p_B) - p_B u_B(q_R - q_B) + x_B(y_R^{(1)} - y_B) = 0,$$

i.e.

$$p_R^{(1)} - 1.25q_R^{(1)} + 0.2788 = 0 - 1.36p_R^{(1)} - 1.632q_R^{(1)} + 3.2986,$$

from which $p_R^{(1)} = 1.1009$ and $q_R^{(1)} = 1.1038$. Usual approximation
to $dU = p\,dx + q\,dy$ gives that $u_R^{(1)} = 1.3711$.

$$y_R^{(2)} = \tfrac{1}{2}(2.1009)(x_R^{(2)} - 0.5) = -\tfrac{1}{2}(1.36 + 1.3711)(x_R^{(2)} - 0.6).$$

Therefore $x_R^{(2)} = 0.5565$ and $y_R^{(2)} = 0.0594$.

10. The transverse displacement U of a point at a distance X
from one end of a vibrating string of length L at time T satisfies
the equation $\partial^2 U/\partial T^2 = c^2\,\partial^2 U/\partial X^2$. Show this can be reduced to
the non-dimensional form

$$\frac{\partial^2 u}{\partial x^2} = \frac{\partial^2 u}{\partial t^2}, \quad 0 \leqslant x \leqslant 1,$$

by putting $x = X/L$, $u = U/L$, and $t = cT/L$.
 A solution of the latter equation satisfies the boundary conditions

$$u = 0 \quad \text{at} \quad x = 0 \text{ and } 1, \quad t \geqslant 0,$$

and the initial conditions

$$u = \tfrac{1}{2}x(1-x) \quad \text{and} \quad \frac{\partial u}{\partial t} = 0, \quad \text{for } 0 \leqslant x \leqslant 1 \text{ when } t = 0.$$

Use the method of characteristics to calculate the non-
dimensional velocities and displacements at time $t = 0.3$ of the
points on the string defined by $x = 0(0.1)0.5$.
 Use the method of separation of the variables to show that the
analytical solution to this problem is

$$u = \frac{2}{\pi^3} \sum_1^\infty \frac{1}{n^3}(1 - \cos n\pi)\cos n\pi t \sin n\pi x.$$

Compare the two solutions at $x = 0.5$, $t = 0.3$.

Solution

The characteristics are $t \pm x = $ constant, on which $p \pm q = $ constant, where $p = \partial u/\partial x$, $q = \partial u/\partial t = $ velocity. Construct the characteristics through C and B as shown in Fig. 4.13. From the initial conditions, $p_A = 0.3$, $q_A = 0$, $p_D = 0.1$, $q_D = 0$.

Along DC, $p_D + q_D = 0.1 = p_C + q_C$.
Along AB, $p_A + q_A = 0.3 = p_B + q_B = p_B$.
Along BC, $p_B - q_B = p_B = p_C - q_C$.

Hence $p_C = 0.2$ and $q_C = -0.1 = $ speed of C. The displacement u can be found by step-wise integration of p with respect to x along $t = 0.3$, i.e. from $\mathrm{d}u = (\partial u/\partial x)\,\mathrm{d}x$, which can be approximated by

$$u_C - u_F = \tfrac{1}{2}(p_F + p_C)(x_C - x_F) = \tfrac{1}{2}(p_F + 0.2)(0.1).$$

Along EF, $p_E + q_E = 0.2 = p_F + q_F = p_F$.
As $u_F = 0$, $u_C = 0.02$. Similarly for the other points.

	$x =$	0.1	0.2	0.3	0.4	0.5
Numerical solution	q	-0.1	-0.2	-0.3	-0.3	-0.3
	u	0.02	0.04	0.06	0.075	0.08
Analytical solution	q	-0.1	-0.2	-0.3	-0.3	-0.3
	u	0.02	0.04	0.06	0.075	0.08

11. The equations of motion for steady two-dimensional isen-

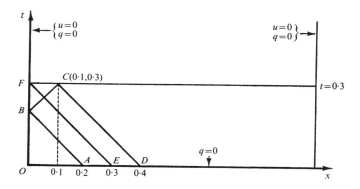

Fig. 4.13

tropic flow of compressible fluid can be written as

$$u\frac{\partial u}{\partial x}+v\frac{\partial u}{\partial y}=-\frac{a^2}{\rho}\frac{\partial\rho}{\partial x},$$

$$u\frac{\partial v}{\partial x}+v\frac{\partial v}{\partial y}=-\frac{a^2}{\rho}\frac{\partial\rho}{\partial y},$$

and the continuity equation as

$$u\frac{\partial\rho}{\partial x}+v\frac{\partial\rho}{\partial y}+\rho\left(\frac{\partial u}{\partial x}+\frac{\partial v}{\partial y}\right)=0,$$

where u, v are the Cartesian components of velocity, ρ the density and a the local speed of sound. Eliminate ρ between these equations and show that for irrotational flow, defined by $u=-\partial\phi/\partial x$, $v=-\partial\phi/\partial y$, the eliminant reduces to

$$(a^2-u^2)\frac{\partial^2\phi}{\partial x^2}-2uv\frac{\partial^2\phi}{\partial x\,\partial y}+(a^2-v^2)\frac{\partial^2\phi}{\partial y^2}=0,$$

where ϕ is the velocity potential.

Deduce that the fluid must be moving supersonically if it is possible for two different flows to exist side by side with continuous values for u and v across the dividing curve C but with a discontinuity in one of the space-derivatives of u or v. (C defines a characteristic.)

Prove that the differentials of u and v along C are related by the equation $dv/du=-1/m$, where m is the slope of C.

Solution

Substitute for $\partial\rho/\partial x$, $\partial\rho/\partial y$ from the first two equations into the third, etc. Then

$$0=(a^2-u^2)\phi_{xx}-2uv\phi_{xy}+(a^2-v^2)\phi_{yy},$$

$$du=u_x\,dx+u_y\,dy=-\phi_{xx}\,dx-\phi_{yx}\,dy,$$

$$dv=v_x\,dx+v_y\,dy=-\phi_{yx}\,dx-\phi_{yy}\,dy,$$

where the suffixes denote differentiations. Solve in the form

$$\phi_{xx}/\Delta_1=-\phi_{xy}/\Delta_2=\phi_{yy}/\Delta_3=-1/\Delta_4.$$

When $\Delta_1=\Delta_2=\Delta_3=\Delta_4=0$, any one of the second-order ϕ derivatives can be discontinuous. $\Delta_4=0$ gives a quadratic for the slope of C, namely

$$(a^2 - u^2)\left(\frac{dy}{dx}\right)^2 + 2uv\frac{dy}{dx} + (a^2 - v^2) = 0.$$

These directions are real and different when $u^2 v^2 > (a^2 - u^2) \times (a^2 - v^2)$, giving $u^2 + v^2 > a^2$.

$\Delta_2 = 0$ gives $-\dfrac{dy}{dx} = \dfrac{(a^2 - v^2)}{(a^2 - u^2)}\dfrac{dv}{du}$. Substitution into the quadratic gives

$$(a^2 - v^2)\left(\frac{dv}{du}\right)^2 - 2uv\frac{dv}{du} + (a^2 - u^2) = 0.$$

Hence the result because the roots of this quadratic are the negative reciprocals of the roots of the quadratic for dy/dx.

12. (a) Prove that the differential relationship between $p = \partial U/\partial x$ and $q = \partial U/\partial y$ along the coincident characteristics of the parabolic equation $\partial^2 U/\partial x^2 = \partial U/\partial y$ is the parabolic equation itself.

Solution

Eliminate r and s, say, between eqns (4.32), (4.33), and (4.34) for this problem, etc.

(b) Show that the change of independent variables defined by $\xi = x + y$, $\eta = x - y$ transforms the equation $\partial^2 U/\partial x^2 = \partial^2 U/\partial y^2$ to $\partial^2 U/\partial \xi \, \partial \eta = 0$. Hence, deduce that the solution of the wave-equation satisfying the initial conditions, $U = f(x)$ and $\partial U/\partial y = g(x)$ on $y = 0$, is

$$U = \tfrac{1}{2}\left\{ f(x+y) + f(x-y) + \int_{x-y}^{x+y} g(t)\, dt \right\}.$$

Also deduce that the solution at (x_0, y_0) depends only on the data on the segment of the initial line between the characteristics $x + y = \text{constant}$, $x - y = \text{constant}$, that pass through (x_0, y_0).

Solution

Integration of $\partial^2 U/\partial \xi \, \partial \eta = 0$ gives $U = \phi(x+y) + \psi(x-y)$. Hence $\partial U/\partial y = \phi'(x+y) - \psi'(x-y)$. The initial conditions along $y = 0$

give $f(x) = \phi(x) + \psi(x)$; $g(x) = \phi'(x) - \psi'(x)$. Integrate the last equation to get

$$\phi(x) - \psi(x) + \text{constant} = \int_0^x g(x)\, dx.$$

Solve for $\phi(x)$ and $\psi(x)$ in terms of f and g then replace x in $\phi(x)$ by $x + y$ and x in $\psi(x)$ by $x - y$, etc.

13. The function U satisfies the equation

$$\frac{\partial^2 U}{\partial x^2} = \frac{\partial^2 U}{\partial t^2},$$

the boundary conditions

$$U = 0 \quad \text{at} \quad x = 0 \text{ and } 1, \quad t \geq 0,$$

and the initial conditions

$$U = \tfrac{1}{8} \sin \pi x, \quad \frac{\partial U}{\partial t} = 0, \quad \text{when} \quad t = 0, \quad 0 \leq x \leq 1.$$

Use the explicit finite-difference formula of eqn (4.54) and a central-difference approximation for the derivative condition, to calculate a solution for $x = 0(0.1)1$ and $t = 0(0.1)0.5$.

Derive the analytical solution

$$U = \tfrac{1}{8} \sin \pi x \cos \pi t$$

and compare with the numerical solution at several points.

Solution $(r = 1)$

$$u_{i,j+1} = u_{i-1,j} + u_{i+1,j} - u_{i,j-1}, \quad (j \geq 1).$$

$$\frac{\partial U}{\partial t} \simeq \frac{u_{i,1} - u_{i,-1}}{2\delta t} = 0, \text{ and } j = 0 \text{ in the previous equation, gives}$$

$$u_{i,1} = \tfrac{1}{2}(u_{i-1,0} + u_{i+1,0}).$$

The problem is symmetric with respect to $x = \tfrac{1}{2}$. The following values of u were obtained by working to $5D$ and rounding to $4D$. The analytical solution equals the finite-difference solution to this degree of accuracy.

x = 0	0.1	0.2	0.3	0.4	0.5
t = 0.1 0	0.0367	0.0699	0.0962	0.1131	0.1189
0.2 0	0.0312	0.0594	0.0818	0.0962	0.1011
0.3 0	0.0227	0.0432	0.0594	0.0699	0.0735
0.4 0	0.0119	0.0227	0.0312	0.0368	0.0386
0.5 0	0	0	0	0	0

14. The equation $(\partial^2 U/\partial t^2) = (\partial^2 U/\partial x^2)$, $0 < x < 1$, $t > 0$, is approximated at the point (ih, jk) by the difference equation

$$\frac{1}{k^2}\,\delta_t^2 u_{i,j} = \frac{1}{h^2}\left(\tfrac{1}{2}\delta_x^2 u_{i,j+1} + \tfrac{1}{2}\delta_x^2 u_{i,j-1}\right).$$

Given that U has known initial values throughout $0 \le x \le 1$, $t = 0$, known boundary values at $x = 0$ and 1, $t > 0$, and that $Nh = 1$, use the matrix method of analysis to prove that the equations are unconditionally stable in the sense that errors do not increase exponentially with increasing t.

Solution

The equations in matrix form are $\mathbf{A}\mathbf{u}_{j+1} = 2\mathbf{u}_j - \mathbf{A}\mathbf{u}_{j-1} - \mathbf{b}_j$, where $\mathbf{b}_j$ is a column vector of known values,

$$\mathbf{A} = \begin{bmatrix} (1+r) & -\tfrac{1}{2}r & & \\ -\tfrac{1}{2}r & (1+r) & -\tfrac{1}{2}r & \\ & \cdot & \cdot & \\ & & -\tfrac{1}{2}r & (1+r) \end{bmatrix}, \quad \text{an } (N-1)\times(N-1) \text{ matrix,}$$

and $r = k^2/h^2$. Hence a perturbation $\mathbf{e}_0$ of the initial values satisfies $\mathbf{e}_{j+1} = 2\mathbf{A}^{-1}\mathbf{e}_j - \mathbf{e}_{j-1}$, which, with $\dot{\mathbf{e}}_j = \mathbf{e}_j$ can be expressed as

$$\begin{bmatrix} \mathbf{e}_{j+1} \\ \mathbf{e}_j \end{bmatrix} = \begin{bmatrix} 2\mathbf{A}^{-1} & -\mathbf{I} \\ \mathbf{I} & \mathbf{0} \end{bmatrix} \begin{bmatrix} \mathbf{e}_j \\ \mathbf{e}_{j-1} \end{bmatrix},$$

i.e. as $\mathbf{v}_{j+1} = \mathbf{P}\mathbf{v}_j$. The eigenvalues λ of P are given by

$$\det \begin{vmatrix} (2\lambda_k^{-1} - \lambda) & -1 \\ 1 & -\lambda \end{vmatrix} = 0,$$

where

$$\lambda_k = (1+r) - 2(\tfrac{1}{2}r)\cos\frac{k\pi}{N}, \quad k = 1(1)(N-1).$$

This leads to $\lambda = \{1 \pm \sqrt{-1}\sqrt{(\lambda_k^2 - 1)}\}/\lambda_k$. As $\lambda_k = 1 + 2r \times \sin^2(k\pi/2N) > 1$, $|\lambda| = 1$ for all k.

15. The simultaneous equations

$$\frac{\partial p}{\partial x} = \frac{\partial q}{\partial t}, \quad \frac{\partial q}{\partial x} = \frac{\partial p}{\partial t},$$

which define $\partial^2 U/\partial x^2 = \partial^2 U/\partial t^2$, are represented by the following finite-difference formulae. Prove that both are stable for $\delta t/\delta x \leq 1$.

(a)
$$\frac{1}{2\delta x}(p_{i+1,j} - p_{i-1,j}) = \frac{1}{\delta t}\{q_{i,j+1} - \tfrac{1}{2}(q_{i+1,j} + q_{i-1,j})\},$$

$$\frac{1}{2\delta x}(q_{i+1,j} - q_{i-1,j}) = \frac{1}{\delta t}\{p_{i,j+1} - \tfrac{1}{2}(p_{i+1,j} + p_{i-1,j})\}.$$

(b)
$$\frac{1}{\delta x}(p_{i+\frac{1}{2},j} - p_{i-\frac{1}{2},j}) = \frac{1}{\delta t}(q_{i,j+1} - q_{i,j}),$$

$$\frac{1}{\delta x}(q_{i,j+1} - q_{i-1,j+1}) = \frac{1}{\delta t}(p_{i-\frac{1}{2},j+1} - p_{i-\frac{1}{2},j}).$$

Solution

(a) Substitution of $p_{i,j} = A e^{\beta i \delta x \sqrt{-1}}\xi^i$ and $q_{i,j} = B e^{\beta i \delta x \sqrt{-1}}\xi^i$ into the equations, and elimination of A/B gives

$$\xi = \cos\beta\,\delta x \pm (\rho\sin\beta\,\delta x)\sqrt{-1} \quad \text{where} \quad \rho = \frac{\delta t}{\delta x}.$$

Hence

$$|\xi|^2 = \cos^2\beta\,\delta x + \rho^2\sin^2\beta\,\delta x \leq 1 \quad \text{for} \quad \rho \leq 1.$$

Similarly for (b).

16. Solve the problem in Exercise 10 using the finite-difference equations of Exercise 15(a), with $\delta x = \delta t = 0.1$.

Solution

$$p_{i,j+1} = \tfrac{1}{2}(p_{i+1,j} + p_{i-1,j} + q_{i+1,j} - q_{i-1,j}),$$
$$q_{i,j+1} = \tfrac{1}{2}(q_{i+1,j} + q_{i-1,j} + p_{i+1,j} - p_{i-1,j}).$$

Along $t = 0$, $p = \partial u/\partial x = \frac{1}{2}(1 - 2x)$, $q = \partial u/\partial t = 0$. Along $x = 0$, $u = q = 0$, so $\partial q/\partial t = 0 = \partial p/\partial x$, giving $p_{1,j} = p_{-1,j}$. Put $i = 0$ in the equations above and eliminate $q_{-1,j}$, giving

$$p_{0,j+1} = p_{1,j} + q_{1,j},$$

which is the relationship along the characteristics, since $q_{0,j+1} = 0$. Integrate parallel to Ox using $\delta u = p\,\delta x \simeq \frac{1}{2}(p_i + p_{i+1})\,\delta x$. The values of p, q, and u when $t = 0.3$ are as follows.

$x = 0$	0.1	0.2	0.3	0.4	0.5
p	0.2	0.2	0.2	0.2	0.1	0
q	0	−0.1	−0.2	−0.3	−0.3	−0.3
u	0	0.02	0.04	0.06	0.075	0.08

17. The equation $(\partial^2 U/\partial t^2) = (\partial^2 U/\partial x^2)$ is approximated at the point (ih, jk) by the implicit difference scheme

$$\frac{1}{k^2}\delta_t^2 u_{i,j} = \frac{1}{h^2}(\tfrac{1}{4}\delta_x^2 u_{i,j+1} + \tfrac{1}{2}\delta_x^2 u_{i,j} + \tfrac{1}{4}\delta_x^2 u_{i,j-1}).$$

(a) Use the von Neumann method of analysis to prove that the equations are unconditionally stable.

(b) Use the matrix method to establish unconditional stability for fixed mesh lengths given that the boundary values are known at $x = 0$ and 1, $0 \leqslant x \leqslant 1$, and that $Nh = 1$.

(c) Prove that the principal part of the local truncation error at the point (ih, jk) is $-\frac{1}{12}h^2k^2(1 + 2r^2)(\partial^4 U/\partial x^4)_{i,j}$, where $r = k/h$.

Solution

(a) Equations are

$$-\tfrac{1}{4}r^2 u_{i-1,j+1} + (1 + \tfrac{1}{2}r^2)u_{i,j+1} - \tfrac{1}{4}r^2 u_{i+1,j+1}$$
$$= \tfrac{1}{2}r^2 u_{i-1,j} + (2 - r^2)u_{i,j} + \tfrac{1}{2}r^2 u_{i+1,j} + \tfrac{1}{4}r^2 u_{i-1,j-1}$$
$$+ (-1 - \tfrac{1}{2}r^2)u_{i,j-1} + \tfrac{1}{4}r^2 u_{i+1,j-1}.$$

The error function $E_{i,j} = e^{\sqrt{(-1)}\beta x}\xi^j$ is a solution if

$$(1 + r^2\sin^2\tfrac{1}{2}\beta h)\xi^2 - (2 - 2r^2\sin^2\tfrac{1}{2}\beta h)\xi + (1 + r^2\sin^2\tfrac{1}{2}\beta h) = 0.$$

It is easily shown that $|\xi| = 1$.

(b) In matrix form the equations are $\mathbf{u}_{j+1} = \mathbf{A}^{-1}\mathbf{B}\mathbf{u}_j + \mathbf{A}^{-1}\mathbf{C}\mathbf{u}_{j-1} + \mathbf{A}^{-1}\mathbf{b}_j$, where $\mathbf{b}_j$ is a column vector of known constants,

$$\mathbf{A} = (1 + \tfrac{1}{2}r^2)\mathbf{I} - \tfrac{1}{4}r^2\mathbf{E},$$
$$\mathbf{B} = (2 - r^2)\mathbf{I} + \tfrac{1}{2}r^2\mathbf{E},$$
$$\mathbf{C} = (-1 - \tfrac{1}{2}r^2)\mathbf{I} + \tfrac{1}{4}r^2\mathbf{E} = -\mathbf{A}$$

and the matrix $\mathbf{E}$ has 1's along each diagonal immediately above and below the main diagonal, zeros elsewhere. A perturbation $\mathbf{e}_0$ of the initial values will satisfy $\mathbf{e}_{j+1} = \mathbf{A}^{-1}\mathbf{B}\mathbf{e}_j + \mathbf{A}^{-1}\mathbf{C}\mathbf{e}_{j-1}$. Hence

$$\begin{bmatrix} \mathbf{e}_{j+1} \\ \mathbf{e}_j \end{bmatrix} = \begin{bmatrix} \mathbf{A}^{-1}\mathbf{B} & \mathbf{A}^{-1}\mathbf{C} \\ \mathbf{I} & \mathbf{O} \end{bmatrix} \begin{bmatrix} \mathbf{e}_j \\ \mathbf{e}_{j-1} \end{bmatrix},$$

i.e. $\mathbf{v}_{j+1} = \mathbf{P}\mathbf{v}_j$. The matrices $\mathbf{A}$, $\mathbf{B}$, and $\mathbf{C}$ have the same system of linearly independent eigenvectors as $\mathbf{E}$. So have $\mathbf{A}^{-1}\mathbf{B}$ and $\mathbf{A}^{-1}\mathbf{C}$. Therefore the eigenvalues λ of $\mathbf{P}$ are given by

$$\det\begin{bmatrix} \alpha_k^{-1}\beta_k - \lambda & \alpha_k^{-1}\gamma_k \\ 1 & -\lambda \end{bmatrix} = 0,$$

$k = 1(1)(N-1)$, where

$$\alpha_k = 1 + r^2 \sin\frac{k\pi}{2N}, \quad \beta_k = 2 - 2r^2 \sin^2\frac{k\pi}{2N},$$

$$\text{and} \quad \gamma_k = -1 - r^2 \sin^2\frac{k\pi}{2N}$$

are the eigenvalues of $\mathbf{A}$, $\mathbf{B}$, and $\mathbf{C}$ respectively. It is easily shown that $|\lambda| = 1$.

(c) Expand

$$T_{i,j} = \frac{1}{k^2}\delta_t^2 U_{i,j} - \frac{1}{h^2}(\tfrac{1}{4}\delta_x^2 U_{i,j+1} + \tfrac{1}{2}\delta_x^2 U_{i,j} + \tfrac{1}{4}\delta_x^2 U_{i,j-1})$$

about the point (i, j).
(See Chapter 2, Exercise 11.)

18. (a) Verify that the general solution of the equation

$$\partial^2 U/\partial x^2 = \partial^2 U/\partial t^2,$$

namely

$$U = \phi(x + t) + \psi(x - t),$$

is the exact solution of the explicit finite-difference scheme (4.54) for $\delta x = \delta t$.

(b) The solution of the equation $\partial^2 U/\partial x^2 = \partial^2 U/\partial t^2$ satisfies the initial conditions $U = f(x)$ and $\partial U/\partial t = g(x)$ on $t = 0$. When the equation is approximated by the explicit finite-difference scheme (4.54) with $r = 1$, namely,

$$u_{i,j+1} = u_{i+1,j} + u_{i-1,j} - u_{i,j-1},$$

and the derivative condition is approximated by a forward-difference, prove that;

(i) $$|e_{i,1}| \leqslant \tfrac{1}{2}h^2 M_2,$$

where $e = U - u$, $\delta t = \delta x = h$, and M_2 is the modulus of the largest value of $\partial^2 U/\partial t^2$ in the first time-interval;

(ii) $$e_{i,j+1} = e_{i+1,j} + e_{i-1,j} - e_{i,j-1} + \tfrac{1}{6}h^4 \eta M_4.$$

where M_4 is the modulus of the largest of $\partial^4 U/\partial t^4$ and $\partial^4 U/\partial x^4$ throughout the solution domain and $|\eta| \leqslant 1$.

Hence prove that

$$|e_{i,j}| \leqslant \tfrac{1}{2}jh^2 M_2 + \tfrac{1}{12}j(j-1)h^4 M_4, \quad 1 < j < K,$$

and deduce that u converges to U as h tends to zero, where Kh is finite.

Solution

(b) $g_i = (u_{i,1} - u_{i,0})/h$ gives $u_{i,1} = f_i + hg_i$, assuming no initial errors. By Taylor's expansion,

$$U_{i,1} = U_{i,0} + h\, \partial U_{i,0}/\partial t + \tfrac{1}{2}h^2\, \partial^2 U_{i,\theta}/\partial t^2, \quad (0 < \theta < 1),$$
$$= f_i + hg_i + \tfrac{1}{2}h^2\, \partial^2 U_{i,\theta}/\partial t^2.$$

Hence

$$|e_{i,1}| = |U_{i,1} - u_{i,1}| \leqslant \tfrac{1}{2}h^2 M_2.$$

Substitution of $u_{i,j} = U_{i,j} - e_{i,j}$ into the finite-difference equation and expansion in terms of $U_{i,j}$ by Taylor's theorem gives

$$e_{i,j+1} = e_{i+1,j} + e_{i-1,j} - e_{i,j-1} + \tfrac{1}{24}h^4$$
$$\times \left(\frac{\partial^4 U_{i,j+\theta_1}}{\partial t^4} + \frac{\partial^4 U_{i,j+\theta_2}}{\partial t^4} - \frac{\partial^4 U_{i+\theta_3,j}}{\partial x^4} - \frac{\partial^4 U_{i+\theta_4,j}}{\partial x^4} \right),$$

where $|\theta_s| < 1$, $(s = 1, 2, 3, 4)$. Hence

$$e_{i,j+1} = e_{i+1,j} + e_{i-1,j} - e_{i,j-1} + \tfrac{1}{6}h^4\eta M_4.$$

Draw the straight line characteristics through the point $(i, j+1)$ at $\pm 45°$ to Ox until they meet the line $j = 1$, and mark the points within this triangle contributing terms to $e_{i,j+1}$ when working backwards using the last equation so as to express $e_{i,j+1}$ entirely in terms of errors along $j = 1$. It will be seen there is one point at $(i, j+1)$, two points along $t = jh$, three along $t = (j-1)h$, etc., and $j + 1$ along $t = h$. As the $(j+1)$ points along $t = h$ each contribute an error $\leqslant \tfrac{1}{2}h^2 M_2$, and the $1 + 2 + \ldots + j = \tfrac{1}{2}j(j+1)$ points between $t = 2h$ and $(j+1)h$ each contribute an error $\leqslant \tfrac{1}{6}h^4 M_4$, it follows that

$$|e_{i,j+1}| \leqslant \tfrac{1}{2}(j+1)h^2 M_2 + \tfrac{1}{12}j(j+1)h^4 M_4.$$

Changing j into $(j-1)$ completes the proof. Since $jh = t$,

$$|e_{i,j}| \leqslant \tfrac{1}{2}thM_2 + \tfrac{1}{12}t^2h^2 M_4.$$

As h tends to zero this error tends to zero for finite values of t.

5 Elliptic equations and systematic iterative methods

Introduction

Elliptic partial differential equations arise usually from equilibrium or steady-state problems and their solutions, in relation to the calculus of variations, frequently maximize or minimize an integral representing the energy of the system. The best known elliptic equations are Poisson's equation, $\partial^2\phi/\partial x^2 + \partial^2\phi/\partial y^2 = f(x, y)$, often written as $\nabla^2\phi = f(x, y)$, and Laplace's equation

$$\partial^2\phi/\partial x^2 + \partial^2\phi/\partial y^2 = \nabla^2\phi = 0.$$

Poisson's equation, e.g., summarizes the St. Venant theory of torsion, the slow motion of incompressible viscous fluid, and the inverse-square law theories of electricity, magnetism and gravitating matter at points where the charge density, pole strength or mass density respectively, are non-zero. Laplace's equation arises in the theories associated with the steady flow of heat or electricity in homogeneous conductors, with the irrotational flow of incompressible fluid, and with potential problems in electricity, magnetism and gravitating matter at points devoid of these entities.

The domain of integration of a two-dimensional elliptic equation is always an area S bounded by a closed curve C. The boundary condition usually specifies either the value of the function or the value of its normal derivative at every point on C, or a mixture of both. Unlike hyperbolic equations both conditions cannot be given arbitrarily at any one point. Green's theorem provides an easy proof of this for Poisson's equation by showing that a specified function value on C determines a single-valued and differentiable solution uniquely throughout S. From this it follows that the normal derivative at every point on C is uniquely determined. Let ϕ satisfy $\partial^2\phi/\partial x^2 + \partial^2\phi/\partial y^2 = f(x, y)$ at every point of S and be specified at every point on C. Assume there can be two different solutions ϕ_1 and ϕ_2 inside S. Put

$U = \phi_1 - \phi_2$. Then $U = 0$ on C and

$$\partial^2 U/\partial x^2 + \partial^2 U/\partial y^2 = \nabla^2 \phi_1 - \nabla^2 \phi_2 = f(x, y) - f(x, y) = 0$$

at every point of S. Green's theorem in two dimensions is

$$\int_S \left\{ \left(\frac{\partial U}{\partial x}\right)^2 + \left(\frac{\partial U}{\partial y}\right)^2 \right\} dS = \int_C U \frac{\partial U}{\partial n} ds - \int_S U \left(\frac{\partial^2 U}{\partial x^2} + \frac{\partial^2 U}{\partial y^2}\right) dS,$$

where ds is an element of the boundary curve C and dS an element of S. The restrictions on U make the right side zero so $\partial U/\partial x = \partial U/\partial y = 0$ because the left side is the sum of squares. Hence U is constant throughout S. Since it is zero on C it must be zero throughout S, so $\phi_1 = \phi_2$ proving that the solution is unique. As the solution is known to be continuous the result follows. When $\partial \phi/\partial n$ is specified at every point on C the solution is no longer unique. If ϕ is a solution then so is $\phi + A$ where A is any constant. These results are valid even when the boundary values are discontinuous, from which it follows that discontinuities in boundary values are not propagated into the solution.

For Laplace's equation it can also be proved that the solution throughout S is bounded by the extreme values of ϕ on C and has no maxima or minima.

The following examples are typical elliptic problems but have been worked only to illustrate the use of finite-difference methods.

Example 5.1

The torsion problem

The shear stresses and displacements at points within a long uniform cylinder in torsion can be calculated from a stress function Φ that is constant round the perimeter C of a right cross-section by the XOY plane, and which satisfies the equation

$$\frac{\partial^2 \Phi}{\partial X^2} + \frac{\partial^2 \Phi}{\partial Y^2} + 2 = 0$$

at every point $P(X, Y)$ of the cross-section.

The problem can be written in non-dimensional form by put-

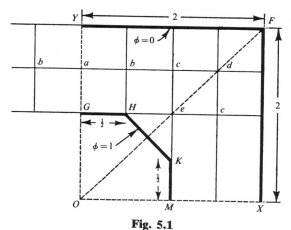

Fig. 5.1

ting

$$x = X/L, \quad y = Y/L \quad \text{and} \quad \phi = \Phi/L^2,$$

where L is some representative length.

Consider a thick cylindrical tube whose cross-section is symmetrical with respect to OX, OY, Fig. 5.1, and such that one-quarter of it is the area between the boundaries YFX and $GHKM$. Let its dimensions, in units of L, be as shown; let $\phi = 0$ on YFX and $\phi = 1$ on $GHKM$. Take a square mesh of side $h = \frac{1}{2}$ and denote the different approximate values of ϕ at the mesh points by a, b, c, d, e, taking into account the symmetry of the problem. The simplest finite-difference approximation to the non-dimensional equation

$$\frac{\partial^2 \phi}{\partial x^2} + \frac{\partial^2 \phi}{\partial y^2} + 2 = 0,$$

is the five-point equation

$$\frac{1}{h^2}(\phi_{i+1,j} - 2\phi_{i,j} + \phi_{i-1,j} + \phi_{i,j+1} - 2\phi_{i,j} + \phi_{i,j-1}) + 2 = 0,$$

where $x = ih$, $y = jh$. With $h = \frac{1}{2}$ we get

$$2(\phi_{i+1,j} + \phi_{i,j+1} + \phi_{i-1,j} + \phi_{i,j-1} - 4\phi_{i,j}) + 1 = 0. \qquad (5.1)$$

Application of eqn (5.1) to each mesh point leads to

$$2(2b + 1 - 4a) + 1 = 0,$$
$$2(c + a + 1 - 4b) + 1 = 0,$$
$$2(d + b + e - 4c) + 1 = 0,$$
$$2(2c - 4d) + 1 = 0,$$
$$2(2c + 2 - 4e) + 1 = 0.$$

The solution of these five equations for the five pivotal values of ϕ is easily shown to be

$$a = 0.737, \quad b = 0.724, \quad c = 0.658, \quad d = 0.454, \quad e = 0.954.$$

As a solution of the differential equation this is not very accurate because the mesh size is large. In particular the value of e contains a large error because the interior angles at H and K exceed $180°$.

Example 5.2

Derivative boundary conditions in a heat-conduction problem

The current in electrical windings gives rise to an internal generation of heat that is dissipated by radiation from the boundaries. The steady temperature distribution in a winding with a rectangular cross-section may therefore be approximated by the solution of the equation for steady heat flow across a rectangle, with internal generation of heat and radiation from its perimeter into the surrounding medium. As the thermal conductivity K parallel to Ox differs from its value λK parallel to Oy the equation for the temperature U is

$$\frac{\partial U^2}{\partial x^2} + \lambda \frac{\partial^2 U}{\partial y^2} = -\frac{q}{K},$$

where q is the number of units of heat generated per unit area per second. This equation can be made non-dimensional as in Chapter 2.

Purely for illustrative purposes, let U be the solution of the equation

$$\frac{\partial^2 U}{\partial x^2} + 3\frac{\partial^2 U}{\partial y^2} = -16 \tag{5.2}$$

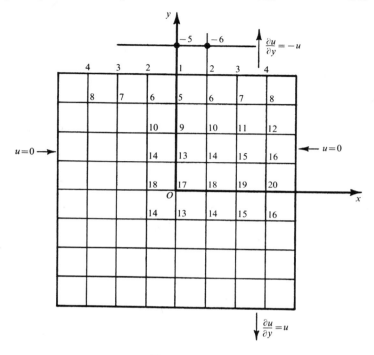

Fig. 5.2

inside the square $x = \pm 1$, $y = \pm 1$ (Fig. 5.2) satisfying the boundary conditions

$$U = 0 \quad \text{on} \quad x = 1, \quad \frac{\partial U}{\partial y} = -U \quad \text{on} \quad y = 1,$$

and be symmetric with respect to Ox and Oy, i.e.

$$\frac{\partial U}{\partial x} = 0 \quad \text{on} \quad Oy, \quad \frac{\partial U}{\partial y} = 0 \quad \text{on} \quad Ox.$$

Take a square mesh of side $h = \frac{1}{4}$ and label the mesh points as indicated, equal numbers denoting equal values of U. Denote the mirror image of the point 6 (say) in $y = 1$ by -6, and the value of U at point 6 by U_6, etc. One finite-difference approximation of equation (5.2) is

$$\frac{1}{h^2}(u_{i+1,j}-2u_{i,j}+u_{i-1,j})+\frac{3}{h^2}(u_{i,j+1}-2u_{i,j}+u_{i,j-1})=-16.$$

When $h=\frac{1}{4}$,

$$u_{i+1,j}+3u_{i,j+1}+u_{i-1,j}+3u_{i,j-1}-8u_{ij}=-1,$$

and this equation can be represented very conveniently by the pattern of numbers

$$\begin{bmatrix} & 3 & \\ 1 & -8 & 1 \\ & 3 & \end{bmatrix} u = -1. \tag{5.3}$$

At the boundary point 1, on imaging the heat-conducting area to be extended to the first row of external mesh points,

$$u_2+3u_{-5}+u_2+3u_5-8u_1=-1,$$

and from the boundary condition, using central differences,

$$(u_{-5}-u_5)/2h = 2(u_{-5}-u_5)=-u_1.$$

Elimination of u_{-5} gives the equation

$$-9\tfrac{1}{2}u_1+2u_2+6u_5=-1.$$

Similarly, points 2, 3, and 4 yield the equations

$$u_1-9\tfrac{1}{2}u_2+u_3+6u_6=-1,$$

$$u_2-9\tfrac{1}{2}u_3+u_4+6u_7=-1,$$

$$u_3-9\tfrac{1}{2}u_4+6u_8=-1.$$

Point 5 gives the equation

$$3u_1-8u_5+2u_6+3u_9=-1,$$

and point 6 the equation

$$3u_2+u_5-8u_6+u_7+3u_{10}=-1.$$

Similar equations can easily be written down for points 7 to 20, giving twenty linear algebraic equations for twenty unknowns. Their solution is not a trivial computation and would normally be carried out on a digital computer. On writing out the equations in detail it will be found that their matrix form is

$$\mathbf{Au} = \mathbf{b},$$

where $\mathbf{u}$ is a column vector with components $u_1, u_2, \ldots, u_{20}$, $\mathbf{b}$ a column vector with each component -1, and $\mathbf{A}$ a matrix which can be expressed in partitioned form as

$$\begin{bmatrix} (\mathbf{B} - 1\frac{1}{2}\mathbf{I}) & 6\mathbf{I} & & & \\ 3\mathbf{I} & \mathbf{B} & 3\mathbf{I} & & \\ & 3\mathbf{I} & \mathbf{B} & 3\mathbf{I} & \\ & & 3\mathbf{I} & \mathbf{B} & 3\mathbf{I} \\ & & & 6\mathbf{I} & \mathbf{B} \end{bmatrix}$$

where

$$\mathbf{B} = \begin{bmatrix} -8 & 2 & & \\ 1 & -8 & 1 & \\ & 1 & -8 & 1 \\ & & 1 & -8 \end{bmatrix}$$

Their solution is

$$u = 3.067, \quad u_2 = 2.909, \quad u_3 = 2.411, \quad u_4 = 1.496,$$

$$u_5 = 3.720, \quad u_6 = 3.527, \quad u_7 = 2.917, \quad u_8 = 1.801,$$

$$u_9 = 4.169, \quad u_{10} = 3.949, \quad u_{11} = 3.258, \quad u_{12} = 2.000,$$

$$u_{13} = 4.431, \quad u_{14} = 4.195, \quad u_{15} = 3.455, \quad u_{16} = 2.113,$$

$$u_{17} = 4.518, \quad u_{18} = 4.276, \quad u_{19} = 3.520, \quad u_{20} = 2.150.$$

Finite-differences in polar co-ordinates

Finite-difference problems involving circular boundaries can usually be solved more conveniently in polar co-ordinates than Cartesian co-ordinates because they avoid the use of awkward differentiation formulae near the curved boundary.

As an example consider Laplace's equation,

$$\frac{\partial^2 U}{\partial r^2} + \frac{1}{r}\frac{\partial U}{\partial r} + \frac{1}{r^2}\frac{\partial^2 U}{\partial \theta^2} = 0.$$

Define the mesh points in the $r - \theta$ plane by the points of intersection of the circles $r = ih$, $(i = 1, 2, \ldots)$, and the straight

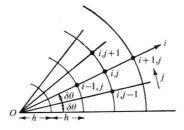

Fig. 5.3

lines $\theta = j\,\delta\theta$, $(j = 0, 1, 2, \ldots)$, Fig. 5.3. Laplace's equation at the point (i, j) may then be approximated by

$$\frac{(u_{i+1,j} - 2u_{i,j} + u_{i-1,j})}{h^2} + \frac{1}{ih} \frac{(u_{i+1,j} - u_{i-1,j})}{2h}$$

$$+ \frac{1}{(ih)^2} \frac{(u_{i,j+1} - 2u_{i,j} + u_{i,j-1})}{(\delta\theta)^2} = 0,$$

giving

$$\left(1 - \frac{1}{2i}\right)u_{i-1,j} + \left(1 + \frac{1}{2i}\right)u_{i+1,j} - 2\left\{1 + \frac{1}{(i\,\delta\theta)^2}\right\}u_{i,j}$$

$$+ \frac{1}{(i\,\delta\theta)^2}\, u_{i,j-1} + \frac{1}{(i\,\delta\theta)^2}\, u_{i,j+1} = 0.$$

If these equations are written out in detail for $i = 1, 2, \ldots, n$ and $j = 1, 2, \ldots, m$, and the boundary values are assumed to be known for $i = 0$, $i = (n+1)$, $j = 0$ and $j = (m+1)$, it will be found that their matrix form is

$$\mathbf{Au} = \mathbf{b},$$

where $\mathbf{b}$ is a column vector determined by the boundary values, $\mathbf{u}$ a column vector whose transpose is

$$(u_{1,1}, u_{1,2}, \ldots, u_{1,m}, u_{2,1}, \ldots, u_{2,m}, \ldots, u_{n,1}, u_{n,2}, \ldots, u_{n,m})$$

and $\mathbf{A}$ a matrix which can be written in partitioned form as

$$
\begin{bmatrix}
\mathbf{B}_I & (1+\tfrac{1}{2})\mathbf{I} & & & & \\
(1-\tfrac{1}{4})\mathbf{I} & \mathbf{B}_2 & (1+\tfrac{1}{4})\mathbf{I} & & & \\
& (1-\tfrac{1}{6})\mathbf{I} & \mathbf{B}_3 & (1+\tfrac{1}{6})\mathbf{I} & & \\
& & & \left(1 - \dfrac{1}{2(n-1)}\right)\mathbf{I} & \mathbf{B}_{n-1} & \left(1 + \dfrac{1}{2(n-1)}\right)\mathbf{I} \\
& & & & \left(1 - \dfrac{1}{2n}\right)\mathbf{I} & \mathbf{B}_n
\end{bmatrix}
$$

where each **B** and **I** are $m \times m$ matrices and

$$
\mathbf{B}_p =
\begin{bmatrix}
-2\left\{1+\dfrac{1}{(p\,\delta\theta)^2}\right\} & \dfrac{1}{(p\,\delta\theta)^2} & & & & \\[2ex]
\dfrac{1}{(p\,\delta\theta)^2} & -2\left\{1+\dfrac{1}{(p\,\delta\theta)^2}\right\} & \dfrac{1}{(p\,\delta\theta)^2} & & & \\[2ex]
& \dfrac{1}{(p\,\delta\theta)^2} & -2\left\{1+\dfrac{1}{(p\,\delta\theta)^2}\right\} & \dfrac{1}{(p\,\delta\theta)^2} & & \\[2ex]
& & & \ddots & & \\[2ex]
& & & \dfrac{1}{(p\,\delta\theta)^2} & -2\left\{1+\dfrac{1}{(p\,\delta\theta)^2}\right\} & \dfrac{1}{(p\,\delta\theta)^2} \\[2ex]
& & & & \dfrac{1}{(p\,\delta\theta)^2} & -2\left\{1+\dfrac{1}{(p\,\delta\theta)^2}\right\}
\end{bmatrix}
$$

Formulae for derivatives near a curved boundary when using a square mesh

When the boundary is curved and intersects the rectangular mesh at points that are not mesh points the formulae previously used to approximate first- and second-order derivatives at points near the boundary can no longer be used. This section is concerned with the finite-difference approximations to the derivatives at a point such as O, close to the boundary curve C on which the values of U are assumed to be known, Fig. 5.4. Let the mesh be a square

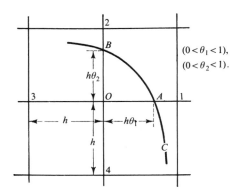

Fig. 5.4

of side h. By Taylor's theorem

$$U_A = U_0 + h\theta_1 \frac{\partial U_0}{\partial x} + \tfrac{1}{2}h^2\theta_1^2 \frac{\partial^2 U_0}{\partial x^2} + O(h^3),$$

$$U_3 = U_0 - h\frac{\partial U_0}{\partial x} + \tfrac{1}{2}h^2 \frac{\partial^2 U_0}{\partial x^2} + O(h^3).$$

Elimination of $\partial^2 U_0/\partial x^2$ gives

$$\frac{\partial U_0}{\partial x} = \frac{1}{h}\left\{\frac{1}{\theta_1(1+\theta_1)} U_A - \frac{(1-\theta_1)}{\theta_1} U_0 - \frac{\theta_1}{(1+\theta_1)} U_3\right\},$$

with a leading error of order h^2.

Similarly, the elimination of $\partial U_0/\partial x$ leads to

$$\frac{\partial^2 U_0}{\partial x^2} = \frac{1}{h^2}\left\{\frac{2}{\theta_1(1+\theta_1)} U_A + \frac{2}{(1+\theta_1)} U_3 - \frac{2}{\theta_1} U_0\right\},$$

with a leading error of order h, which is of lower accuracy than the usual formula employing only pivotal values.

Hence Poisson's equation $\partial^2 U/\partial x^2 + \partial^2 U/\partial y^2 = f(x, y)$ at the point O is approximated by

$$\frac{2u_A}{\theta_1(1+\theta_1)} + \frac{2u_B}{\theta_2(1+\theta_2)} + \frac{2u_3}{(1+\theta_1)} + \frac{2u_4}{(1+\theta_2)} - 2\left(\frac{1}{\theta_1} + \frac{1}{\theta_2}\right)u_0 = h^2 f_0.$$

Finite-difference formulae approximating normal derivatives at points on the boundary C, in terms of internal pivotal values, are extremely awkward and can be found in reference 8.

Improvement of the accuracy of solutions

Finer mesh

Although no useful general results concerning the magnitude of the discretization error as a function of the mesh lengths have yet been established, it seems reasonable to assume that this error will usually decrease as the mesh lengths are reduced. One hopes therefore that the sequence of solutions obtained using finer and finer meshes will eventually give a solution that differs from its immediate predecessor by less than some assigned tolerance. With this approach however, the size of the matrix of coefficients increases rapidly and becomes, after a number of refinements, too large for storage in the immediate access store of a computer.

Deferred approach to the limit

This method, suggested by Richardson, is extremely useful when there is a reliable estimate of the discretization error as a function of the mesh length. It is of dubious value, however, near curved boundaries, near corners with interior angles exceeding 180°, and near boundaries on which specified function values are not smooth.

Let U be the exact solution of the differential equation and u_1, u_2 the approximate solutions at the same mesh-point for mesh lengths h_1 and h_2 respectively. When the leading term in the discretization error is proportional to h^p,

$$U - u_1 = Ah_1^p \quad \text{and} \quad U - u_2 = Ah_2^p.$$

Hence

$$U = \frac{h_2^p u_1 - h_1^p u_2}{h_2^p - h_1^p}.$$

For the five-point formula for Laplace's equation, namely

$$u_{i+1,j} + u_{i-1,j} + u_{i,j+1} + u_{i,j-1} - 4u_{i,j} = 0,$$

the discretization error for a rectangular region with smooth known boundary-values is proportional to h^2. In this case, for $h_1 = 2h_2$,

$$U = u_2 + \tfrac{1}{3}(u_2 - u_1).$$

When the value of p is not known an estimate can be found from three approximate solutions at the same grid-point. Taking

$$h_3 = \tfrac{1}{2}h_2 = \tfrac{1}{4}h_1$$

gives

$$\frac{u_2 - u_1}{u_3 - u_2} = 2^p.$$

Deferred-correction method

In this method we calculate an initial approximate solution, difference it to obtain correction terms, add the correction terms to the initial finite-difference equations, then solve the amended equations for a more accurate solution.

As an example consider Laplace's equation. To fourth-order central-differences

$$h^2 \frac{\partial^2 U}{\partial x^2} = \delta_x^2 u - \tfrac{1}{12} \delta_x^4 u,$$

and

$$h^2 \frac{\partial^2 U}{\partial y^2} = \delta_y^2 u - \tfrac{1}{12} \delta_y^4 u,$$

where $\delta_x^2 u$ represents the second-order differences obtained by differencing parallel to Ox, i.e.

$$\delta_x^2 u_{i,j} = \delta_x u_{i+\frac{1}{2},j} - \delta_x u_{i-\frac{1}{2},j} = u_{i+1,j} - 2u_{i,j} + u_{i-1,j}.$$

Hence $h^2(\partial^2 U/\partial x^2 + \partial^2 U/\partial y^2) = 0$ can be approximated by

$$\delta_x^2 u + \delta_y^2 u = \tfrac{1}{12}(\delta_x^4 u + \delta_y^4 u). \tag{5.4}$$

The initial approximation is given by solving $\delta_x^2 u + \delta_y^2 u = 0$, which is, of course, the usual five-point formula

$$u_{i-1,j} + u_{i+1,j} + u_{i,j-1} + u_{i,j+1} - 4u_{i,j} = 0.$$

The numbers $\delta_x^4 u_{i,j}$, $\delta_y^4 u_{i,j}$ are then derived from

$$\delta_x^4 u_{i,j} = \delta_x^2(\delta_x^2 u_{i,j}) = u_{i+2,j} - 4u_{i+1,j} + 6u_{i,j} - 4u_{i-1,j} + u_{i-2,j}, \text{ etc.,}$$

and used to produce the equations (5.4) for an improved solution.

More accurate finite-difference formulae

Accuracy can also be improved by representing the differential equation by a higher-order finite-difference approximation designed to reduce the truncation error. This increases the number of pivotal values in each difference equation, but for Laplace's and Poisson's equations the increase is not large for truncation errors of order six and four respectively. A number of such formulae have been devised by W. G. Bickley and A. Thom and are given in reference 8. Their mode of derivation is indicated below by the establishment of a nine-point finite-difference approximation to Poisson's equation that has a truncation error of order h^4.

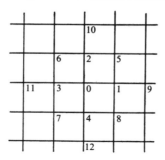

Fig. 5.5

Take a square mesh of side h and label the relative mesh-point positions as in Fig. 5.5.

Denote the Laplacian $\partial^2 U/\partial x^2 + \partial^2 U/\partial y^2$ by $\nabla^2 U$, and let

$$\xi = h\frac{\partial}{\partial x}, \quad \eta = h\frac{\partial}{\partial y}, \quad \mathscr{D}^2 = \frac{\partial^2}{\partial x \partial y},$$

so that

$$\xi^2 + \eta^2 = h^2\nabla^2, \quad \xi\eta = h^2\mathscr{D}^2,$$

$$\xi^4 + \eta^4 = (\xi^2 + \eta^2)^2 - 2\xi^2\eta^2 = h^4(\nabla^4 - 2\mathscr{D}^4).$$

As Taylor's series can be written as

$$U(x+h) = \left(1 + h\frac{d}{dx} + \ldots + \frac{h^n}{n!}\frac{d^n}{dx^n} + \ldots\right)U(x)$$

$$= (e^{h(d/dx)})U(x),$$

it follows that

$$U_1 = e^\xi U_0, \quad U_2 = e^\eta U_0, \quad U_3 = e^{-\xi} U_0, \quad U_5 = e^{\xi+\eta} U_0, \text{ etc.}$$

Because Poisson's equation is symmetric in its derivatives, define the following symmetric sums;

$$S_1 = U_1 + U_2 + U_3 + U_4, \quad S_2 = U_5 + U_6 + U_7 + U_8,$$

and

$$S_3 = U_9 + U_{10} + U_{11} + U_{12}.$$

On substituting for $U_1, U_2, \ldots$ in terms of U_0 and expanding the exponential operators in powers of ξ and η, it can be shown that

$$S_1 = 4U_0 + h^2\nabla^2 U_0 + \tfrac{1}{12}h^4(\nabla^4 - 2\mathscr{D}^4)U_0 + \tfrac{1}{360}h^6(\nabla^6 - 3\mathscr{D}^2\nabla^2)U_0 \\ + \ldots,$$

$$S_2 = 4U_0 + 2h^2\nabla^2 U_0 + \tfrac{1}{6}h^4(\nabla^4 + 4\mathscr{D}^4)U_0 + \tfrac{1}{180}h^6(\nabla^6 + 12\mathscr{D}^4\nabla^2)U_0 \\ + \ldots,$$

$$S_3 = 4U_0 + 4h^2\nabla^2 U_0 + \tfrac{4}{3}h^4(\nabla^4 - 2\mathscr{D}^4)U_0 + \tfrac{8}{45}h^6(\nabla^6 - 3\mathscr{D}^4\nabla^2)U_0 + \ldots$$

$$(5.5)$$

Elimination of $\mathscr{D}^4 U_0$ between S_1 and S_2 gives

$$\nabla^2 U_0 = \frac{4S_1 + S_2 - 20U_0}{6h^2} - \tfrac{1}{12}h^2\nabla^4 U_0 + O(h^4).$$

For Poisson's equation the second term on the right is known because of $\nabla^2 U = f$, so $\nabla^4 U = \nabla^2 f$. For Laplace's equation the coefficients of h^2 and h^4 both vanish. (See Exercise 7.) Hence the nine-point formula $4S_1 + S_2 - 20u_0 = 0$ is a more accurate finite-difference representation of Laplace's equation than $S_1 - 4u_0 = 0$ because its truncation error is of order h^6. (See Exercise 8.) The error in $S_1 - 4u_0 = 0$ is of order h^2. The most convenient way of exhibiting this nine-point formula approximating Poisson's equation $\nabla^2 u = f$, is by the 'molecular' display

$$\begin{bmatrix} 1 & 4 & 1 \\ 4 & -20 & 4 \\ 1 & 4 & 1 \end{bmatrix} u = 6h^2 f + \tfrac{1}{2}h^4\nabla^2 f.$$

Analysis of the discretization error of the five-point difference approximation to Poisson's equation over a rectangle

The following analysis is based on that given in reference 27. Consider the Dirichlet problem

$$\frac{\partial^2 U}{\partial x^2} + \frac{\partial^2 U}{\partial y^2} = f(x, y), \quad (x, y) \in D, \qquad (5.6)$$

$$U(x, y) = g(x, y), \quad (x, y) \in C, \qquad (5.7)$$

where $g(x, y)$ is known on the boundary curve C of the rectangular solution domain D defined by $0 < x < a,\ 0 < y < b$. (Termed a Dirichlet problem because U is known on C.) Define the mesh points as the points of intersection of the straight lines $x_i = ih$, $i = 0(1)m$ and $y_j = jh$, $j = 0(1)n$, where $mh = a$ and $nh = b$. Denote the set of mesh points in D by D_h and those on C by C_h and let G_h denote the set of all these mesh points, called the union of D_h and C_h and written as $G_h = D_h \cup C_h$, i.e. G_h is the set of all mesh points in $G = D \cup C$.

The five-point difference approximation to eqn (5.6) at the point (i, j) is

$$\frac{1}{h^2}(u_{i+1,j} + u_{i-1,j} + u_{i,j+1} + u_{i,j-1} - 4u_{i,j}) = f_{i,j},$$

so the difference equations approximating the problem may be written as

$$Lu_{i,j} = f_{i,j}, \quad (i, j) \in D_h, \tag{5.8}$$

and

$$u_{i,j} = g_{ij}, \quad (i, j) \in C_h, \tag{5.9}$$

where the five-point difference operator L is defined by

$$Lu_{i,j} = \frac{1}{h^2}(u_{i+1,j} + u_{i-1,j} + u_{i,j+1} + u_{i,j-1} - 4u_{i,j}), \quad (i, j) \in D_h.$$

Our problem is to express, if possible, the discretization error $e_{i,j} = U_{i,j} - u_{i,j}$ at the (i, j)th mesh point of D_h in terms of h. Operating on this error with L leads to

$$
\begin{aligned}
Le_{i,j} &= LU_{i,j} - Lu_{i,j}, \quad (i, j) \in D_h, \\
&= LU_{i,j} - f_{i,j}, \text{ by eqn (5.8)}, \\
&= T_{i,j}, \tag{5.9a}
\end{aligned}
$$

where $T_{i,j}$ is the local truncation error of the difference approximation $Lu - f = 0$ at the point (i, j).

In Exercise 5 it is proved that

$$T_{i,j} = \tfrac{1}{12}h^2\left\{\left(\frac{\partial^4 U}{\partial x^4}\right)_{\xi,y_j} + \left(\frac{\partial^4 U}{\partial y^4}\right)_{x_i,\eta}\right\},$$

where $x_i - h < \xi < x_i + h$, $y_j - h < \eta < y_j + h$, and it is assumed that U has continuous derivatives of all orders up to and including the fourth in $G = D \cup C$. If

$$M_4 = \max\left\{\max_G\left|\frac{\partial^4 U}{\partial x^4}\right|, \max_G\left|\frac{\partial^4 U}{\partial y^4}\right|\right\},$$

then

$$\max_{D_h} |T_{i,j}| \leqslant \tfrac{1}{6}h^2 M_4$$

and hence, by (5.9a),

$$\max_{D_h} |Le_{i,j}| = \max_{D_h} |T_{i,j}| \leqslant \tfrac{1}{6}h^2 M_4. \tag{5.10}$$

Clearly, we need a result relating $e_{i,j}$ to $Le_{i,j}$. This is provided by a theorem, proved in an addendum to this section, which states that if v is any function defined on the set of mesh points G_h in the rectangular region $0 \leqslant x \leqslant a$, $0 \leqslant y \leqslant b$, then

$$\max_{D_n} |v| \leqslant \max_{C_h} |v| + \tfrac{1}{4}(a^2 + b^2)\max_{D_h} |Lv|.$$

Applying this theorem to the discretization error $e_{i,j}$ gives that

$$\max_{D_h} |e_{i,j}| \leqslant \max_{C_h} |e_{i,j}| + \tfrac{1}{4}(a^2 + b^2)\max_{D_h} |Le_{i,j}|.$$

But $e_{i,j} = 0$ on the boundary because $U_{i,j} = u_{i,j} = g_{i,j}$, $(i, j) \in C_h$, by eqs (5.7) and (5.9). Hence, by eqn (5.10), we can conclude that

$$\max_{D_h} |e_{i,j}| \leqslant \tfrac{1}{24}(a^2 + b^2)h^2 M_4.$$

This is an extremely useful result because it proves that $u_{i,j}$ converges to $U_{i,j}$ as h tends to zero, in spite of the fact that M_4 is unknown for most problems. It also proves that the discretization error is proportional to h^2 so Richardson's 'deferred approach to the limit' method can be used effectively to improve the accuracy of the solution of the difference equations.

Elliptic problems with irregular boundaries

When the closed boundary curve C of a simply-connected open bounded domain D is irregular so that it intersects the sides of the square meshes as in Fig. 5.4, the analysis of the discretization error associated with either the five-point or the nine-point difference approximation to Poisson's equation with known boundary values is much more difficult than for a rectangle. The following result has, however, been obtained for the five-point difference replacement of Laplace's equation. If

$$\frac{\partial^2 U}{\partial x^2} + \frac{\partial^2 U}{\partial y^2} = 0, \quad (x, y) \in D,$$

and

$$U(x, y) = g(x, y), \quad (x, y) \in C,$$

where $g(x, y)$ is continuous and known on C and U has continuous fourth derivatives in $G = D \cup C$, then

$$\max |U(P) - u(P)| \leqslant \tfrac{1}{12} h^2 r^2 M_4 + h^2 M_2, \qquad (5.11)$$

where P is any internal mesh point of D, h is the side of the square mesh, r is the radius of a circle enclosing G, and M_k is defined by

$$M_k = \max\left\{\max_G \left|\frac{\partial^k U}{\partial x^k}\right|, \max_G \left|\frac{\partial^k U}{\partial y^k}\right|\right\}.$$

For this result to hold it is necessary to approximate Laplace's equation at a point such as O, Fig. 5.4, by the difference equation derived in that section, or to calculate u by interpolating linearly on u_3 and u_A or u_4 and u_B by

$$u_0 = \frac{\theta_1}{1 + \theta_1} u_3 + \frac{1}{1 + \theta_1} u_A \quad \text{or} \quad u_0 = \frac{\theta_2}{1 + \theta_2} u_4 + \frac{1}{1 + \theta_2} u_B.$$

Equation (5.11) shows that $u(P)$ converges to $U(P)$ as $h \to 0$ provided M_4 is bounded. This condition is not satisfied if, for example, the boundary contains corners with internal angles in excess of $180°$.

Addendum

If v is any function defined on the set of mesh points $G_h = D_h \cup C_h$ in the rectangular region $0 \leqslant x \leqslant a$, $0 \leqslant y \leqslant b$, then

$$\max_{D_h} |v| \leqslant \max_{C_h} |v| + \tfrac{1}{4}(a^2 + b^2)\max_{D_h} |Lv|,$$

where

$$Lv_{i,j} = \frac{1}{h^2}(v_{i+1,j} + v_{i-1,j} + v_{i,j+1} + v_{i,j-1} - 4v_{i,j}).$$

Proof

Define the function $\phi_{i,j}$ by the equation

$$\phi_{i,j} = \tfrac{1}{4}(x_i^2 + y_j^2) = \tfrac{1}{4}(i^2 + j^2)h^2, \quad (i, j) \in G_h.$$

Clearly,

$$0 \le \phi_{i,j} \le \tfrac{1}{4}(a^2 + b^2) \quad \text{for all} \quad (i, j) \in G_h. \tag{5.12}$$

It also follows that at all points $(i, j) \in D_h$,

$$L\phi_{i,j} = \tfrac{1}{4}\{(i+1)^2 + j^2 + (i-1)^2 + j^2 + i^2$$
$$+ (j+1)^2 + i^2 + (j-1)^2 - 4i^2 - 4j^2\} = 1. \tag{5.13}$$

Now define the functions w^+ and w^- by

$$w^+ = v + N\phi \quad \text{and} \quad w^- = -v + N\phi, \tag{5.14}$$

where

$$N = \max_{D_h} |Lv_{i,j}|.$$

Operating on eqns (5.14) with L and using (5.13) leads to

$$Lw_{i,j}^{\pm} = \pm Lv_{i,j} + N, \quad (i, j) \in D_h,$$
$$\ge 0 \text{ in virtue of the definition of } N.$$

In Exercise 9 it is proved that if $Lw_{i,j} \ge 0$ for all $(i, j) \in D_h$, then

$$\max_{D_h} w_{i,j} \le \max_{C_h} w_{i,j}.$$

Applying this result to w^+ and w^- gives that

$$\max_{D_h} w_{i,j}^{\pm} \le \max_{C_h} w_{i,j}^{\pm} = \max_{C_h}(\pm v_{i,j} + N\phi_{i,j}) \text{ by } (5.14).$$
$$\le \max_{C_h}(\pm v_{i,j}) + \tfrac{1}{4}(a^2 + b^2)N \text{ by } (5.12). \tag{5.14a}$$

Since $w_{i,j}^{\pm} = \pm v_{i,j} + N\phi_{i,j}$ and $N\phi_{i,j} \ge 0$, it is seen that $\pm v_{i,j} \le w_{i,j}^{\pm}$ for all $(i, j) \in G_h$. Hence by (5.14a)

$$\max_{D_h}(\pm v_{i,j}) \le \max_{C_h}(\pm v_{i,j}) + \tfrac{1}{4}(a^2 + b^2)N,$$

i.e.

$$\max_{D_h} |v_{i,j}| \le \max_{C_h} |v_{i,j}| + \tfrac{1}{4}(a^2 + b^2)\max_{D_h} |Lv_{i,j}|.$$

Comments on the solution of finite-difference equations approximating boundary-value problems

A finite-difference method for approximating a boundary value problem leads to a system of algebraic simultaneous equations. For linear boundary-value problems these equations are always linear but their number is generally large and, for this reason, their solution is a major problem in itself. Methods of solution belong essentially to either the class of direct methods or the class of iterative methods.

Direct methods solve the system of equations in a known number of arithmetic operations, and errors in the solution arise entirely from rounding errors introduced during the computation. Basically, these direct methods are elimination methods of which the best known examples are the systematic Gaussian elimination method and the triangular decomposition method which factorizes the matrix $\mathbf{A}$ of the equations $\mathbf{Ax} = \mathbf{b}$ into $\mathbf{A} = \mathbf{LU}$, where $\mathbf{L}$ and $\mathbf{U}$ are lower and upper triangular matrices respectively. In the latter method, once the decomposition has been determined, the solution is calculated from $\mathbf{LUx} = \mathbf{b}$ by putting $\mathbf{Ux} = \mathbf{y}$ and then solving $\mathbf{Ly} = \mathbf{b}$ for $\mathbf{y}$ by forward substitution and $\mathbf{Ux} = \mathbf{y}$ for $\mathbf{x}$ by back-substitution. (See Exercise 20.) With both methods it is usually necessary to employ partial pivoting with scaling to control the growth of rounding errors. The stable non-pivoting elimination procedure given on p. 24 for equations with tridiagonal matrices is exceptional. The ideas behind these methods are simple but the details associated with both their efficient computer implementation and the analysis of errors are considerable and can be found in references 30, 32, and 9.

It is of interest to note that when the arithmetic of these methods is carried out on a large modern computer the rounding errors introduced during the computation often have less effect on the solution than the rounding errors in the coefficients and constants of the equations. Say the coefficients and constants of n linear algebraic equations are all less than one in modulus and are known with certainty to only q decimal places, i.e. they contain errors $\leqslant \frac{1}{2} 10^{-q}$. Then it has been proved (reference 31) that if the arithmetic of the Gaussian elimination method with partial pivoting is carried out with numbers having more than $(q + \log_{10} n)$ decimal places, i.e. with more than $\log_{10} n$ guarding figures, the solution obtained to q decimal places is the exact solution of a set of n linear equations whose coefficients and constants differ from those of the original equations by less than $\frac{1}{2} 10^{-q}$. (In binary arithmetic, the arithmetic most frequently used on a computer, 10^{-q} is replaced by 2^{-q} and $\log_{10} n$ by $\log_2 n$.) In other words, the solution is as accurate as the data warrants. This does not imply that the number of correct significant figures in the solution is the same as in the data. It might be less. If it is, the loss of accuracy occurs *not* through the method of solution but because the equations are ill-conditioned in the sense that small changes in the coefficients produce large changes in the solution. The number of meaningful digits in the solution cannot, in fact, exceed the number of significant figures in the first q decimal places of the smallest pivot.

Although Gaussian elimination and **LU** decomposition are mathematically equivalent, the overall errors of the **LU** method can be made smaller than those of the Gaussian elimination method if inner products are accumulated in double length arithmetic, an operation provided for on some computers. Without double length accumulation the two methods are numerically equivalent as well as mathematically equivalent.

If the accuracy of a computed solution needs to be improved the following method is normally recommended. Denote the computed solution of $\mathbf{Ax} = \mathbf{b}$ by $\mathbf{x}^{(1)}$. Compute the residual $\mathbf{r}^{(1)} = \mathbf{b} - \mathbf{Ax}^{(1)}$ *using double length arithmetic*, i.e. to $2t$ decimal or binary digits instead of the standard t digits. Put $\mathbf{x} = \mathbf{x}^{(1)} + \mathbf{d}$, where $\mathbf{d}$ is the exact correction term to be added to the computed solution. As $\mathbf{A}(\mathbf{x}^{(1)} + \mathbf{d}) = \mathbf{b}$ in follows that $\mathbf{Ad} = \mathbf{b} - \mathbf{Ax}^{(1)} = \mathbf{r}^{(1)}$. The solution of this equation will not be $\mathbf{d}$ exactly but a computed approximation $\mathbf{d}^{(1)}$. Therefore, an improved solution will be $\mathbf{x}^{(2)} = \mathbf{x}^{(1)} + \mathbf{d}^{(1)}$, and the procedure can be repeated. In order to keep the solution of the equations to single length arithmetic, i.e. t digits, the residuals could be multiplied by 10^t or 2^t and the corresponding correction terms $\mathbf{d}^{(1)}$, $\mathbf{d}^{(2)}, \ldots$, multiplied by 10^{-t} or 2^{-t}. (See references 30 and 9.) It should be noted that this method of iterative refinement requires only a small amount of extra arithmetic because the solution of $\mathbf{Ad} = \mathbf{r}^{(1)}$ involves only forward and backward substitutions since $\mathbf{A} = \mathbf{LU}$ has already been found.

If the matrix of coefficients can be entered into the fast store of the computer then direct methods, in general, are quicker and more accurate than iterative methods. (See p. 260.) If also the matrix of coefficients has some special property or structure it is usually possible to increase the number of equations that can be solved by very efficient programming. In particular, Martin and Wilkinson, reference 32, have written such a programme for equations with symmetric positive definite matrices, and the NAG library of routines has a programme for banded matrices. Direct methods are certainly preferable to iterative methods when:

(i) Several sets of equations with the same coefficient matrix but different right-hand sides have to be solved.

(ii) The matrix is nearly singular. In this case small residuals do not imply small errors in the solution. This is easily seen since $\mathbf{A}^{-1}\mathbf{r}^{(1)} = \mathbf{A}^{-1}\mathbf{b} - \mathbf{x}^{(1)} = \mathbf{x} - \mathbf{x}^{(1)}$. Therefore $(\mathbf{x} - \mathbf{x}^{(1)})$ could have large components when the components of the residual vector are small because some of the elements of $\mathbf{A}^{-1}$ will be large if $\mathbf{A}$ is nearly singular.

The economic use of direct methods has been extended by Hockney 1968, reference 12, and George 1970. Stone 1968, reference 26, has also formulated a very economical iterative method that is more allied to direct methods than to the classical iterative methods and each is dealt with briefly at the end of this book.

As seen in this and earlier chapters, the matrices associated with difference equations approximating partial differential equations are band matrices, i.e. matrices with non-zero elements lying between two sub-diagonals parallel to the main diagonal. They are also usually 'sparse', i.e. the number of zero elements in the matrix is much greater than the number of non-zero elements. For this type of matrix standard Gaussian elimination is inefficient in the sense that it 'fills-in' the zero elements within the band with non-zero numbers that have to be stored in the computer and used at subsequent stages of the elimination process.

With iterative methods however, no arithmetic is associated with zero coefficients so considerably fewer numbers have to be stored in the computer. As a consequence, they can be used to solve systems of equations that are too large for the use of direct methods. Programming and data handling are also much simpler than for direct methods. Another advantage, not possessed by direct methods, is their frequent extension to the solution of sets of non-linear equations. The efficient use of iterative methods is very dependent however upon the direct calculation or estimation of the value or values of some acceleration parameter, and upon the coefficient matrix being well-conditioned, otherwise convergence will be slow and the volume of arithmetic enormous. With optimum acceleration parameters the volume of arithmetic of iterative methods for large sets of equations may actually be less than for direct methods.

In this book, only the classical Jacobi, Gauss–Seidel, and SOR methods are considered. A far more comprehensive treatment of all present-day iterative methods is given in Young's book, reference 33.

Systematic iterative methods for solving large linear systems of algebraic equations

An iterative method for solving equations is one in which a first approximation is used to calculate a second approximation which in turn is used to calculate a third and so on. The iterative procedure is said to be convergent when the differences between the exact solution and the successive approximations tend to zero as the number of iterations increase. In general, the exact solution is never obtained in a finite number of steps, but this does not matter. What is important is that the successive iterates converge fairly rapidly to values that are correct to a specified accuracy.

As mentioned previously, one would consider using iterative methods when a direct method requires more computer fast storage space than is available and the matrix of coefficients is sparse but well-conditioned, a situation that often arises with the difference equations approximating elliptic boundary value problems.

For simplicity consider the four equations

$$a_{11}x_1 + a_{12}x_2 + a_{13}x_3 + a_{14}x_4 = b_1$$
$$a_{21}x_1 + a_{22}x_2 + a_{23}x_3 + a_{24}x_4 = b_2$$
$$a_{31}x_1 + a_{32}x_2 + a_{33}x_3 + a_{34}x_4 = b_3$$
$$a_{41}x_1 + a_{42}x_2 + a_{43}x_3 + a_{44}x_4 = b_4,$$

$$(5.15)$$

where $a_{ii} \neq 0$, $i = 1(1)4$. These may be written as

$$x_1 = \frac{1}{a_{11}}(b_1 - a_{12}x_2 - a_{13}x_3 - a_{14}x_4)$$

$$x_2 = \frac{1}{a_{22}}(b_2 - a_{21}x_1 - a_{23}x_3 - a_{24}x_4)$$

$$x_3 = \frac{1}{a_{33}}(b_3 - a_{31}x_1 - a_{32}x_2 - a_{34}x_4) \tag{5.16}$$

$$x_4 = \frac{1}{a_{44}}(b_4 - a_{41}x_1 - a_{42}x_2 - a_{43}x_3).$$

Jacobi method

Denote the first approximation to x_i by $x_i^{(1)}$, the second by $x_i^{(2)}$, etc., and assume that n of them have been calculated, i.e. $x_i^{(n)}$ is known for $i = 1(1)4$. Then the Jacobi iterative method expresses the $(n+1)$th iterative values exclusively in terms of the nth iterative values and the iteration corresponding to eqns (5.16) is defined by

$$x_1^{(n+1)} = \frac{1}{a_{11}}\{b_1 - a_{12}x_2^{(n)} - a_{13}x_3^{(n)} - a_{14}x_4^{(n)}\}$$

$$x_2^{(n+1)} = \frac{1}{a_{22}}\{b_2 - a_{21}x_1^{(n)} - a_{23}x_3^{(n)} - a_{24}x_4^{(n)}\}$$

$$x_3^{(n+1)} = \frac{1}{a_{33}}\{b_3 - a_{31}x_1^{(n)} - a_{32}x_2^{(n)} - a_{34}x_4^{(n)}\} \tag{5.17}$$

$$x_4^{(n+1)} = \frac{1}{a_{44}}\{b_4 - a_{41}x_1^{(n)} - a_{42}x_2^{(n)} - a_{43}x_3^{(n)}\}.$$

In the general case for m equations,

$$x_i^{(n+1)} = \frac{1}{a_{ii}}\left\{b_i - \sum_{j=1}^{i-1} a_{ij}x_j^{(n)} - \sum_{j=i+1}^{m} a_{ij}x_j^{(n)}\right\}, \quad i = 1(1)m.$$

Gauss–Seidel method

In this method the $(n+1)$th iterative values are used as soon as they are available and the iteration corresponding to eqns

(5.16) is defined by

$$x_1^{(n+1)} = \frac{1}{a_{11}} \{b_1 - a_{12}x_2^{(n)} - a_{13}x_3^{(n)} - a_{14}x_4^{(n)}\}$$

$$x_2^{(n+1)} = \frac{1}{a_{22}} \{b_2 - a_{21}x_1^{(n+1)} - a_{23}x_3^{(n)} - a_{24}x_4^{(n)}\}$$

(5.18)

$$x_3^{(n+1)} = \frac{1}{a_{33}} \{b_3 - a_{31}x_1^{(n+1)} - a_{32}x_2^{(n+1)} - a_{34}x_4^{(n)}\}$$

$$x_4^{(n+1)} = \frac{1}{a_{44}} \{b_4 - a_{41}x_1^{(n+1)} - a_{42}x_2^{(n+1)} - a_{43}x_3^{(n+1)}\}.$$

In the general case for m equations,

$$x_i^{(n+1)} = \frac{1}{a_{ii}} \left\{ b_i - \sum_{j=1}^{i-1} a_{ij}x_j^{(n+1)} - \sum_{j=i+1}^{m} a_{ij}x_j^{(n)} \right\}, \quad i = 1(1)m.$$

Successive over-relaxation method

Add and subtract $x_i^{(n)}$ to the right-hand side of the ith Gauss–Seidel equation (5.18), $i = 1(1)4$, and write the set as

$$x_1^{(n+1)} = x_1^{(n)} + \left[\frac{1}{a_{11}} \{b_1 - a_{11}x_1^{(n)} - a_{12}x_2^{(n)} - a_{13}x_3^{(n)} - a_{14}x_4^{(n)}\} \right]$$

$$x_2^{(n+1)} = x_2^{(n)} + \left[\frac{1}{a_{22}} \{b_2 - a_{21}x_1^{(n+1)} - a_{22}x_2^{(n)} - a_{23}x_3^{(n)} - a_{24}x_4^{(n)}\} \right]$$

(5.19)

$$x_3^{(n+1)} = x_3^{(n)} + \left[\frac{1}{a_{33}} \{b_3 - a_{31}x_1^{(n+1)} - a_{32}x_2^{(n+1)} - a_{33}x_3^{(n)} - a_{34}x_4^{(n)}\} \right]$$

$$x_4^{(n+1)} = x_4^{(n)} + \left[\frac{1}{a_{44}} \{b_4 - a_{41}x_1^{(n+1)} - a_{42}x_2^{(n+1)} - a_{43}x_3^{(n+1)} - a_{44}x_4^{(n)}\} \right].$$

It is then seen that the expressions in the square brackets are the corrections or changes made to $x_i^{(n)}$, $i = 1(1)4$, by one Gauss–Seidel iteration. (Also called the 'displacements' in many books.) If successive corrections are all one-signed, as they usually are for the approximating difference equations of elliptic problems (see p. 273), it would be reasonable to expect convergence to be accelerated if each equation of (5.19) was given a larger correction term than is defined by (5.19). This idea leads to the successive over-relaxation or SOR iteration which is defined by the equations

$$x_1^{(n+1)} = x_1^{(n)} + \frac{\omega}{a_{11}} \{b_1 - a_{11}x_1^{(n)} - a_{12}x_2^{(n)} - a_{13}x_3^{(n)} - a_{14}x_4^{(n)}\}$$

$$
\begin{array}{ccc}
. & . & . \\
. & . & . \\
\end{array}
$$

$$x_4^{(n+1)} = x_4^{(n)} + \frac{\omega}{a_{44}} \{b_1 - a_{41}x_1^{(n+1)} - a_{42}x_2^{(n+1)} - a_{43}x_3^{(n+1)} - a_{44}x_4^{(n)}\}.$$

The factor ω, called the acceleration parameter or relaxation factor, generally lies in the range $1 < \omega < 2$. The determination of the optimum value of ω for maximum rate of convergence will be considered later for a particular class of matrices. The value $\omega = 1$ gives the Gauss–Seidel iteration.

In the general case for m equations the SOR iteration is defined by

$$x_i^{(n+1)} = x_i^{(n)} + \frac{\omega}{a_{ii}} \left\{ b_i - \sum_{j=1}^{i-1} a_{ij}x_j^{(n+1)} - \sum_{j=i}^{m} a_{ij}x_j^{(n)} \right\}, \quad i = 1(1)m.$$

This scheme can easily be remembered by noting that it may be written as

$$x_i^{(n+1)} = \frac{\omega}{a_{ii}} \left\{ b_i - \sum_{j=1}^{i-1} a_{ij}x_j^{(n+1)} - \sum_{j=i+1}^{m} a_{ij}x_j^{(n)} \right\} - (\omega - 1)x_i^{(n)}$$

$$= \omega(\text{R.H.S. of the Gauss–Seidel iteration equations})$$

$$- (\omega - 1)x_i^{(n)}. \tag{5.20}$$

Example 5.3

The function U satisfies the equation

$$\frac{\partial U}{\partial t} = \frac{\partial^2 U}{\partial x^2}, \quad 0 < x < 1, \quad t > 0,$$

the initial condition $U = 1$ when $t = 0$, $0 < x < 1$, and the boundary conditions $U = 0$ at $x = 0$ and 1, $t \geq 0$.

Approximate the differential equation by the Crank–Nicolson equations taking $\delta x = 0.1$ and $r = \delta t/(\delta x)^2 = 1$. Solve them to 4D for one time-step by the Jacobi, Gauss–Seidel, and SOR iterative methods, taking $\omega = 1.064$ and the first approximation values as equal to the initial values.

The Crank–Nicolson equations are

$$-u_{i-1,j+1} + 4u_{i,j+1} - u_{i+1,j+1} = u_{i-1,j} + u_{i+1,j}.$$

Denote the unknowns $u_{i,j+1}$, $i = 1(1)9$, by u_i. The equations for the unknowns along the first time-level, taking into account the initial conditions, the boundary conditions and the symmetry about $x = 0.5$ are then

$$4u_1 - u_2 = 1 \quad \text{i.e.,} \quad u_1 = \tfrac{1}{4}(u_2 + 1)$$

$$-u_1 + 4u_2 - u_3 = 2 \qquad u_2 = \tfrac{1}{4}(u_1 + u_3 + 2)$$

$$-u_2 + 4u_3 - u_4 = 2 \qquad u_3 = \tfrac{1}{4}(u_2 + u_4 + 2)$$

$$-u_3 + 4u_4 - u_5 = 2 \qquad u_4 = \tfrac{1}{4}(u_3 + u_5 + 2)$$

$$-2u_4 + 4u_5 = 2 \qquad u_5 = \tfrac{1}{2}(u_4 + 1).$$

The Jacobi iteration equations are

$$u_1^{(n+1)} = \tfrac{1}{4}(u_2^{(n)} + 1),$$

$$u_2^{(n+1)} = \tfrac{1}{4}(u_1^{(n)} + u_3^{(n)} + 2),$$

$$u_3^{(n+1)} = \tfrac{1}{4}(u_2^{(n)} + u_4^{(n)} + 2),$$

$$u_4^{(n+1)} = \tfrac{1}{4}(u_3^{(n)} + u_5^{(n)} + 2)$$

and

$$u_5^{(n+1)} = \tfrac{1}{2}(u_4^{(n)} + 1).$$

Taking the first approximation values $u_i^{(0)}$, $i = 1(1)5$, as equal to the initial values of 1 leads to the first iteration values

$$u_1^{(1)} = \tfrac{1}{4}(u_2^{(0)} + 1) = \tfrac{1}{4}(1 + 1) = 0.5,$$

$$u_2^{(1)} = \tfrac{1}{4}(u_1^{(0)} + u_3^{(0)} + 2) = \tfrac{1}{4}(1 + 1 + 2) = 1,$$

$$u_3^{(1)} = \tfrac{1}{4}(u_2^{(0)} + u_4^{(0)} + 2) = \tfrac{1}{4}(1 + 1 + 2) = 1,$$

$$u_4^{(1)} = \tfrac{1}{4}(1 + 1 + 2) = 1$$

and

$$u_5^{(1)} = \tfrac{1}{2}(u_4^{(0)} + 1) = \tfrac{1}{2}(1 + 1) = 1.$$

The solution to $4D$ is obtained after 11 iterations and is displayed in Table 5.1.

TABLE 5.1 (Jacobi)

$u_{i,0} = 0$	1.0	1.0	1.0	1.0	1.0
$n = 0$	0	1.0	1.0	1.0	1.0	1.0
$n = 1$	0	0.5	1.0	1.0	1.0	1.0
$n = 2$	0	0.5	0.875	1.0	1.0	1.0
$n = 3$	0	0.4688	0.875	0.9688	1.0	1.0
$n = 4$	0	0.4688	0.8594	0.9688	0.9922	1.0
$n = 10$	0	0.4641	0.8564	0.9614	0.9890	0.9946
$n = 11$	0	0.4641	0.8564	0.9613	0.9890	0.9945

The Gauss–Seidel iteration equations are

$$u_1^{(n+1)} = \tfrac{1}{4}(u_2^{(n)} + 1),$$
$$u_2^{(n+1)} = \tfrac{1}{4}(u_1^{(n+1)} + u_3^{(n)} + 2),$$
$$u_3^{(n+1)} = \tfrac{1}{4}(u_2^{(n+1)} + u_4^{(n)} + 2),$$
$$u_4^{(n+1)} = \tfrac{1}{4}(u_3^{(n+1)} + u_5^{(n)} + 2),$$

and

$$u_5^{(n+1)} = \tfrac{1}{2}(u_4^{(n+1)} + 1).$$

Therefore

$$u_1^{(1)} = \tfrac{1}{4}(u_2^{(0)} + 1) = \tfrac{1}{4}(1 + 1) = 0.5,$$

$$u_2^{(1)} = \tfrac{1}{4}(u_1^{(1)} + u_3^{(0)} + 2) = \tfrac{1}{4}(0.5 + 1 + 2) = 0.875,$$

$$u_3^{(1)} = \tfrac{1}{4}(u_2^{(1)} + u_4^{(0)} + 2) = \tfrac{1}{4}(0.875 + 1 + 2) = 0.96875,$$

$$u_4^{(1)} = \tfrac{1}{4}(u_3^{(1)} + u_5^{(0)} + 2) = \tfrac{1}{4}(0.96875 + 1 + 2) = 0.99219,$$

and

$$u_5^{(1)} = \tfrac{1}{2}(u_4^{(1)} + 1) = \tfrac{1}{2}(0.99219 + 1) = 0.99609.$$

The solution to $4D$ is obtained after five iterations and is given in Table 5.2.

TABLE 5.2 (Gauss–Seidel)

$i = 0$	1	2	3	4	5
$i_{i,0} = 0$	1.0	1.0	1.0	1.0	1.0
$n = 0$	0	1.0	1.0	1.0	1.0	1.0
$n = 1$	0	0.5	0.875	0.9688	0.9922	0.9961
$n = 2$	0	0.4688	0.8594	0.9629	0.9898	0.9949
$n = 3$	0					
$n = 4$	0					
$n = 5$	0	0.4641	0.8564	0.9613	0.9890	0.9945

The SOR iteration equations by (5.20) are

$$u_1^{(n+1)} = \frac{\omega}{4}(u_2^{(n)} + 1) - (\omega - 1)u_1^{(n)} = 0.266(u_2^{(n)} + 1) - 0.064u_1^{(n)},$$

$$u_2^{(n+1)} = 0.266(u_1^{(n+1)} + u_3^{(n)} + 2) - 0.064u_2^{(n)},$$

$$u_3^{(n+1)} = 0.266(u_2^{(n+1)} + u_4^{(n)} + 2) - 0.064u_3^{(n)},$$

$$u_4^{(n+1)} = 0.266(u_3^{(n+1)} + u_5^{(n)} + 2) - 0.064u_4^{(n)},$$

$$u_5^{(n+1)} = 0.532(u_4^{(n+1)} + 1) - 0.064u_5^{(n)}.$$

Therefore

$$u_1^{(1)} = 0.266(u_2^{(0)} + 1) - 0.064u_1^{(0)} = 0.266(1+1) - 0.064 = 0.468.$$

$$u_2^{(1)} = 0.266(u_1^{(1)} + u_3^{(0)} + 2) - 0.064u_2^{(0)} = 0.266(0.468 + 3) - 0.064$$
$$= 0.85849.$$

$$u_3^{(1)} = 0.266(0.85849 + 3) - 0.064 = 0.96236.$$

$$u_4^{(1)} = 0.266(0.96236 + 3) - 0.064 = 0.98999.$$

$$u_5^{(1)} = 0.532(0.98999 + 1) - 0.064 = 0.99467.$$

The solution to 4D is obtained after three iterations and is shown in Table 5.3.

TABLE 5.3 (S.O.R.)

$i = 0$	1	2	3	4	5
$u_{i,0} = 0$	1.0	1.0	1.0	1.0	1.0
$n = 0$ 0	1.0	1.0	1.0	1.0	1.0
$n = 1$ 0	0.4680	0.8585	0.9624	0.9900	0.9947
$n = 2$ 0	0.4644	0.8566	0.9616	0.9890	0.9945
$n = 3$ 0	0.4641	0.8564	0.9613	0.9890	0.9945

Systematic iterative methods in matrix form

Equations (5.15) in matrix form are

$$\begin{bmatrix} a_{11} & a_{12} & a_{13} & a_{14} \\ a_{21} & a_{22} & a_{23} & a_{24} \\ a_{31} & a_{32} & a_{33} & a_{34} \\ a_{41} & a_{42} & a_{43} & a_{44} \end{bmatrix} \begin{bmatrix} x_1 \\ x_2 \\ x_3 \\ x_4 \end{bmatrix} = \begin{bmatrix} b_1 \\ b_2 \\ b_3 \\ b_4 \end{bmatrix}$$

or, more briefly,

$$\mathbf{Ax} = \mathbf{b}.$$

For our purposes it will be found convenient to express the matrix $\mathbf{A}$ as the sum of its main diagonal elements, its strictly lower triangular elements, and its strictly upper triangular elements and to write it as $\mathbf{A} = \mathbf{D} - \mathbf{L} - \mathbf{U}$, where

$$\mathbf{D} = \begin{bmatrix} a_{11} & 0 & 0 & 0 \\ 0 & a_{22} & 0 & 0 \\ 0 & 0 & a_{33} & 0 \\ 0 & 0 & 0 & a_{44} \end{bmatrix} \quad -\mathbf{L} = \begin{bmatrix} 0 & 0 & 0 & 0 \\ a_{21} & 0 & 0 & 0 \\ a_{31} & a_{32} & 0 & 0 \\ a_{41} & a_{42} & a_{43} & 0 \end{bmatrix},$$

$$-\mathbf{U} = \begin{bmatrix} 0 & a_{12} & a_{13} & a_{14} \\ 0 & 0 & a_{23} & a_{24} \\ 0 & 0 & 0 & a_{34} \\ 0 & 0 & 0 & 0 \end{bmatrix}.$$

Equations (5.15) can then be written as

$$\mathbf{Dx} = (\mathbf{L} + \mathbf{U})\mathbf{x} + \mathbf{b}$$

and a study of eqns (5.17) shows that the corresponding Jacobi iteration can be expressed as

$$\mathbf{Dx}^{(n+1)} = (\mathbf{L} + \mathbf{U})\mathbf{x}^{(n)} + \mathbf{b}, \tag{5.21}$$

giving

$$\mathbf{x}^{(n+1)} = \mathbf{D}^{-1}(\mathbf{L} + \mathbf{U})\mathbf{x}^{(n)} + \mathbf{D}^{-1}\mathbf{b}.$$

The matrix $\mathbf{D}^{-1}(\mathbf{L} + \mathbf{U})$ is called the point Jacobi iteration matrix. The word 'point' refers to the fact that the algebraic equations approximate a differential equation at a number of mesh points and the iterative procedure expresses the next iterative value at only one mesh point in terms of known iterative values at other mesh points.

Similarly, by eqns (5.18), the Gauss–Seidel iteration is defined by

$$\mathbf{Dx}^{(n+1)} = \mathbf{Lx}^{(n+1)} + \mathbf{Ux}^{(n)} + \mathbf{b}. \tag{5.22}$$

Hence

$$(\mathbf{D} - \mathbf{L})\mathbf{x}^{(n+1)} = \mathbf{Ux}^{(n)} + \mathbf{b},$$

giving that
$$\mathbf{x}^{n+1} = (\mathbf{D} - \mathbf{L})^{-1}\mathbf{U}\mathbf{x}^{(n)} + (\mathbf{D} - \mathbf{L})^{-1}\mathbf{b},$$

which shows that the point Gauss–Seidel iteration matrix is $(\mathbf{D} - \mathbf{L})^{-1}\mathbf{U}$.

The correction or displacement vector for the Gauss–Seidel iteration is, by eqn (5.22),

$$\mathbf{x}^{(n+1)} - \mathbf{x}^{(n)} = \mathbf{D}^{-1}(\mathbf{L}\mathbf{x}^{(n+1)} + \mathbf{U}\mathbf{x}^{(n)} + \mathbf{b}) - \mathbf{x}^{(n)}$$
$$= \mathbf{D}^{-1}(\mathbf{L}\mathbf{x}^{(n+1)} + \mathbf{U}\mathbf{x}^{(n)} + \mathbf{b} - \mathbf{D}\mathbf{x}^{(n)}).$$

Hence the SOR iteration is defined by

$$\mathbf{x}^{(n+1)} - \mathbf{x}^{(n)} = \omega\mathbf{D}^{-1}(\mathbf{L}\mathbf{x}^{(n+1)} + \mathbf{U}\mathbf{x}^{(n)} + \mathbf{b} - \mathbf{D}\mathbf{x}^{(n)}),$$

giving

$$(\mathbf{I} - \omega\mathbf{D}^{-1}\mathbf{L})\mathbf{x}^{(n+1)} = \{(1 - \omega)\mathbf{I} + \omega\mathbf{D}^{-1}\mathbf{U}\}\mathbf{x}^{(n)} + \omega\mathbf{D}^{-1}\mathbf{b}.$$

Therefore,

$$\mathbf{x}^{(n+1)} = (\mathbf{I} - \omega\mathbf{D}^{-1}\mathbf{L})^{-1}\{(1 - \omega)\mathbf{I} + \omega\mathbf{D}^{-1}\mathbf{U}\}\mathbf{x}^{(n)}$$
$$+ (\mathbf{I} - \omega\mathbf{D}^{-1}\mathbf{L})^{-1}\omega\mathbf{D}^{-1}\mathbf{b}, \quad (5.23)$$

showing that the point *SOR* iteration matrix $\mathbf{H}(\omega)$ is

$$\mathbf{H}(\omega) = (\mathbf{I} - \omega\mathbf{D}^{-1}\mathbf{L})^{-1}\{(1 - \omega)\mathbf{I} + \omega\mathbf{D}^{-1}\mathbf{U}\}.$$

A necessary and sufficient condition for the convergence of iterative methods

Each of the three iterative methods described above can be written as
$$\mathbf{x}^{(n+1)} = \mathbf{G}\mathbf{x}^{(n)} + \mathbf{e}, \quad (5.24)$$

where $\mathbf{G}$ is the iteration matrix and $\mathbf{c}$ a column vector of known values. This equation was derived from the original equations by rearranging them into the form

$$\mathbf{x} = \mathbf{G}\mathbf{x} + \mathbf{c}, \quad (5.25)$$

i.e. the unique solution of the m linear equations $\mathbf{A}\mathbf{x} = \mathbf{b}$ is the solution of eqn (5.25). The error $\mathbf{e}^{(n)}$ in the nth approximation to the exact solution is defined by $\mathbf{e}^{(n)} = \mathbf{x} - \mathbf{x}^{(n)}$ so it follows by the subtraction of eqn (5.24) from eqn (5.25) that

$$\mathbf{e}^{(n+1)} = \mathbf{G}\mathbf{e}^{(n)}.$$

Therefore

$$\mathbf{e}^{(n)} = \mathbf{G}\mathbf{e}^{(n-1)} = \mathbf{G}^2\mathbf{e}^{(n-2)} = \ldots = \mathbf{G}^n\mathbf{e}^{(0)}. \qquad (5.26)$$

The sequence of iterative values $\mathbf{x}^{(1)}, \mathbf{x}^{(2)}, \ldots, \mathbf{x}^{(n)}, \ldots$ will converge to $\mathbf{x}$ as n tends to infinity if

$$\lim_{n \to \infty} \mathbf{e}^{(n)} = \mathbf{O}.$$

Since $\mathbf{x}^{(0)}$ and therefore $\mathbf{e}^{(0)}$ is arbitrary it follows that the iteration will converge if and only if

$$\lim_{n \to \infty} \mathbf{G}^n = \mathbf{O}.$$

Assume now that the matrix $\mathbf{G}$ of order m has m linearly independent eigenvectors $\mathbf{v}_s$, $s = 1(1)m$. Then these eigenvectors can be used as a basis for our m-dimensional vector space and the arbitrary error vector $\mathbf{e}^{(0)}$, with its m components, can be expressed uniquely as a linear combination of them, namely,

$$\mathbf{e}^{(0)} = \sum_{s=1}^{m} c_s \mathbf{v}_s,$$

where the c_s are scalars. Hence

$$\mathbf{e}^{(1)} = \mathbf{G}\mathbf{e}^{(0)} = \sum_{s=1}^{m} c_s \mathbf{G}\mathbf{v}_s.$$

But $\mathbf{G}\mathbf{v}_s = \lambda_s \mathbf{v}_s$ by the definition of an eigenvalue, where λ_s is the eigenvalue corresponding to $\mathbf{v}_s$. Hence

$$\mathbf{e}^{(1)} = \sum_{1}^{m} c_s \lambda_s \mathbf{v}_s.$$

Similarly,

$$\mathbf{e}^{(n)} = \sum_{1}^{m} c_s \lambda_s^n \mathbf{v}_s. \qquad (5.27)$$

Therefore $\mathbf{e}^{(n)}$ will tend to the null vector as n tends to infinity, for arbitrary $\mathbf{e}^{(0)}$, if and only if $|\lambda_s| < 1$ for all s. In other words, the iteration will converge for arbitrary $\mathbf{x}^{(0)}$ if and only if the spectral radius $\rho(\mathbf{G})$ of $\mathbf{G}$ is less than one.

As a corollary to this result a sufficient condition for convergence is that $\|\mathbf{G}\| < 1$ because $\rho(\mathbf{G}) \leqslant \|\mathbf{G}\|$.

As an example, consider the Jacobi iteration matrix $\mathbf{D}^{-1}(\mathbf{L}+\mathbf{U})$. If the ith equation of $\mathbf{Ax}=\mathbf{b}$ is

$$a_{i1}x_1 + a_{i2}x_2 + \ldots + a_{ii}x_i + \ldots + a_{im}x_m = b_i$$

then the ith row of $\mathbf{D}^{-1}(\mathbf{L}+\mathbf{U})$ is

$$\frac{a_{i1}}{a_{ii}} \quad \frac{a_{i2}}{a_{ii}} \quad . \quad . \quad \frac{a_{i,i-1}}{a_{ii}} \quad 0 \quad \frac{a_{i,i+1}}{a_{ii}} \quad . \quad . \quad \frac{a_{im}}{a_{ii}}.$$

Let the sum of the moduli of these elements be the greatest row sum of the moduli of the elements of the iteration matrix. Then if we take our matrix norm as the infinity norm, the Jacobi iteration will converge if

$$|a_{i1}| + |a_{i2}| + \ldots + |a_{i,i-1}| + 0 + |a_{i,i+1}| + |a_{im}| < |a_{ii}|.$$

This states that the Jacobi method applied to the equations $\mathbf{Ax}=\mathbf{b}$ will converge if $\mathbf{A}$ is a strictly diagonally dominant matrix, i.e., if in each row of $\mathbf{A}$ the modulus of the diagonal element exceeds the sum of the moduli of the off-diagonal elements.

Rate of convergence

Assume that the iteration matrix $\mathbf{G}$ has m linearly independent eigenvectors $\mathbf{v}_s$ corresponding to the eigenvalues λ_s and that $|\lambda_1| > |\lambda_2| \geqslant |\lambda_3| \geqslant \ldots \geqslant |\lambda_m|$. By eqn (5.27) the error vector $\mathbf{e}^{(n)}$ can be expressed as

$$\mathbf{e}^{(n)} = \lambda_1^n \left\{ c_1\mathbf{v}_1 + \left(\frac{\lambda_2}{\lambda_1}\right)^n c_2\mathbf{v}_2 + \ldots + \left(\frac{\lambda_m}{\lambda_1}\right)^n c_m\mathbf{v}_m \right\}.$$

For large values of n this shows that

$$\mathbf{e}^{(n)} \simeq \lambda_1^n c_1 \mathbf{v}_1.$$

Similarly,

$$\mathbf{e}^{(n+1)} \simeq \lambda_1^{n+1} c_1 \mathbf{v}_1,$$

so

$$\mathbf{e}^{(n+1)} \simeq \lambda_1 \mathbf{e}^{(n)}. \tag{5.28}$$

If the ith component of $\mathbf{e}^{(n)}$ is denoted by $e_i^{(n)}$ it is seen that

$$\frac{|e_i^{(n)}|}{|e_i^{(n+1)}|} \simeq \frac{1}{|\lambda_1|} = \frac{1}{\rho(\mathbf{G})}.$$

For simplicity, assume that $e_i^{(n)} = 10^{-2}$ and $e_i^{(n+1)} = 10^{-4}$. Then $e_i^{(n)}/e_i^{(n+1)} = 10^2$ and the logarithm of this to base 10 is 2. More generally, $\log_{10}(1/\rho) = -\log \rho$ gives an indication of the number of decimal digits by which the error is eventually decreased by each convergent iteration. Since, for convergence, $0 < \rho < 1$, the number of decimal digits of accuracy gained per iteration increases as ρ decreases. Alternatively, for large n, $\mathbf{e}^{(n)} \simeq \lambda_1 \mathbf{e}^{(n-1)}$, therefore

$$\mathbf{e}^{(n+p)} \simeq \lambda_1 \mathbf{e}^{(n+p-1)} \simeq \ldots \simeq \lambda_1^p \mathbf{e}^{(n)}, \quad p = 1, 2, \ldots$$

If we want to reduce the size of the error by 10^{-q}, say, then the number of iterations needed to do this will be the least value of p for which

$$|\lambda_1^p| = \rho^p \leqslant 10^{-q}.$$

Taking logs and remembering that $\log \rho$ is negative for a convergent iteration leads to

$$p \geqslant q/(-\log_{10} \rho),$$

which shows that p decreases as $(-\log \rho)$ increases. Clearly, the number $(-\log \rho)$ provides a measure for the comparison of the rates of convergence of different iterative methods when n is sufficiently large. For this reason $(-\log_e \rho)$ is defined to be *the asymptotic rate of convergence* and is denoted by $R_\infty(\mathbf{G})$.

The *average rate of convergence* $R_n(\mathbf{G})$ after n iterations is defined by

$$R_n(\mathbf{G}) = -\frac{1}{n} \log_e \|\mathbf{G}^n\|_2,$$

where $\|\mathbf{G}^n\|_2$ is the 2-norm or spectral norm of the matrix $\mathbf{G}^n$ which is given by the square root of the spectral radius of $[\mathbf{G}^n]^H[\mathbf{G}^n]$. In reference 33 it is proved that the asymptotic rate of convergence

$$R_\infty(\mathbf{G}) = \lim_{n \to \infty} R_n(\mathbf{G})$$

and that the number of iterations required to reduce the error $\|\mathbf{e}^{(n)}\|$ to $\|\mathbf{e}^{(0)}\|/\alpha$, for sufficiently large n, is $\geqslant -(\log_e \alpha)/R_n$.

Methods for accelerating the convergence of iterative processes

Both of the following methods are applicable to any iterative process when λ_1 is real.

Lyusternik's method

By eqn (5.28),

$$e^{(n+1)} = \lambda_1 e^{(n)} + \delta^{(n)}, \quad \lambda_1 \text{ real}, \tag{5.29}$$

where the components of $\delta^{(n)}$ are small and $|\lambda_1| < 1$. Also, by definition, the exact solution x and $e^{(n)}$ are related by

$$x = x^{(n)} + e^{(n)} \tag{5.30}$$
$$= x^{(n+1)} + e^{(n+1)}$$
$$= x^{(n+1)} + \lambda_1 e^{(n)} + \delta^{(n)}. \tag{5.31}$$

Eliminating $e^{(n)}$ between (5.30) and (5.31) gives that

$$x = \frac{x^{(n+1)} - \lambda_1 x^{(n)}}{1 - \lambda_1} + \frac{\delta^{(n)}}{1 - \lambda_1},$$

so if $\|\delta^{(n)}\|$ is small compared with $(1 - \lambda_1)$, a good approximation to the solution will be given by

$$x \simeq \frac{x^{(n+1)} - \lambda_1 x^{(n)}}{1 - \lambda_1},$$

which can also be written as

$$x \simeq x^{(n)} + \frac{x^{(n+1)} - x^{(n)}}{(1 - \lambda_1)}.$$

The last equation shows that small differences between successive iterates do not necessarily imply a close approximation to the solution. If, for example,

$$\max_i |x_i^{(n+1)} - x_i^{(n)}| = \varepsilon \quad \text{and} \quad \lambda_1 = 0.99$$

then

$$\max_i |x_i^{(n+1)} - x_i^{(n)}|/(1 - \lambda_1) = 100\varepsilon,$$

where $x_i^{(n)}$ is the ith component of $x^{(n)}$. To obtain a solution with a maximum error of ε in any component the iteration would have to be continued until

$$\max_i |x_i^{(n+1)} - x_i^{(n)}| \simeq 0 \cdot 01\varepsilon.$$

An alternative derivation of this result is given in Exercise 11.

For most problems λ_1 will not be known analytically in which case its value must be estimated. One straightforward way of doing this is as follows.

For sufficiently large n,

$$\mathbf{e}^{(n)} \simeq \lambda_1 \mathbf{e}^{(n-1)},$$

therefore

$$\mathbf{e}^{(n+1)} - \mathbf{e}^{(n)} \simeq \lambda_1 (\mathbf{e}^{(n)} - \mathbf{e}^{(n-1)}),$$

i.e.

$$\{(\mathbf{x} - \mathbf{x}^{(n+1)}) - (\mathbf{x} - \mathbf{x}^{(n)})\} \simeq \lambda_1 \{(\mathbf{x} - \mathbf{x}^{(n)}) - (\mathbf{x} - \mathbf{x}^{(n-1)})\}$$

showing that

$$\mathbf{x}^{(n+1)} - \mathbf{x}^{(n)} \simeq \lambda_1 (\mathbf{x}^{(n)} - \mathbf{x}^{(n-1)}),$$

i.e.

$$\mathbf{d}^{(n)} \simeq \lambda_1 \mathbf{d}^{(n-1)}. \tag{5.32}$$

Hence

$$\|\mathbf{d}^{(n)}\| \simeq |\lambda_1| \, \|\mathbf{d}^{(n-1)}\|,$$

so

$$|\lambda_1| = \rho \simeq \frac{\|\mathbf{d}^{(n)}\|}{\|\mathbf{d}^{(n-1)}\|},$$

where the norm of $\mathbf{d}^{(n)}$ can be defined by

$$\|\mathbf{d}^{(n)}\| = \max_i |x_i^{(n+1)} - x_i^{(n)}|, \quad i = 1(1)m,$$

or

$$\|\mathbf{d}^{(n)}\| = |x_1^{(n+1)} - x_1^{(n)}| + |x_2^{(n+1)} - x_2^{(n)}| + \ldots + |x_m^{(n+1)} - x_m^{(n)}|$$

or

$$\|\mathbf{d}^{(n)}\| = \{(x_1^{(n+1)} - x_1^{(n)})^2 + (x_2^{(n+1)} - x_2^{(n)})^2 + \ldots + (x_m^{(n+1)} - x_m^{(n)})^2\}^{\frac{1}{2}}.$$

Equation (5.32) justifies, incidentally, the basis of the SOR iterative method because it proves that when λ_1 is positive the corresponding components of successive correction or displacement vectors are one signed. More refined methods for estimating λ_1 are given in reference 33.

Aitken's method

$$\mathbf{e}^{(n)} \simeq \lambda_1 \mathbf{e}^{(n-1)}, \quad \text{i.e.} \quad \mathbf{x} - \mathbf{x}^{(n)} \simeq \lambda_1 (\mathbf{x} - \mathbf{x}^{(n-1)})$$

and

$$\mathbf{e}^{(n+1)} \simeq \lambda_1 \mathbf{e}^{(n)}, \quad \text{i.e.} \quad \mathbf{x} - \mathbf{x}^{(n+1)} \simeq \lambda_1 (\mathbf{x} - \mathbf{x}^{(n)}).$$

Take the ith components of these approximations and eliminate λ_1 by simple division. Then

$$\frac{x_i - x_i^{(n)}}{x_i - x_i^{(n+1)}} \simeq \frac{x_i - x_i^{(n-1)}}{x_i - x_i^{(n)}}$$

and solving this for x_i gives that

$$x_i \simeq \frac{x_i^{(n+1)} x_i^{(n-1)} - (x_i^{(n)})^2}{x_i^{(n+1)} - 2x_i^{(n)} + x_i^{(n-1)}} = x_i^{(n+1)} - \frac{\{x_i^{(n+1)} - x_i^{(n)}\}^2}{x_i^{(n+1)} - 2x_i^{(n)} + x_i^{(n-1)}}.$$

This method of acceleration avoids the explicit estimation of λ_1. It is of interest to note however that the two methods are identical if, in Lyusternick's method, λ_1 is approximated by

$$|x_i^{(n+1)} - x_i^{(n)}|/|x_i^{(n)} - x_i^{(n-1)}| \quad \text{for each } i.$$

As an illustrative example of these methods the following table gives the 9th, 10th, and 11th Jacobi iteration values arising from the five-point approximation of $\partial^2 \phi/\partial x^2 + \partial^2 \phi/\partial y^2 + 2 = 0$ over the rectangular domain $0 < x < 2$, $0 < y < 1$, where $\phi = 0$ on the boundary. (The torsion problem.) For a square mesh of side $\frac{1}{4}$ there are only eight unknowns because of symmetry. The starting values were zero. The solution ϕ to 4D required 65 iterations. Lyusternik's approximation is denoted by L, i.e. $L_i = \phi_i^{(10)} + (\phi_i^{(11)} - \phi_i^{(10)})/(1 - \lambda_1)$, where

$$\lambda_1 = \max_i |\phi_i^{(11)} - \phi_i^{(10)}|/\max_i |\phi_i^{(10)} - \phi_i^{(9)}| = 0.8281.$$

Better L values could have been obtained by calculating λ_1 from the average of $|\phi_i^{(11)} - \phi_i^{(10)}|/|\phi_i^{(10)} - \phi_i^{(9)}|$, $i = 1(1)8$. This average is $0 \cdot 8146$ which is closer to the theoretical value of $\frac{1}{2}(\cos \pi/8 + \cos \pi/4) = 0 \cdot 8155$ than is $0 \cdot 8281$. (See p. 285). The improved values by Aitken's method are denoted by A.

In practice, these methods would be used to improve early iterative values for subsequent iterations.

TABLE 5.4

$\phi^{(9)}$	0.3533	0.4959	0.5486	0.5613	0.4505	0.6430	0.7154	0.7335
$\phi^{(10)}$	0.3616	0.5112	0.5681	0.5827	0.4624	0.6644	0.7434	0.7634
$\phi^{(11)}$	0.3684	0.5235	0.5843	0.5999	0.4791	0.6821	0.7660	0.7880
ϕ	0.3980	0.5782	0.6555	0.6769	0.5138	0.7592	0.8668	0.8968
L	0.4012	0.5828	0.6623	0.6828	0.5177	0.7674	0.8749	0.9071
A	0.4002	0.5733	0.6618	0.6722	0.5093	0.7655	0.8597	0.9070

The eigenvalues of the Jacobi and SOR iteration matrices

Assume that the m linear equations

$$Ax = b$$

are such that matrix $A = D - L - U$ is non-singular and has non-zero diagonal elements a_{ii}, $i = 1(1)m$.

Jacobi iteration

The eigenvalues μ of the Jacobi iteration matrix $D^{-1}(L+U)$ are the μ roots of

$$\det\{\mu I - D^{-1}(L+U)\} = \det D^{-1}(\mu D - L - U)$$
$$= \det D^{-1} \det(\mu D - L - U) = 0,$$

where $\det D^{-1} = 1/\det D = 1/a_{11}a_{22} \ldots a_{mm} \neq 0$.
Hence

$$\det(\mu D - L - U) = 0.$$

Example 5.4

Calculate the eigenvalues of the Jacobi iteration matrix corresponding to

$$A = D - L - U = \begin{bmatrix} -4 & 1 & 0 \\ 1 & -4 & 1 \\ 0 & 1 & -4 \end{bmatrix}.$$

$$\det(\mu D - L - U) = \det \begin{bmatrix} -4u & 1 & 0 \\ 1 & -4\mu & 1 \\ 0 & 1 & -4\mu \end{bmatrix} = 0 = -8\mu(8\mu^2 - 1).$$

Hence $\mu = 0$ or $\pm 1/2\sqrt{2}$, showing that the Jacobi iteration would converge since $|\mu| < 1$.

SOR iteration

The eigenvalues λ of the *SOR* iteration matrix

$$\mathbf{H} = (\mathbf{I} - \omega\mathbf{D}^{-1}\mathbf{L})^{-1}\{(1-\omega)\mathbf{I} + \omega\mathbf{D}^{-1}\mathbf{U}\}$$

are the λ roots of $\det(\lambda\mathbf{I} - \mathbf{H}) = 0$. Now

$$\begin{aligned}
\lambda\mathbf{I} - \mathbf{H} &= \lambda(\mathbf{I} - \omega\mathbf{D}^{-1}\mathbf{L})^{-1}(\mathbf{I} - \omega\mathbf{D}^{-1}\mathbf{L})\\
&\quad - (\mathbf{I} - \omega\mathbf{D}^{-1}\mathbf{L})^{-1}\{(1-\omega)\mathbf{I} + \omega\mathbf{D}^{-1}\mathbf{U}\}\\
&= (\mathbf{I} - \omega\mathbf{D}^{-1}\mathbf{L})^{-1}\{\lambda(\mathbf{I} - \omega\mathbf{D}^{-1}\mathbf{L}) - (1-\omega)\mathbf{I} - \omega\mathbf{D}^{-1}\mathbf{U}\}\\
&= (\mathbf{I} - \omega\mathbf{D}^{-1}\mathbf{L})^{-1}\mathbf{D}^{-1}\{(\lambda + \omega - 1)\mathbf{D} - \lambda\omega\mathbf{L} - \omega\mathbf{U}\}.
\end{aligned}$$

Therefore

$$\det(\lambda\mathbf{I} - \mathbf{H}) = \det\{(\lambda + \omega - 1)\mathbf{D} - \lambda\omega\mathbf{L} - \omega\mathbf{U}\}/\det(\mathbf{I} - \omega\mathbf{D}^{-1}\mathbf{L})\det\mathbf{D}.$$

But $\det(\mathbf{I} - \omega\mathbf{D}^{-1}\mathbf{L}) = $ determinant of a unit lower triangular matrix $= 1$. Also

$$\det\mathbf{D} = a_{11}a_{22}\ldots a_{mm} \neq 0.$$

Hence the eigenvalues λ are the λ roots of

$$\det\{(\lambda + \omega - 1)\mathbf{D} - \lambda\omega\mathbf{L} - \omega\mathbf{U}\} = 0. \tag{5.33}$$

This result holds for any set of linear equations satisfying $\det\mathbf{A} \neq 0$ and $\det\mathbf{D} \neq 0$.

The eigenvalues λ of the Gauss–Seidel iteration matrix are given by putting $\omega = 1$, i.e. they are the λ roots of

$$\det(\lambda\mathbf{D} - \lambda\mathbf{L} - \mathbf{U}) = 0.$$

Example 5.5

Calculate the eigenvalues of the *SOR* iteration matrix corresponding to matrix $\mathbf{A}$ of Example 5.4.

The eigenvalues are the λ roots of

$$\det\{\lambda + \omega - 1)\mathbf{D} - \lambda\omega\mathbf{L} - \omega\mathbf{U}\} = 0 = \det\{\alpha\mathbf{D} - \lambda\omega\mathbf{L} - \omega\mathbf{U}\},$$

where $\alpha = \lambda + \omega - 1$. As

$$\det(\alpha\mathbf{D} - \lambda\omega\mathbf{L} - \omega\mathbf{U}) = \det\begin{bmatrix} -4\alpha & \omega & 0 \\ \lambda\omega & -4\alpha & \omega \\ 0 & \lambda\omega & -4\alpha \end{bmatrix}$$

$$= -2\alpha(8\alpha^2 - \lambda\omega^2) = 0,$$

therefore $\alpha = \lambda + \omega - 1 = 0$, implying $\lambda = 1 - \omega$, or $8(\lambda + \omega - 1)^2 = \lambda\omega^2$, a quadratic for λ in terms of ω.

The optimum acceleration parameter for the SOR method

A problem of paramount importance associated with the SOR method is the determination of a suitable value for the acceleration parameter ω. Ideally, we want the optimum value ω_b of ω which minimizes the spectral radius of the SOR iteration matrix and thereby maximizes the rate of convergence of the method. At the present time no formula exists for the determination of ω_b for an arbitrary set of linear equations. Fortunately, it can be calculated for many of the difference equations approximating first- and second-order partial differential equations because their matrices are often of a special type called property (A) or 2-cyclic matrices, and the significance of this was first revealed by Young in reference 34. He proved that when a 2-cyclic matrix is put into what he termed a consistently ordered form $\mathbf{A}$, which can be done by a simple re-ordering of the rows and corresponding columns of the original 2-cyclic matrix, then the eigenvalues λ of the point SOR iteration matrix $\mathbf{H}(\omega)$ associated with $\mathbf{A}$ are related to the eigenvalues μ of the point Jacobi iteration matrix $\mathbf{B}$ associated with $\mathbf{A}$ by the equation

$$(\lambda + \omega - 1)^2 = \lambda \omega^2 \mu^2. \tag{5.34}$$

From this it can be proved that

$$\omega_b = \frac{2}{1 + \sqrt{\{1 - \rho^2(\mathbf{B})\}}}, \tag{5.35}$$

where $\rho(\mathbf{B})$ is the spectral radius of the Jacobi iteration matrix.

Varga, reference 29, subsequently generalized these concepts for p-cyclic matrices, $p \geq 2$, but in this book only 2-cyclic matrices will be considered because p-cyclic matrices, $p \geq 3$, occur very rarely in the mathematics of the physical sciences. Young's definitions of 2-cyclic matrices and consistent ordering will be dealt with later. For our present purposes it is sufficient to know that block tridiagonal matrices $\mathbf{A}$ of the form

$$\mathbf{A} = \begin{bmatrix} \mathbf{D}_1 & \mathbf{A}_1 & & & & \\ \mathbf{B}_1 & \mathbf{D}_2 & \mathbf{A}_2 & & & \\ & \mathbf{B}_2 & \mathbf{D}_3 & \mathbf{A}_3 & & \\ & & \cdot & \cdot & \cdot & \\ & & & \mathbf{B}_{k-2} & \mathbf{D}_{k-1} & \mathbf{A}_{k-1} \\ & & & & \mathbf{B}_{k-1} & \mathbf{D}_k \end{bmatrix} \tag{5.36}$$

where the $\mathbf{D}_i$, $i = 1(1)k$, are square diagonal matrices not necessarily of the same order, are 2-cyclic and consistently ordered. A special case corresponding to $k = 2$ is

$$\mathbf{A} = \begin{bmatrix} \mathbf{D}_1 & \mathbf{A}_1 \\ \mathbf{B}_1 & \mathbf{D}_2 \end{bmatrix}. \tag{5.37}$$

To avoid interruption of the development of the proof of (5.34) it is convenient at this point to prove the following theorem.

Theorem 1

If the block tridiagonal matrix (5.36) is written as $\mathbf{A} = \mathbf{D} - \mathbf{L} - \mathbf{U}$, where

$$\mathbf{D} = \begin{bmatrix} \mathbf{D}_1 & & & \\ & \mathbf{D}_2 & & \\ & & \cdot & \\ & & & \mathbf{D}_k \end{bmatrix}, \quad -\mathbf{L} = \begin{bmatrix} 0 & & & \\ \mathbf{B}_1 & 0 & & \\ & \mathbf{B}_2 & & \\ & & \mathbf{B}_{k-1} & 0 \end{bmatrix}$$

and

$$-\mathbf{U} = \begin{bmatrix} \mathbf{0} & \mathbf{A}_1 & & \\ & \mathbf{0} & \mathbf{A}_2 & \\ & & \cdot & \\ & & \mathbf{0} & \mathbf{A}_{k-1} \\ & & & \mathbf{0} \end{bmatrix},$$

then $\det(\mathbf{D} - \mathbf{L} - \mathbf{U}) = \det(\mathbf{D} - \alpha\mathbf{L} - \alpha^{-1}\mathbf{U})$, \quad (5.38)

where α is any non-zero number.

Proof

Let

$$\mathbf{C} = \begin{bmatrix} \mathbf{I}_1 & & & & \\ & \alpha\mathbf{I}_2 & & & \\ & & \alpha^2\mathbf{I}_3 & & \\ & & & \cdot & \\ & & & & \alpha^{k-1}\mathbf{I}_k \end{bmatrix}$$

so that

$$\mathbf{C}^{-1} = \begin{bmatrix} \mathbf{I}_1 & & & & \\ & \alpha^{-1}\mathbf{I}_2 & & & \\ & & \alpha^{-2}\mathbf{I}_3 & & \\ & & & \cdot & \\ & & & & \alpha^{1-k}\mathbf{I}_k \end{bmatrix},$$

where $\mathbf{I}_i$ denotes the unit matrix of the same order as $\mathbf{D}_i$, $i = 1(1)k$. Then

$$\mathbf{CAC}^{-1} = \begin{bmatrix} \mathbf{D}_1 & \alpha^{-1}\mathbf{A}_1 & & & \\ \alpha\mathbf{B}_1 & \mathbf{D}_2 & \alpha^{-1}\mathbf{A}_2 & & \\ & \cdot & \cdot & \cdot & \\ & \alpha\mathbf{B}_{k-2} & \mathbf{D}_{k-1} & \alpha^{-1}\mathbf{A}_{k-1} \\ & & \alpha\mathbf{B}_{k-1} & \mathbf{D}_k \end{bmatrix} = \mathbf{D} - \alpha\mathbf{L} - \alpha^{-1}\mathbf{U}.$$

Hence,

$$\det(\mathbf{D} - \alpha\mathbf{L} - \alpha^{-1}\mathbf{U}) = \det \mathbf{C} \det \mathbf{A}/\det \mathbf{C} = \det \mathbf{A} = \det(\mathbf{D} - \mathbf{L} - \mathbf{U}).$$

This theorem clearly holds even when each $\mathbf{D}_i$ is a full square matrix because the diagonal blocks of $\mathbf{CAC}^{-1}$ are independent of α.

The relationship between the eigenvalues of the Jacobi and SOR point iteration matrices corresponding to a block tridiagonal coefficient matrix

Let $\mathbf{H}(\omega)$ and $\mathbf{B}$ represent respectively the *SOR* and Jacobi point iteration matrices corresponding to the non-singular block tridiagonal matrix $\mathbf{A}$ of eqn (5.36) with non-zero diagonal elements. Then the non-zero eigenvalues λ of $\mathbf{H}(\omega)$ are related to the eigenvalues μ of $\mathbf{B}$ by the equation

$$(\lambda + \omega - 1)^2 = \lambda\omega^2\mu^2, \quad \omega \neq 0. \tag{5.39}$$

Proof

By eqn (5.33) the eigenvalues λ of $\mathbf{H}(\omega)$ are the λ roots of

$$\det\{(\lambda + \omega - 1)\mathbf{D} - \lambda\omega\mathbf{L} - \omega\mathbf{U}\} = 0.$$

But it is given that $\mathbf{A} = \mathbf{D} - \mathbf{L} - \mathbf{U}$ is block tridiagonal, so the matrix $\{(\lambda + \omega - 1)\mathbf{D} - \lambda\omega\mathbf{L} - \omega\mathbf{U}\}$ is also block triadiagonal. Hence, by Theorem 1,

$$\det\{(\lambda + \omega - 1)\mathbf{D} - \alpha\lambda\omega\mathbf{L} - \alpha^{-1}\omega\mathbf{U}\} = 0, \quad \alpha \neq 0, \quad \omega \neq 0.$$

Put $\alpha\lambda\omega = \alpha^{-1}\omega$, i.e. $\alpha^2 = 1/\lambda$, and choose $\alpha = \lambda^{-\frac{1}{2}}$, $\lambda \neq 0$. Then

$$\det\{(\lambda + \omega - 1)\mathbf{D} - \lambda^{\frac{1}{2}}\omega(\mathbf{L} + \mathbf{U})\} = 0$$
$$= \lambda^{m/2}\omega^m \det\{\lambda^{-\frac{1}{2}}\omega^{-1}(\lambda + \omega - 1)\mathbf{D} - (\mathbf{L} + \mathbf{U})\}.$$

Therefore

$$\det\{\lambda^{-\frac{1}{2}}\omega^{-1}(\lambda + \omega - 1)\mathbf{D} - (\mathbf{L} + \mathbf{U}) = 0. \tag{5.40}$$

Since the eigenvalues μ of the Jacobi iteration matrix corresponding to $(\mathbf{D} - \mathbf{L} - \mathbf{U})$ are the μ roots of the equation

$$\det\{\mu\mathbf{D} - (\mathbf{L} + \mathbf{U})\} = 0,$$

it follows that eqn (5.40) states that

$$\mu = \lambda^{-\frac{1}{2}}\omega^{-1}(\lambda + \omega - 1).$$

This relation is usually written as

$$(\lambda + \omega - 1)^2 = \lambda\omega^2\mu^2, \quad \omega \neq 0, \lambda \neq 0. \tag{5.41}$$

Additional comments are made on this result after the section on consistent ordering.

It is easily seen that λ cannot be zero when $\omega \neq 1$ because if it was, eqn (5.33) would give

$$\det\{(\omega - 1)\mathbf{D} - \omega\mathbf{U}\} = 0 = (\omega - 1)^m a_{11}a_{22}\ldots a_{mm}.$$

But $a_{ii} \neq 0$, $i = 1(1)m$, by hypothesis. This equation can be satisfied only by $\omega = 1$. In other words, the Gauss–Seidel iteration matrix will have some zero eigenvalues but not the *SOR*, $\omega \neq 1$.

Theorem 2

The non-zero eigenvalues of the Jacobi iteration matrix corresponding to matrix (5.36) occur in pairs $\pm\mu_i$.

Consider the m linear equations

$$\mathbf{Ax} = \mathbf{b}$$

where the $m \times m$ matrix $\mathbf{A}$ has the block tridiagonal form (5.36) and each $\mathbf{D}_i$ is a square diagonal matrix, i.e. $\mathbf{D}$ is an $m \times m$ diagonal matrix. Then by eqn (5.21) the Jacobi iteration matrix $\mathbf{B}$ corresponding to $\mathbf{A}$ is $\mathbf{D}^{-1}(\mathbf{L}+\mathbf{U}) = \mathbf{E}+\mathbf{F}$, where $\mathbf{D}^{-1}\mathbf{L} = \mathbf{E}$ and $\mathbf{D}^{-1}\mathbf{U} = \mathbf{F}$ are of the form

$$
\mathbf{E} = \begin{bmatrix} 0 & & & & \\ \mathbf{E}_1 & 0 & & & \\ & \mathbf{E}_2 & 0 & & \\ & & \cdot & \cdot & \\ & & & \mathbf{E}_{k-1} & 0 \end{bmatrix}
$$

$$
\text{and } \mathbf{F} = \begin{bmatrix} 0 & \mathbf{F}_1 & & & \\ & 0 & \mathbf{F}_2 & & \\ & & \cdot & \cdot & \\ & & & 0 & \mathbf{F}_{k-1} \\ & & & & 0 \end{bmatrix} \tag{5.42}
$$

The characteristic polynomial of $\mathbf{B}$ is

$$
P(\mu) = \det[\mu\mathbf{I} - \mathbf{B}] = \det[\mu\mathbf{I} - \mathbf{E} - \mathbf{F}],
$$

where $\mu\mathbf{I} - \mathbf{B}$ has the block tridiagonal form of (5.36). Therefore, by theorem 1,

$$
\det[\mu\mathbf{I} - \mathbf{E} - \mathbf{F}] = \det[\mu\mathbf{I} - \alpha\mathbf{E} - \alpha^{-1}\mathbf{F}], \tag{5.43}
$$

where $\alpha \neq 0$ is any number. With $\alpha = -1$, eqn (5.43) gives that

$$
\det[\mu\mathbf{I} - \mathbf{E} - \mathbf{F}] = \det[\mu\mathbf{I} + \mathbf{E} + \mathbf{F}] = (-1)^m \det[-\mu\mathbf{I} - \mathbf{E} - \mathbf{F}],
$$

i.e.

$$
P(\mu) = (-1)^m P(-\mu).
$$

It follows therefore that if m is odd, $P(\mu) = -P(-\mu)$, so $P(\mu)$ is an odd polynomial of μ. Similarly, $P(\mu)$ is an even polynomial when m is even. Hence $P(\mu)$, for some integer $r \geq 0$, will either have the form

$$
P(\mu) = \mu^{2r} g(\mu^2), \quad \text{when } m \text{ is even,}
$$

or the form $\tag{5.43a}$

$$
P(\mu) = \mu^{2r+1} f(\mu^2), \quad \text{when } m \text{ is odd,}
$$

where $g(x)$ and $f(x)$ denote polynomials in x such that $g(0) \neq 0$ and $f(0) \neq 0$.

By eqns (5.43a) the non-zero roots of $P(\mu) = 0$ are given either by $g(\mu^2) = 0$ or by $f(\mu^2) = 0$, thus proving that the non-zero eigenvalues of **B** occur in pairs $\pm\mu_i$. Moreover, for some integer $r \geq 0$, **B** will have $2r$ (m even), or $(2r+1)$, (m odd), eigenvalues equal to zero.

Theoretical determination of the optimum relaxation parameter ω_b

The following results hold when some of the eigenvalues μ_i of the Jacobi iteration matrix are complex (reference 29), but the analysis is difficult so we shall assume each μ_i is real, as is the case when matrix **A** is real and symmetric.

Theorem 3

When the matrix **A** is block tridiagonal of the form (5.36), with non-zero diagonal elements, and all the eigenvalues of the Jacobi iteration matrix **B** associated with **A** are real and such that $0 < \rho(\mathbf{B}) < 1$, then

$$\omega_b = \frac{2}{1 + \sqrt{\{1 - \rho^2(\mathbf{B})\}}}$$

and

$$\rho\{\mathbf{H}(\omega_b)\} = \omega_b - 1.$$

Furthermore, the SOR method applied to the equations $\mathbf{Ax} = \mathbf{b}$ converges for all ω in the range $0 < \omega < 2$.

Proof

Write eqn (5.41) as

$$\frac{1}{\omega}(\lambda + \omega - 1) = \pm\lambda^{\frac{1}{2}}\mu$$

and let

$$y_1(\lambda) = \frac{1}{\omega}(\lambda + \omega - 1) = \frac{1}{\omega}\lambda + 1 - \frac{1}{\omega}$$

and

$$y_2(\lambda) = \pm\lambda^{\frac{1}{2}}\mu.$$

Then the pair of eigenvalues of $\mathbf{H}(\omega)$ that correspond to the pair

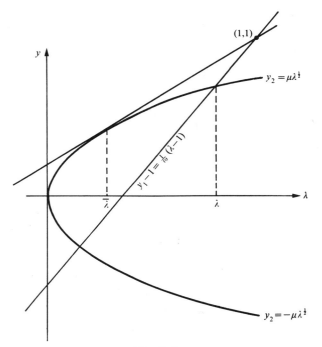

Fig. 5.6

of non-zero eigenvalues $\pm\mu$ of **B** are the λ values of the points where the straight line

$$y_1 - 1 = \frac{1}{\omega}(\lambda - 1)$$

intersects the parabola

$$y_2^2 = \lambda\mu^2.$$

This straight line passes through the point $(1, 1)$ and its slope $1/\omega$ decreases as ω increases. From Fig. 5.6 it is seen that the largest abscissa of the two points of intersection decreases with increasing ω until the line is a tangent to the parabola. The λ-values of the points of intersection are the λ roots of (5.41) which can be rearranged as

$$\lambda^2 + 2\lambda\{(\omega - 1) - \tfrac{1}{2}\omega^2\mu^2\} + (\omega - 1)^2 = 0.$$

For tangency,

$$\{(\omega - 1) - \tfrac{1}{2}\omega^2\mu^2\}^2 - (\omega - 1)^2 = 0,$$

giving

$$\omega = \bar{\omega} = \frac{2\{1 \pm (1 - \mu^2)^{\frac{1}{2}}\}}{\mu^2}. \tag{5.44}$$

The positive sign in (5.44) gives the value of ω corresponding to the other tangent from the point $(1, 1)$ and for which the λ-value of the point of contact exceeds 1, i.e., a non-convergent case. The negative sign gives

$$\bar{\omega} = \frac{2}{1 + \sqrt{(1 - \mu^2)}},$$

and it is easily shown that the abscissa of the point of tangency is

$$\bar{\lambda} = \bar{\omega} - 1.$$

For values of $\omega > \bar{\omega}$ the λ roots of eqn (5.41) are complex, each with modulus $|\lambda| = \omega - 1$ which increases with increasing ω. Since the SOR method diverges for $|\lambda| \geq 1$ it follows that $\omega \geq 2$ gives a divergent iteration, i.e., the iteration is convergent for $0 < \omega < 2$.

Hence we have shown that for a fixed eigenvalue μ of $\mathbf{B}$ the value of ω which minimizes the corresponding λ eigenvalue of largest modulus is $\bar{\omega} = 2/\{1 + \sqrt{(1 - \mu^2)}\}$.

Now the largest parabola, i.e. the one which contains all the others inside its interior, is given by $\pm\mu_1$, where $|\mu_1| = \rho(\mathbf{B})$, and for this parabola $\bar{\lambda}$ is larger than for the other eigenvalues of $\mathbf{B}$. Since the maximum rate of convergence of the SOR method is determined by the minimum value of the eigenvalue of largest modulus of $\mathbf{H}(\omega)$, $\lambda = \lambda_1$, say, it follows that

$$\bar{\lambda}_1 = \bar{\omega}_1 - 1,$$

where

$$\bar{\omega}_1 = \frac{2}{1 + \sqrt{(1 - \mu_1^2)}},$$

and λ_1 corresponds to $\pm\mu_1$. In other words the optimum value of ω for maximum rate of convergence is given by

$$\omega_b = \frac{2}{1 + \sqrt{\{1 - \rho^2(\mathbf{B})\}}},$$

where $\rho(\mathbf{B})$ is the spectral radius of the Jacobi iteration matrix.

The calculation of ω_b

For Poisson's equation over a rectangle of sides ph and qh, with Dirichlet boundary conditions, it is proved in Exercise 18 that $\rho(\mathbf{B})$ for the five-point difference approximation, using a square mesh of side h, is

$$\rho(\mathbf{B}) = \frac{1}{2}\left(\cos\frac{\pi}{p} + \cos\frac{\pi}{q}\right).$$

In general, however, either the spectral radius of the Jacobi iteration matrix, or the spectral radius $\rho\{\mathbf{H}(1)\} = \rho^2(\mathbf{B})$ of the Gauss–Seidel iteration matrix must be estimated as on p. 273 or by one of the methods given in references 29 and 33.

The Gauss–Seidel iteration matrix H(1)

This section is merely the special case $\omega = 1$ of the preceding section. In other words, the pair of eigenvalues of the Gauss–Seidel iteration matrix $\mathbf{H}(1)$ that correspond to the pair of non-zero eigenvalues $\pm\mu_i$ of $\mathbf{B}$ are the λ values of the points where

$$y_1 = \lambda,$$

i.e. the straight line through the origin at $45°$, intersects the parabola

$$y_2^2 = \lambda\mu_i^2.$$

These are the roots of $y_1^2 = y_2^2$, giving that

$$\lambda(\lambda - \mu_i^2) = 0, \quad \text{i.e.} \quad \lambda_i = 0 \quad \text{or} \quad \lambda_i = \mu_i^2.$$

This shows that to each non-zero pair of eigenvalues $\pm\mu_i$ of $\mathbf{B}$ there corresponds the pair $\lambda_i = 0$ and $\lambda_i = \mu_i^2$ of $\mathbf{H}(1)$. If $\mu_i = 0$ then $\lambda_i = 0$. It follows from this that when the matrix $\mathbf{A}$ of the equations has the form (5.36) then the Gauss–Seidel method converges if and only if the Jacobi method converges. If both methods converge then

$$\rho\{\mathbf{H}(1)\} = \{\rho(\mathbf{B})\}^2 < 1,$$

and by the definition of the asymptotic rate of convergence R_∞, $R_\infty\{\mathbf{H}(1)\} = 2R_\infty(\mathbf{B})$, i.e. the Gauss–Seidel point iterative method converges twice as fast as the Jacobi point iterative method after a large number of iterations. (See Worked Example 5.3.)

Re-ordering of equations and unknowns

As will be seen later it is sometimes necessary to re-order a set of equations, together with the unknowns, so that the matrix of the re-ordered set has some desired form. In all that follows the unknown corresponding to the ith mesh point will be denoted by x_i and the subscript will not be changed in any re-ordering. For illustrative purposes, say the equations corresponding to the first, second, third, and fourth mesh points are

$$a_{11}x_1 + a_{12}x_2 \qquad\quad + a_{14}x_4 = b_1,$$

$$a_{21}x_1 + a_{22}x_2 + a_{23}x_3 \qquad\quad = b_2,$$

$$a_{32}x_2 + a_{33}x_3 + a_{34}x_4 = b_3,$$

$$a_{41}x_1 + \qquad\quad a_{43}x_3 + a_{44}x_4 = b_4,$$

respectively, which may be written as

$$
\begin{bmatrix}
a_{11} & a_{12} & 0 & a_{14} \\
a_{21} & a_{22} & a_{23} & 0 \\
0 & a_{32} & a_{33} & a_{34} \\
a_{41} & 0 & a_{43} & a_{44}
\end{bmatrix}
\begin{bmatrix}
x_1 \\ x_2 \\ x_3 \\ x_4
\end{bmatrix}
=
\begin{bmatrix}
b_1 \\ b_2 \\ b_3 \\ b_4
\end{bmatrix},
$$

i.e. as

$$\mathbf{A}\mathbf{x}_A = \mathbf{b}_A.$$

If we now write down the equations at the mesh points four, two, three, and one, *in that order*, the column vector of unknowns being in the *same order*, namely, x_4, x_2, x_3, x_1 we obtain

$$
\begin{bmatrix}
a_{44} & 0 & a_{43} & a_{41} \\
0 & a_{22} & a_{23} & a_{21} \\
a_{34} & a_{32} & a_{33} & 0 \\
a_{14} & a_{12} & 0 & a_{11}
\end{bmatrix}
\begin{bmatrix}
x_4 \\ x_2 \\ x_3 \\ x_1
\end{bmatrix}
=
\begin{bmatrix}
b_4 \\ b_2 \\ b_3 \\ b_1
\end{bmatrix},
\tag{5.45}
$$

which may be represented by

$$\mathbf{C}\mathbf{x}_C = \mathbf{b}_C.$$

It is easily verified that the matrix $\mathbf{C}$ can be derived from matrix $\mathbf{A}$ by interchanging rows one and four of $\mathbf{A}$ to give matrix $\mathbf{B}$ then interchanging columns one and four of $\mathbf{B}$ to give $\mathbf{C}$, or by interchanging columns one and four of $\mathbf{A}$ to give $\mathbf{B}_1$ then rows one and four of $\mathbf{B}_1$ to give $\mathbf{C}$. In matrix algebra, rows one and four of $\mathbf{A}$ can be interchanged by premultiplying $\mathbf{A}$ by the elementary permutation matrix $\mathbf{P}_{14}$, where

$$\mathbf{P}_{14} = \begin{bmatrix} 0 & 0 & 0 & 1 \\ 0 & 1 & 0 & 0 \\ 0 & 0 & 1 & 0 \\ 1 & 0 & 0 & 0 \end{bmatrix}$$

is obtained by interchanging either rows one and four or columns one and four of the unit matrix of order four. Similarly, columns one and four of $\mathbf{B}$ can be interchanged by postmultiplying $\mathbf{B}$ with $\mathbf{P}_{14}$. Hence $\mathbf{C} = \mathbf{P}_{14}\mathbf{A}\mathbf{P}_{14}$. As $\mathbf{x}_C = \mathbf{P}_{14}\mathbf{x}_A$ and $\mathbf{b}_C = \mathbf{P}_{14}\mathbf{b}_A$, the re-ordered equations and unknowns (5.45) may be written as

$$(\mathbf{P}_{14}\mathbf{A}\mathbf{P}_{14})\mathbf{P}_{14}\mathbf{x}_A = \mathbf{P}_{14}\mathbf{b}_A. \tag{5.46}$$

Since $\mathbf{P}_{14} = \mathbf{P}_{41} = \mathbf{P}_{14}^T = \mathbf{P}_{41}^T = \mathbf{P}_{14}^{-1} = \mathbf{P}_{41}^{-1}$, it follows that $\mathbf{C} = \mathbf{P}_{14}\mathbf{A}\mathbf{P}_{14}^T = \mathbf{P}_{14}\mathbf{A}\mathbf{P}_{14}^{-1}$ is an orthogonal similarity transformation of $\mathbf{A}$. Therefore the eigenvalues of $\mathbf{C}$ are those of $\mathbf{A}$. This result would not hold if the unknowns were not ordered in the same way as the equations. It should also be noticed that the coefficient a_{ii} of x_i in the re-ordered equations is a diagonal element of $\mathbf{P}_{14}\mathbf{A}\mathbf{P}_{14}$. Clearly, all possible re-orderings of the equations and unknowns can be obtained by a succession of transformations (5.46). If, for example, the equations and unknowns of (5.46) are to be written down at the mesh points four, three, two and one, in that order, the corresponding matrix equation is

$$(\mathbf{P}_{23}\mathbf{P}_{14}\mathbf{A}\mathbf{P}_{14}\mathbf{P}_{23})\mathbf{P}_{23}\mathbf{P}_{14}\mathbf{x}_A = \mathbf{P}_{23}\mathbf{P}_{14}\mathbf{b}_A,$$

i.e.

$$(\mathbf{P}\mathbf{A}\mathbf{P}^T)\mathbf{P}\mathbf{x}_A = \mathbf{P}\mathbf{b}_A,$$

where $\mathbf{P} = \mathbf{P}_{23}\mathbf{P}_{14}$ is called a permutation matrix and satisfies $\mathbf{P} = \mathbf{P}^T = \mathbf{P}^{-1}$.

Point iterative methods and re-orderings

In all point iterative methods the equation corresponding to the ith mesh point is 'solved' initially for x_i in terms of the other components of the solution vector $\mathbf{x}$. The iteration then expresses the unknown component $x_i^{(n+1)}$ of the $(n+1)$th iteration vector in terms of known $(n+1)$ and nth iterative values of the remaining components of **x**.

With the point Jacobi method the $(n+1)$th iterative values are expressed exclusively in terms of nth iterative values so the order in which they are calculated, i.e. the order in which the mesh points are scanned, does not affect the values of successive iterates at a particular mesh point. Mathematically, this obvious result is equivalent to proving that the Jacobi iteration matrices associated with all possible orderings of equations and unknowns have the same set of eigenvalues, and the same set of eigenvectors in the sense that they have the same components but in different orders. The error vectors $\mathbf{x} - \mathbf{x}^{(n)}$ for different orderings will then have the same components but in different orders. This is proved in Exercise 19.

With both the Gauss–Seidel and SOR point methods the latest iterates are used as soon as they are available. Any change therefore in the scanning of the mesh points, that is, in the order in which the iteration equations are 'solved' will, in general, affect the rate of convergence of the method. For this reason, any two Gauss–Seidel (or SOR) iteration matrices associated respectively with two different re-orderings of the same set of equations have, in general, different eigenvalues and vectors.

2-cyclic matrices and consistent orderings

The concept of consistent ordering is central to the theory of the SOR iterative method for solving the equations $\mathbf{Ax} = \mathbf{b}$ because at present the calculation of the optimum acceleration parameter is possible only for consistently ordered matrices.

The earliest definition was due to Young, reference 34, and was related to a class of matrices whose members possessed what was termed property (A). (A matrix **A** possesses property (A) if there exists a permutation matrix **P** such that the similarity transformation $\mathbf{PAP}^T$ is block tridiagonal of the form (5.36)). A number of more general definitions have since been formulated. In particular, Varga, reference 29, extended the concept of consistent ordering to the class of p-cyclic matrices which include property (A) matrices as the special case $p = 2$. All these definitions are dealt with in a book by Young, reference 33, which brings the theory of iterative procedures up to date.

In this book Young's original definitions are used because his concept of an ordering vector enables us to determine very easily a permutation matrix **P** that will transform a non-consistently ordered 2-cyclic matrix **A** into a consistently ordered 2-cyclic matrix **PAP**^{T}. This use of ordering vectors has been extended to p-cyclic matrices in references 23 and 33.

The following presentation of 2-cyclic matrices and consistent ordering is based on N. Papamichael's M. Tech. dissertation, Brunel University, 1970, reference 22.

2-cyclic matrices

The reader is warned that the definition looks difficult. When related to an example however it will be seen to be easy.

Definition 1

The $N \times N$ matrix $\mathbf{A} = [a_{i,j}]$ is 2-cyclic if there exist two disjoint subsets S and T of W, the set of the first N positive integers, such that $S \cup T = W$ and if $a_{i,j} \neq 0$ then either $i = j$ or $i \in S$ and $j \in T$ or $i \in T$ and $j \in S$.

Example 5.6

Consider

$$\mathbf{A} = \begin{bmatrix} a_{11} & a_{12} & 0 & 0 & a_{15} \\ a_{21} & a_{22} & a_{23} & 0 & 0 \\ 0 & a_{32} & a_{33} & a_{34} & 0 \\ a_{41} & 0 & a_{43} & a_{44} & 0 \\ a_{51} & 0 & a_{53} & 0 & a_{55} \end{bmatrix}$$

Mentally exclude the diagonal elements as $i = j$.

In the first row $i = 1$. Let $1 \in S$. Then $j = 2, 5 \in T$.

In the second row $i = 2 \in T$, so $j = 1, 3 \in S$.

The final distribution of the first five positive integers is as below.

	S	T
Row 1	$i = 1$	$j = 2, 5$
Row 2	$j = 1, 3$	$i = 2$
Row 3	$i = 3$	$j = 2, 4$
Row 4	$j = 1, 3$	$i = 4$
Row 5	$j = 1, 3$	$i = 5$

Since the sets $S = \{1, 3\}$ and $T = \{2, 4, 5\}$ are disjoint and $S \cup T = \{1, 2, 3, 4, 5\} = W$, it follows that matrix $\mathbf{A}$ is 2-cyclic. If the matrix was p-cyclic there would be p disjoint non-empty subsets.

Example 5.7

Consider

$$\mathbf{A} = \begin{bmatrix} a_{11} & a_{12} & a_{13} & 0 \\ 0 & a_{22} & a_{23} & a_{24} \\ a_{31} & 0 & a_{33} & a_{34} \\ a_{41} & a_{42} & 0 & a_{44} \end{bmatrix}$$

As before,

	S	T
Row 1	$i = 1$	$j = 2, 3$
Row 2	$j = 3, 4$	$i = 2$

At the second step of the process the sets S and T cease to be disjoint. Therefore matrix $\mathbf{A}$ is not 2-cyclic.

Theorem 1

A matrix $\mathbf{A}$ is 2-cyclic if and only if there exists a row vector $\boldsymbol{\gamma} = (\gamma_1, \gamma_2, \ldots, \gamma_N)$ with integral components such that if $a_{ij} \neq 0$ and $i \neq j$ then $|\gamma_i - \gamma_j| = 1$.

Proof

Assume that matrix A is 2-cyclic. Referring to definition 1 let $\gamma_i = 1$ if $i \in S$ and $\gamma_i = 0$ if $i \in T$. If $a_{ij} \neq 0$ and $i \neq j$ then either $i \in S$ and $j \in T$ and hence $\gamma_i = 1$, $\gamma_j = 0$, or else $i \in T$ and $j \in S$ and hence $\gamma_i = 0$ and $\gamma_j = 1$. In either case $|\gamma_i - \gamma_j| = 1$.

Conversely, we must show that if $a_{ij} \neq 0$, $i \neq j$ and $|\gamma_i - \gamma_j| = 1$, then matrix $\mathbf{A}$ is 2-cyclic. Let

   (i) S denote a set of integers i for which γ_i is odd,
   (ii) T denote a set of integers i for which γ_i is even. For
       example, if

$$\begin{array}{ll} \gamma_1 = 3 & \gamma_3 = 4 \\ \gamma_2 = 1 & \gamma_4 = 2 \\ \gamma_5 = 5 & \end{array}$$

then the sets in (i) and (ii) are $S = \{1, 2, 5\}$ and $T = \{3, 4\}$. If $i \in S$, γ_i is odd so it follows that γ_j is even since $|\gamma_i - \gamma_j| = 1$ and the difference of two odd numbers is even. Hence $j \in T$. Similarly, $i \in T$ implies that $j \in S$. Hence matrix **A** is 2-cyclic.

Definition 2

A vector $\boldsymbol{\gamma}$ with the properties given in theorem 1 is said to be an *ordering vector* for the matrix **A**.

By this definition an ordering vector for a given 2-cyclic matrix will contain $M \geq 2$ distinct components. The actual numerical value of a component of $\boldsymbol{\gamma}$ is not important. Only the difference between any two components is significant so the components of an ordering vector with M distinct components can always be taken as $0, 1, 2, \ldots, M-1$.

If **A** is 2-cyclic then two ordering vector $\boldsymbol{\gamma}^{(1)}$ and $\boldsymbol{\gamma}^{(2)}$, each with two distinct components, may be obtained by setting

$$\gamma_i^{(1)} = \begin{cases} 0, & \text{if} \quad i \in S \\ 1, & \text{if} \quad i \in T \end{cases}$$

and

$$\gamma_i^{(2)} = \begin{cases} 1, & \text{if} \quad i \in S \\ 0, & \text{if} \quad i \in T \end{cases}$$

Example 5.8

Write down two ordering vectors for the 2-cyclic matrix of Example 5.6.

In that example $S = \{1, 3\}$ and $T = \{2, 4, 5\}$. Therefore we can set $\gamma_1 = \gamma_3 = 0$ and $\gamma_2 = \gamma_4 = \gamma_5 = 1$, giving $\boldsymbol{\gamma}^{(1)} = (0, 1, 0, 1, 1)$, or put $\gamma_1 = \gamma_3 = 1$ and $\gamma_2 = \gamma_4 = \gamma_5 = 0$ giving $\boldsymbol{\gamma}^{(2)} = (1, 0, 1, 0, 0)$.

In general, for a given 2-cyclic matrix, there exist ordering vectors with $M > 2$ distinct components. Such vectors can be derived from $\boldsymbol{\gamma}^{(1)}$ or $\boldsymbol{\gamma}^{(2)}$ by replacing some of the γ_i by $(\gamma_i + 2)$ but the choice is restricted by conditions specified in reference 23. For small matrices all possible ordering vectors may be obtained by setting $\boldsymbol{\gamma} = (0, \gamma_1, \gamma_2, \ldots, \gamma_{N-1})$ then applying $|\gamma_i - \gamma_j| = 1$ to each $a_{ij} \neq 0$, $i \neq j$. This method applied to Example 5.6

yields

$$\gamma^{(1)} = (0, 1, 0, 1, 1), \quad \gamma^{(2)} = (1, 0, 1, 0, 0), \quad \gamma^{(3)} = (1, 0, 1, 2, 2),$$

$$\gamma^{(4)} = (1, 0, 1, 2, 0), \quad \gamma^{(5)} = (1, 0, 1, 0, 2), \quad \gamma^{(6)} = (2, 1, 0, 1, 1),$$

$$\gamma^{(7)} = (1, 2, 1, 0, 2), \quad \gamma^{(8)} = (1, 2, 1, 0, 0), \quad \gamma^{(9)} = (0, 1, 2, 1, 1),$$

$$\gamma^{(10)} = (1, 2, 1, 2, 0).$$

It should be noted that the method indicated gives $\gamma^{(2)}$ as $(0, -1, 0, -1, -1)$ but every component can be made non-negative by adding 1 to each component.

An ordering vector $\gamma = \{\gamma_i\}$ with $M > 2$ distinct components can always be transformed into a vector $\gamma' = \{\gamma_i'\}$ with 2 distinct components by putting $\gamma_i' = \gamma_i \pmod 2$. For example, putting $\gamma_i' = \gamma_i^{(6)} \pmod 2$ transforms $\gamma^{(6)}$ into the vector $(0, 1, 0, 1, 1) = \gamma^{(1)}$.

Consistent orderings

Young defined a consistently ordered matrix as follows.

Definition 3

A 2-cyclic $N \times N$ matrix is consistently ordered if there exists an ordering vector $\gamma = (\gamma_1, \gamma_2, \ldots, \gamma_N)$ such that if $a_{ij} \neq 0$ and $j > i$ then $\gamma_j - \gamma_i = 1$ and if $i > j$ then $\gamma_i - \gamma_j = 1$.

It follows that if the components of γ are in ascending order of magnitude and γ is an ordering vector for the matrix $\mathbf{A}$ then $\mathbf{A}$ is consistently ordered.

The converse is not necessarily true, i.e. a consistently ordered matrix $\mathbf{A}$ might have an ordering vector with successive components not in ascending order of magnitude. (See Example 5.10).

Theorem 2

If the matrix $\mathbf{A}$ is 2-cyclic then there exists a permutation matrix $\mathbf{P}$ such that $\mathbf{P}\mathbf{A}\mathbf{P}^T$ is consistently ordered.

Proof

Let $\boldsymbol{\gamma}$ be an ordering vector for the matrix $\mathbf{A}$. By interchanging the components of $\boldsymbol{\gamma}$ it is always possible to obtain a vector $\boldsymbol{\gamma}'$ with components in ascending order of magnitude.

The matrix $\mathbf{P}$ may be obtained as follows. If the first interchange of the components of $\boldsymbol{\gamma}$ are γ_i and γ_j, write down the elementary permutation matrix $\mathbf{P}_{ij}$. If the next interchange are the components γ_m and γ_n, premultiply $\mathbf{P}_{ij}$ by $\mathbf{P}_{mn}$. The final matrix $\ldots \mathbf{P}_{mn}\mathbf{P}_{ij}$ given by this process is $\mathbf{P}$ and the final ordering vector $\boldsymbol{\gamma}'$ is an ordering vector for the consistently ordered matrix $\mathbf{PAP}^T$.

Clearly $\mathbf{P}$ is not unique and the number of consistently ordered matrices $\mathbf{PAP}^T$ that can be obtained this way from $\boldsymbol{\gamma}$ equals the number of ways $\boldsymbol{\gamma}$ can be transformed into $\boldsymbol{\gamma}'$.

Example 5.9

Consider the matrix $\mathbf{A}$ of Example 5.6 for which one ordering vector $\boldsymbol{\gamma}^{(1)} = (0, 1, 0, 1, 1)$. The vector $\boldsymbol{\gamma}^{(1)'} = (0, 0, 1, 1, 1)$ can be obtained by interchanging elements 2 and 3 of $\boldsymbol{\gamma}^{(1)}$. Therefore $\mathbf{P} = \mathbf{P}_{23}$ and the corresponding consistently ordered matrix $\mathbf{PAP}^T$ is

$$\mathbf{P}_{23}\mathbf{AP}_{23} = \begin{bmatrix} a_{11} & 0 & \vdots & a_{12} & 0 & a_{15} \\ 0 & a_{33} & \vdots & a_{32} & a_{34} & 0 \\ \hline a_{21} & a_{23} & \vdots & a_{22} & 0 & 0 \\ a_{41} & a_{43} & \vdots & 0 & a_{44} & 0 \\ a_{51} & a_{53} & \vdots & 0 & 0 & a_{55} \end{bmatrix}$$

(See 'Re-ordering of equations and unknowns'.) Let $\mathbf{PAP}^T = \mathbf{C} = [c_{ij}]$, then

$$\mathbf{C} = \begin{bmatrix} c_{11} & 0 & \vdots & c_{13} & 0 & c_{15} \\ 0 & c_{22} & \vdots & c_{23} & c_{24} & 0 \\ \hline c_{31} & c_{32} & \vdots & c_{33} & 0 & 0 \\ c_{41} & c_{42} & \vdots & 0 & c_{44} & 0 \\ c_{51} & c_{52} & \vdots & 0 & 0 & c_{55} \end{bmatrix}$$

As expected, $\mathbf{C}$ is 2-cyclic and consistently ordered with ordering vector $\boldsymbol{\gamma}^{(1)'} = (0, 0, 1, 1, 1)$. Another ordering vector for $\mathbf{A}$ is $\boldsymbol{\gamma}^{(2)} = (1, 0, 1, 0, 0)$. Hence $(1, 0, 1, 0, 0) \rightarrow (0, 0, 1, 0, 1)$, $\mathbf{P}_{15}$. Then

$(0, 0, 1, 0, 1) \rightarrow (0, 0, 0, 1, 1)$, $\mathbf{P}_{34}$. Therefore $\mathbf{P} = \mathbf{P}_{34}\mathbf{P}_{15}$ and the corresponding consistently ordered matrix $\mathbf{PAP}^T$ is

$$\mathbf{P}_{34}\mathbf{P}_{15}\mathbf{AP}_{15}\mathbf{P}_{34} = \begin{bmatrix} a_{55} & 0 & 0 & \vdots & a_{53} & a_{51} \\ 0 & a_{22} & 0 & \vdots & a_{23} & a_{21} \\ 0 & 0 & a_{44} & \vdots & a_{43} & a_{41} \\ \hdashline 0 & a_{32} & a_{34} & \vdots & a_{33} & 0 \\ a_{15} & a_{12} & 0 & \vdots & 0 & a_{11} \end{bmatrix}$$

A third ordering vector is $\boldsymbol{\gamma}^{(3)} = (1, 0, 1, 2, 2)$ for which $\mathbf{P} = \mathbf{P}_{12}$. Therefore a consistent ordering associated with $\boldsymbol{\gamma}^{(3)}$ is

$$\mathbf{PAP}^T = \mathbf{P}_{12}\mathbf{AP}_{12} = \begin{bmatrix} a_{22} & \vdots & a_{21} & a_{23} & \vdots & 0 & 0 \\ \hdashline a_{12} & \vdots & a_{11} & 0 & \vdots & 0 & a_{15} \\ a_{32} & \vdots & 0 & a_{33} & \vdots & a_{34} & 0 \\ \hdashline 0 & \vdots & a_{41} & a_{43} & \vdots & a_{44} & 0 \\ 0 & \vdots & a_{51} & a_{53} & \vdots & 0 & a_{55} \end{bmatrix}$$

Consistent orderings and block tridiagonal matrices

All the consistently ordered matrices of Example 5.9 have the block tridiagonal form

$$\begin{bmatrix} \mathbf{D}_1 & \mathbf{A}_1 & & & \\ \mathbf{B}_1 & \mathbf{D}_2 & \mathbf{A}_2 & & \\ & \cdot & \cdot & \cdot & \\ & & \mathbf{B}_{M-2} & \mathbf{D}_{M-1} & \mathbf{A}_{M-1} \\ & & & \mathbf{B}_{M-1} & \mathbf{D}_M \end{bmatrix}, M \geqslant 2, \qquad (5.47)$$

where each $\mathbf{D}_i$, $i = 1(1)M$, is a square diagonal matrix, the forms associated with $\boldsymbol{\gamma}^{(1)}$ and $\boldsymbol{\gamma}^{(2)}$ corresponding to the special case $M = 2$.

Theorem 3

A 2-cyclic consistently ordered matrix $\mathbf{A}$ has the block tridiagonal form (5.47) if and only if it has an ordering vector with components arranged in ascending order of magnitude.

Proof

Assume that the ordering vector $\boldsymbol{\gamma} = \{\gamma_i\}_{i=1}^N$ with $M \geq 2$ distinct components arranged in ascending order of magnitude is an ordering vector for the 2-cyclic consistently ordered $N \times N$ matrix $\mathbf{A}$. Let $\boldsymbol{\gamma}$ have n_0 components equal to 0, n_1 components equal to 1, n_2 components equal to 2, etc., that is,

$$\boldsymbol{\gamma} = (\underbrace{0, 0, \ldots, 0}_{n_0}; \underbrace{1, 1, \ldots, 1}_{n_1}; \underbrace{222 \ldots, 2}_{n_2}; \underbrace{3, \ldots, 3}_{n_3}; \ldots) \tag{5.48}$$

By hypothesis, $\mathbf{A}$ is 2-cyclic and consistently ordered so the vector $\boldsymbol{\gamma}$ is such that if $a_{ij} \neq 0$ and $i > j$ then $\gamma_i - \gamma_j = 1$ and if $j > i$ then $\gamma_j - \gamma_i = 1$.

Now consider, for example, the third row of blocks of Fig. 5.7 from left to right. In the first block, $i = (n_0 + n_1 + 1)(1)(n_0 + n_1 + n_2)$, $\gamma_i = 2$, $j = 0(1)n_0$ and $\gamma_j = 0$. Hence $i > j$ but $\gamma_i - \gamma_j = 2$. Therefore $a_{ij} = 0$.

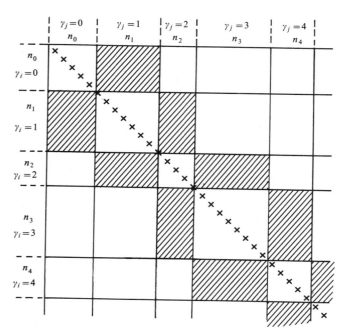

Fig. 5.7

In the second block, $i = (n_0 + n_1 + 1)(1)(n_0 + n_1 + n_2)$, $\gamma_i = 2$, $j = (n_0 + 1)(1)(n_0 + n_1)$ and $\gamma_j = 1$. Hence $i > j$ and $\gamma_i - \gamma_j = 1$ so a_{ij} can be non-zero.

In the third block, irrespective of whether $i > j$ or $j > i$, $\gamma_i = \gamma_j = 2$ so $|\gamma_i - \gamma_j| = 0$, which is contrary to the definition of a 2-cyclic matrix. Hence $a_{ij} = 0$ if $i \neq j$. When $i = j$, a_{ij} need not be zero. Therefore the block on the main diagonal of $\mathbf{A}$ is a square diagonal matrix.

In the fourth block, $j > i$, $\gamma_j = 3$ and $\gamma_i = 2$. Therefore $\gamma_j - \gamma_i = 1$ so a_{ij} can be non-zero.

The same argument applied to blocks $5(1)M$ shows that each of these blocks is a null matrix.

In general, any block which is such that the difference between its row-wise value of γ_i and its column-wise value of γ_j has a modulus greater that 1 must consist of zero elements. In general, every block on the main diagonal of $\mathbf{A}$ is a diagonal matrix.

These observations clearly establish matrix $\mathbf{A}$ as block tridiagonal of the form (5.47).

Conversely, assume that $\mathbf{A}$ has the block form (5.47) where each $\mathbf{D}_i$ is an $n_{i-1} \times n_{i-1}$ diagonal matrix. To prove that $\mathbf{A}$ is 2-cyclic and consistently ordered it is sufficient to find an ordering vector for $\mathbf{A}$ that has components in ascending order of magnitude and is such that $|\gamma_i - \gamma_j| = 1$, $a_{i,j} \neq 0$, $i \neq j$. Such a vector is given by (5.48).

It follows from the proof of Theorem 3 that the consistently ordered block tridiagonal matrix $\mathbf{PAP}^T$ associated with an ordering vector $\boldsymbol{\gamma}$ for $\mathbf{A}$ is determined by the number of distinct components of $\boldsymbol{\gamma}$ in that $\mathbf{PAP}^T$ has the form (5.47) where:

(i) M is equal to the number of distinct components in $\boldsymbol{\gamma}$.

(ii) Each $\mathbf{D}_i$ is a diagonal matrix of order n_{i-1} where n_{i-1} is the number of components in $\boldsymbol{\gamma}$ equal to $(i-1)$.

In particular, one ordering vector for a 2-cyclic matrix $\mathbf{A}$ is $\boldsymbol{\gamma} = \{\gamma_i\}_{i=1}^N$ where

$$\gamma_i = \begin{cases} 0, & \text{if } i \in S \\ 1, & \text{if } i \in T, \end{cases}$$

and S and T are the sets of definition 1. By Theorem 3 the 2-cyclic consistently ordered matrix $\mathbf{PAP}^T$ associated with this $\boldsymbol{\gamma}$ will be

$$\begin{bmatrix} \mathbf{D}_1 & \mathbf{A} \\ \mathbf{B} & \mathbf{D}_2 \end{bmatrix}, \tag{5.49}$$

where the $\mathbf{D}_i$ are square diagonal matrices. Hence another definition of 2-cyclic matrices is as follows.

Definition 4

The $N \times N$ matrix $\mathbf{A}$ is 2-cyclic if there exists an $N \times N$ permutation matrix $\mathbf{P}$ such that $\mathbf{PAP}^T$ has the block form (5.49).

As mentioned before, although a block tridiagonal matrix of the form (5.47) is consistently ordered the converse is not necessarily true.

Example 5.10

Consider the matrix

$$\mathbf{A} = \begin{bmatrix} a_{11} & 0 & 0 & a_{14} \\ a_{21} & a_{22} & a_{23} & 0 \\ 0 & a_{32} & a_{33} & 0 \\ a_{41} & 0 & 0 & a_{44} \end{bmatrix}$$

for which one ordering vector is $\boldsymbol{\gamma} = (0, 1, 2, 1)$. $\mathbf{A}$ is not block tridiagonal. Nevertheless, it is consistently ordered by definition 3 because

$$\gamma_4 - \gamma_1 = \gamma_2 - \gamma_1 = \gamma_3 - \gamma_2 = 1.$$

Additional comments on consistent ordering and the SOR method

(i) The proof of

$$(\lambda + \omega - 1)^2 = \lambda \omega^2 \mu^2, \quad \omega \neq 0, \quad \lambda \neq 0, \tag{5.50}$$

depended on the matrix $\mathbf{A}$ being block tridiagonal. It has been proved however (reference 33) that this relation holds for any 2-cyclic consistently ordered matrix $\mathbf{A}$ and not just for block tridiagonal matrices. (See Exercise 22).

(ii) If the matrix $\mathbf{A}$ of the equations $\mathbf{Ax} = \mathbf{b}$ is 2-cyclic but inconsistently ordered, its 2-cyclic property implies the existence of a permutation matrix $\mathbf{P}$ such that $\mathbf{PAP}^T$ is consistently ordered. The equations and unknowns can then be re-ordered into the form $(\mathbf{PAP}^T)\mathbf{Px} = \mathbf{Pb}$ and this ensures that (5.50) holds. This is the significance of $\mathbf{A}$ being 2-cyclic.

(iii) Relation (5.50) shows that the eigenvalues λ_i of $\mathbf{H}(\omega)$ depend only on the eigenvalues μ_i of the Jacobi iteration matrix $\mathbf{B}$ and the accelerating factor ω. Since the eigenvalues of $\mathbf{B}$ remain unchanged by any transformation of $\mathbf{A}$ to $\mathbf{PAP}^T$, (See Exercise 19), it follows that the SOR iteration matrices associated with two different orderings $\mathbf{P}_1\mathbf{AP}_1^T$ and $\mathbf{P}_2\mathbf{AP}_2^T$ have the same set of eigenvalues. In particular they have the same spectral radii. Therefore the SOR method applied to the equations $(\mathbf{P}_1\mathbf{AP}_1^T)\mathbf{P}_1\mathbf{x} = \mathbf{P}_1\mathbf{b}$ and $(\mathbf{P}_2\mathbf{AP}_2^T)\mathbf{P}_2\mathbf{x} = \mathbf{P}_2\mathbf{b}$ gives the same asymptotic rates of convergence because successive error vectors, for sufficiently large n, are related by the equation $\mathbf{e}^{(n+1)} \simeq \lambda_1\mathbf{e}^{(n)}$, where $|\lambda_1| = \rho(\mathbf{H})$.

(iv) If the SOR iteration matrices associated with two different consistent orderings $\mathbf{P}_1\mathbf{AP}_1^T$ and $\mathbf{P}_2\mathbf{AP}_2^T$ have also the same set of eigenvectors then the arithmetics of the SOR iterations will be identical because of eqn (5.27). This occurs, for example, when two equal components of an ordering vector for a consistently ordered matrix are interchanged. When the matrix $\mathbf{A}$ is block tridiagonal, as in (5.47), the interchange corresponds to the reordering of rows and columns within any diagonal submatrix $\mathbf{D}_j$ of $\mathbf{A}$, $j = 1(1)M$.

(v) The concepts of 2-cyclicity and consistent ordering can be generalized to block partitioned matrices. In particular the nine-point finite-difference equations approximating Poisson's equation are 2-cyclic in block form. See reference 33.

Consistent orderings associated with the five-point difference approximation to Poisson's equation

Consider the Dirichlet problem

$$\frac{\partial^2 U}{\partial x^2} + \frac{\partial^2 U}{\partial y^2} = f(x, y), \quad (x, y) \in D,$$

$$U(x, y) = g(x, y), \quad (x, y) \in C,$$

$$(5.51)$$

where $G = D \cup C$ is the unit square

$$G = \{(x, y): 0 \leqslant x \leqslant 1, 0 \leqslant y \leqslant 1\}.$$

(This notation means that G is the set of points (x, y) such that $0 \leqslant x \leqslant 1$ and $0 \leqslant y \leqslant 1$.) Cover the square G, in the usual man-

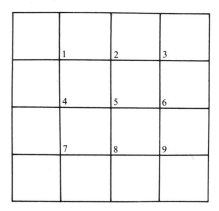

Fig. 5.8

ner, by a square mesh of side h, where $Nh = 1$. Then a numerical solution of (5.51) can be obtained by approximating Poisson's equation at the internal mesh points $(x_i, y_j) = (ih, jh)$, $i, j = 1(1)(N-1)$, by the five-point difference equation

$$u_{i-1,j} + u_{i+1,j} + u_{i,j-1} + u_{i,j+1} - 4u_{i,j} = h^2 f_{i,j}. \qquad (5.52)$$

Take $N = 4$ and label the mesh points as shown in Fig. 5.8.

If the difference equations are taken in the natural order of the points, i.e. in the order 1, 2, 3, 4, 5, 6, 7, 8, 9 and the vector of unknowns is ordered in the same way, the matrix of coefficients is

$$\mathbf{A} = \begin{bmatrix}
-4 & 1 & 0 & 1 & 0 & 0 & 0 & 0 & 0 \\
1 & -4 & 1 & 0 & 1 & 0 & 0 & 0 & 0 \\
0 & 1 & -4 & 0 & 0 & 1 & 0 & 0 & 0 \\
1 & 0 & 0 & -4 & 1 & 0 & 1 & 0 & 0 \\
0 & 1 & 0 & 1 & -4 & 1 & 0 & 1 & 0 \\
0 & 0 & 1 & 0 & 1 & -4 & 0 & 0 & 1 \\
0 & 0 & 0 & 1 & 0 & 0 & -4 & 1 & 0 \\
0 & 0 & 0 & 0 & 1 & 0 & 1 & -4 & 1 \\
0 & 0 & 0 & 0 & 0 & 1 & 1 & 1 & -4
\end{bmatrix} \begin{matrix} 1 \\ 2 \\ 3 \\ 4 \\ 5 \\ 6 \\ 7 \\ 8 \\ 9 \end{matrix}$$

$$(5.53)$$

where the column of numbers on the right-hand side indicates the order of the equations and unknowns. For this matrix the distribution of the elements in the sets S and T of the definition of 2-cyclic matrices is as follows.

	S	T
Row 1	$i = 1$	$j = 2, 4$
Row 2	$j = 1, 3, 5$	$i = 2$
Row 4	$j = 1, 5, 7$	$i = 4$
Row 3	$i = 3$	$j = 2, 6$
Row 5	$i = 5$	$j = 2, 4, 6, 8$
Row 7	$i = 7$	$j = 4, 8$
Row 6	$j = 3, 5, 9$	$i = 6$
Row 8	$j = 5, 7, 9$	$i = 8$
Row 9	$i = 9$	$j = 6, 8$

Thus

$$S = \{1, 3, 5, 7, 9\}, \quad T = \{2, 4, 6, 8\},$$

$S \cup T = W$ and $S \cap T = \phi$. Therefore matrix $\mathbf{A}$ is 2-cyclic.

It follows immediately that two possible ordering vectors for $\mathbf{A}$ are

$$\boldsymbol{\gamma}^{(1)} = (0, 1, 0, 1, 0, 1, 0, 1, 0)$$

and

$$\boldsymbol{\gamma}^{(2)} = (1, 0, 1, 0, 1, 0, 1, 0, 1).$$

A third ordering vector which can either be obtained by means of a theorem given by Papamichael and Smith, reference 23, or as indicated earlier, is

$$\boldsymbol{\gamma}^{(3)} = (0, 1, 2, 1, 2, 3, 2, 3, 4).$$

Although the components of $\boldsymbol{\gamma}^{(3)}$ are not in ascending order of magnitude it is easily checked that matrix $\mathbf{A}$ is consistently ordered with respect to $\boldsymbol{\gamma}^{(3)}$. Therefore matrix $\mathbf{A}$ is consistently ordered but does not have the block tridiagonal form of (5.47) because each diagonal block is not a diagonal submatrix.

A block tridiagonal consistent ordering of $\mathbf{A}$, associated with $\boldsymbol{\gamma}^{(1)}$, may be obtained by scanning the points in the order 1, 3, 5,

7, 9, 2, 4, 6, 8 and gives the matrix

$$
\mathbf{A}_1 =
\begin{bmatrix}
-4 & 0 & 0 & 0 & 0 & 1 & 1 & 0 & 0 \\
0 & -4 & 0 & 0 & 0 & 1 & 0 & 1 & 0 \\
0 & 0 & -4 & 0 & 0 & 1 & 1 & 1 & 1 \\
0 & 0 & 0 & -4 & 0 & 0 & 1 & 0 & 1 \\
0 & 0 & 0 & 0 & -4 & 0 & 0 & 1 & 1 \\
1 & 1 & 1 & 0 & 0 & -4 & 0 & 0 & 0 \\
1 & 0 & 1 & 1 & 0 & 0 & -4 & 0 & 0 \\
0 & 1 & 1 & 0 & 1 & 0 & 0 & -4 & 0 \\
0 & 0 & 1 & 1 & 1 & 0 & 0 & 0 & -4
\end{bmatrix}
\begin{matrix}
1 \\ 3 \\ 5 \\ 7 \\ 9 \\ 2 \\ 4 \\ 6 \\ 8
\end{matrix}
$$

$$(5.54)$$

Similarly, a block tridiagonal consistent ordering associated with $\gamma^{(2)}$ may be obtained by scanning the points in the order 2, 4, 6, 8, 1, 3, 5, 7, 9. It is seen therefore that consistent orderings of $\mathbf{A}$ can be obtained by labelling the mesh points black and white, as on a chessboard, and taking all the 'white' equations before the 'black' equations, or vice-versa, the unknowns being ordered in the same manner as the equations.

Since

$$\mathbf{P}_{67}\mathbf{P}_{34}[\gamma^{(3)}]^T = (0, 1, 1, 2, 2, 3, 3, 4)^T,$$

a block tridiagonal consistent ordering of $\mathbf{A}$ associated with $\gamma^{(3)}$ may be obtained by scanning the points in the order 1, 2, 4, 3, 5, 7, 6, 8, 9 and gives the matrix

$$
\mathbf{A}_2 =
\begin{bmatrix}
-4 & 1 & 1 & 0 & 0 & 0 & 0 & 0 & 0 \\
1 & -4 & 0 & 1 & 1 & 0 & 0 & 0 & 0 \\
1 & 0 & -4 & 0 & 1 & 1 & 0 & 0 & 0 \\
0 & 1 & 0 & -4 & 0 & 0 & 1 & 0 & 0 \\
0 & 1 & 1 & 0 & -4 & 0 & 1 & 1 & 0 \\
0 & 0 & 1 & 0 & 0 & -4 & 0 & 1 & 0 \\
0 & 0 & 0 & 1 & 1 & 0 & -4 & 0 & 1 \\
0 & 0 & 0 & 0 & 1 & 1 & 0 & -4 & 1 \\
0 & 0 & 0 & 0 & 0 & 0 & 1 & 1 & -4
\end{bmatrix}
\begin{matrix}
1 \\ 2 \\ 4 \\ 3 \\ 5 \\ 7 \\ 6 \\ 8 \\ 9
\end{matrix}
$$

Another block tridiagonal consistent ordering of $\mathbf{A}$, associated with $\gamma^{(3)}$, is obtained by scanning the points in the order 1, 4, 2,

7, 5, 3, 8, 6, 9. Hence a consistent ordering of $\mathbf{A}$ can be obtained by scanning the mesh points on successive diagonals in the same direction.

The matrix obtained by scanning successive mesh lines in opposite directions, i.e. by scanning the points in the order 1, 2, 3, 6, 5, 4, 7, 8, 9, is not consistently ordered.

Stone's strongly implicit iterative method

The point SOR method is fully explicit, the pth equation being used to calculate the next iterative value of only the component u_p. A strongly implicit method calculates all of the next iterative values by a direct elimination method. Naturally, one hopes that the successive vectors of iterative values will approximate the exact solution vector very closely after only a few iterations. Let the function U satisfy the elliptic differential equation.

$$a_1(x, y)\frac{\partial^2 U}{\partial x^2} + a_2(x, y)\frac{\partial^2 U}{\partial y^2} + a_3(x, y)\frac{\partial U}{\partial x}$$

$$+ a_4(x, y)\frac{\partial U}{\partial y} + a_5(x, y)U = Q(x, y) \quad (5.55)$$

at every interior point of the rectangle

$$S = \{(x, y): 0 \le x \le a, 0 \le y \le b\}.$$

and have known values on its boundary. Cover the solution domain with rectangular meshes of sides h and k such that $(N+1)h = a$, $(M+1)k = b$, and label the internal mesh points as shown in Fig. 5.9 so as to avoid double subscript notation. Then a five-point difference approximation to (5.55) at the pth mesh point can be written as

$$B_p u_{p-N} + D_p u_{p-1} + E_p u_p + F_p u_{p+1} + H_p u_{p+N} = q_p. \quad (5.56)$$

When the equations at the mesh points $1, 2, 3, \ldots, p, p+1, \ldots, MN$, in that order, are expressed as $\mathbf{Au} = \mathbf{q}$, the matrix $\mathbf{A}$ is as shown in Fig. 5.10, where all the coefficients B_p, $p = (N+1)(1)MN$, lie on the diagonal labelled B, all the coefficients $D_p, p = 2(1)MN$, lie on the diagonal labelled D, etc. The F and D diagonals are immediately above and below the E diagonal. Thus matrix $\mathbf{A}$ is of bandwidth $2N$. For large N the

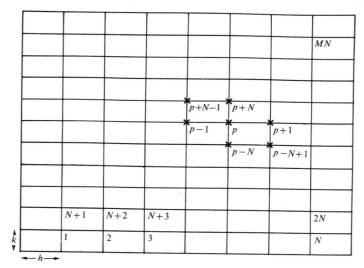

Fig. 5.9

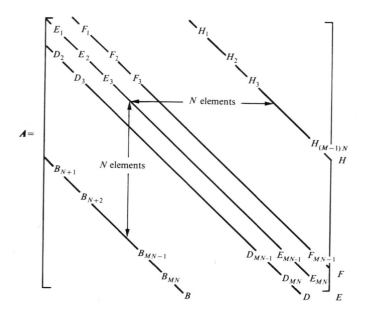

Fig. 5.10

standard **LU** decomposition 'fills-in' a very large number of zeros, the zeros between the B and H diagonals.

Stone's idea was to modify the matrix **A** by the addition of a 'small' matrix **N** so that:

(i) The factorization of $(\mathbf{A}+\mathbf{N})$ into the product $\bar{\mathbf{L}}\bar{\mathbf{U}}$ involves much less arithmetic than the standard **LU** decomposition of **A**.

(ii) The elements of $\bar{\mathbf{L}}$ and $\bar{\mathbf{U}}$ are easily calculated.

(iii) $\|\mathbf{N}\| \ll \|\mathbf{A}\|$.

Assuming this has been done, the development of an iterative procedure to calculate the solution of $\mathbf{Au}=\mathbf{q}$ is then quite straightforward because the equation can be written as

$$(\mathbf{A}+\mathbf{N})\mathbf{u} = (\mathbf{A}+\mathbf{N})\mathbf{u} + (\mathbf{q}-\mathbf{Au})$$

which suggests the iterative procedure

$$(\mathbf{A}+\mathbf{N})\mathbf{u}^{(n+1)} = (\mathbf{A}+\mathbf{N})\mathbf{u}^{(n)} + (\mathbf{q}-\mathbf{Au}^{(n)}). \qquad (5.57)$$

When the right-hand side is known, eqn (5.57) gives an efficient method for solving *directly* for $\mathbf{u}^{(n+1)}$ because the factorization of $(\mathbf{A}+\mathbf{N})$ into $\bar{\mathbf{L}}\bar{\mathbf{U}}$ is efficient. If $\|\mathbf{N}\| \ll \|\mathbf{A}\|$ one would intuitively expect a rapid rate of convergence.

To achieve his objectives Stone decided that $\bar{\mathbf{L}}$ and $\bar{\mathbf{U}}$ would each have only three non-zero elements per row as illustrated in Fig. 5.11. The non-zero elements of $\bar{\mathbf{L}}$ lie on the diagonals b, c, and d, corresponding to the diagonals B, D, and E of **A** respectively, and the non-zero elements of $\bar{\mathbf{U}}$ lie on the diagonals 1, e and f, corresponding to the diagonals E, F, and H of **A** respectively.

The product $\bar{\mathbf{L}}\bar{\mathbf{U}}$ has seven non-zero elements per row, Fig. 5.12, which lie along the diagonals $\bar{B}$, $\bar{C}$, $\bar{D}$, $\bar{E}$, $\bar{F}$, $\bar{G}$ and $\bar{H}$. The

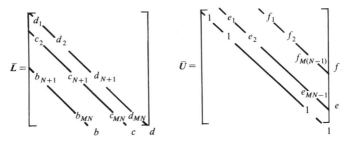

Fig. 5.11

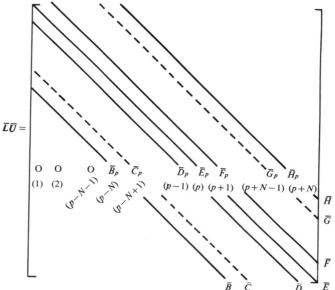

$$\bar{L}\bar{U} =$$

(The bracketed numbers (1), (2), ..., (p),... indicate the position of the first, second,....., pth, ... elements along the pth row.)

Fig. 5.12

diagonals $\bar{B}$, $\bar{D}$, $\bar{E}$, $\bar{F}$ and $\bar{H}$ correspond to the diagonals B, D, E, F and H of Fig. 5.10, and the additional diagonals $\bar{C}$ and $\bar{G}$ are immediately next to the $\bar{B}$ and $\bar{H}$ diagonals.

By Fig. 5.12 it is seen that the pth equation of

$$\bar{L}\bar{U}\mathbf{u} = \mathbf{q}\{ = (\mathbf{A} + \mathbf{N})\mathbf{u}\}$$

is

$$(\bar{B}_p u_{p-N} + \bar{D}_p u_{p-1} + \bar{E}_p u_p + \bar{F}_p u_{p+1} + \bar{H}_p u_{p+N})$$
$$+ (\bar{C}_p u_{p-N+1} + \bar{G}_p u_{p+N-1}) = q_p. \quad (5.58)$$

A comparison of eqns (5.58) and (5.56) shows that the first five terms of (5.58) will coincide with those of (5.56) if

$$\bar{B}_p = B_p, \quad \bar{D}_p = D_p, \quad \bar{E}_p = E_p, \quad \bar{F}_p = F_p \text{ and } \bar{H}_p = H_p.$$
$$(5.59)$$

The matrix $\mathbf{N}$ would then consist of the diagonals $\bar{C}$ and $\bar{G}$. In this case the unknown elements b, c, d, e, and f of $\bar{L}$ and $\bar{U}$, and the unknown elements $\bar{C}$ and $\bar{G}$ of $\mathbf{N}$, would be found from eqns (5.59) and the relationships between the elements of $\bar{\mathbf{L}}\bar{\mathbf{U}}$ and those of $\bar{\mathbf{L}}$ and $\bar{\mathbf{U}}$ which are,

$$\bar{B}_p = b_p \tag{5.60a}$$

$$\bar{D}_p = c_p \tag{5.60b}$$

$$\bar{E}_p = b_p f_{p-N} + c_p e_{p-1} + d_p \tag{5.60c}$$

$$\bar{F}_p = d_p e_p \tag{5.60d}$$

$$\bar{H} = d_p f_p \tag{5.60e}$$

$$\bar{C}_p = b_p e_{p-N} \text{ and } \bar{G}_p = c_p f_{p-1}. \tag{5.60f}$$

Equations (5.59) and (5.60a) to (5.60e) would give b, c, d, e, and f. Equations (5.60f) would then give $\bar{C}_p$ and $\bar{G}_p$. Stone found, however, that this choice of $\mathbf{N}$ did not give a rapidly convergent iteration.

Considering eqn (5.58), in which the term $(\bar{C}_p u_{p-N+1} + \bar{G}_p u_{p+N-1})$ is the pth component of $\mathbf{Nu}$ for the $\mathbf{N}$ considered above, Stone then decided to diminish the magnitude of this term by subtracting from it a closely equivalent expression. By Taylor expansions it is easily shown that

$$u_{p-N+1} = -u_p + u_{p+1} + u_{p-N} + O(hk)$$

and

$$u_{p+N-1} = -u_p + u_{p+N} + u_{p-1} + O(hk).$$

Therefore $u_{p-N+1} - (-u_p + u_{p+1} + u_{p-N}) = O(hk)$.

At this stage Stone also introduced an acceleration parameter α, $0 < \alpha < 1$, and defined the pth component of $\mathbf{Nu}$ to be

$$\bar{C}_p \{u_{p-N+1} - \alpha(-u_p + u_{p+1} + u_{p-N})\}$$

$$+ \bar{G}_p \{u_{p+N-1} - \alpha(-u_p + u_{p+N} + u_{p-1})\}. \tag{5.61}$$

Hence the pth equation of $(\mathbf{A} + \mathbf{N})\mathbf{u} = \mathbf{q}$ is, by eqns (5.56) and (5.61),

$$(B_p - \alpha\bar{C}_p)u_{p-N} + (D_P - \alpha\bar{G}_p)u_{p-1} + \{E_p + \alpha(\bar{C}_p + \bar{G}_p)\}u_p$$

$$+ (F_p - \alpha\bar{C}_p)u_{p+1} + (H_p - \alpha\bar{G}_p)u_{p+N} + \bar{C}_p u_{p-N+1} + \bar{G}_p u_{p+N-1} = q_p.$$

The relationships between the elements of $\bar{\mathbf{L}}\bar{\mathbf{U}} = \mathbf{A} + \mathbf{N}$ and those of $\bar{\mathbf{L}}$ and $\bar{\mathbf{U}}$ are then

$$B_p - \alpha\bar{C}_p = b_p \tag{5.62a}$$

$$D_p - \alpha\bar{G}_p = c_p \tag{5.62b}$$

$$E_p + \alpha(\bar{C}_p + \bar{G}_p) = b_p f_{p-N} + c_p e_{p-1} + d_p \tag{5.62c}$$

$$F_p - \alpha\bar{C}_p = d_p e_p \tag{5.62d}$$

$$H_p - \alpha\bar{G}_p = d_p f_p \tag{5.62e}$$

$$\bar{C}_p = b_p e_{p-N} \text{ and } \bar{G}_p = c_p f_{p-1}. \tag{5.62f}$$

After substituting for $\bar{C}_p$ and $\bar{G}_p$ from (5.62f), the five equations (5.62a) to (5.62e) can be written more usefully as

$$b_p = \frac{B_p}{(1 + \alpha e_{p-N})} \tag{5.63a}$$

$$c_p = \frac{D_p}{(1 + \alpha f_{p-1})} \tag{5.63b}$$

$$d_p = E_p + \alpha(b_p e_{p-N} + c_p f_{p-1}) - b_p f_{p-N} - c_p e_{p-1} \tag{5.63c}$$

$$e_p = \frac{(F_p - \alpha b_p e_{p-N})}{d_p} \tag{5.63d}$$

$$f_p = \frac{(H_p - \alpha c_p f_{p-1})}{d_p}, \tag{5.63e}$$

and give the coefficients in $\bar{\mathbf{L}}$ and $\bar{\mathbf{U}}$ sequentially for $p = 1, 2, \ldots,$ MN. For example, if $N = 3$ and $M = 4$, then

$$B_1 = 0 = B_2 = B_3 = D_1 = F_{12} = H_{12} = H_{11} = H_{10},$$

$$c_1 = 0 = b_1 = b_2 = b_3 = e_{12} = f_{12} = f_{11} = f_{10},$$

and every letter with a zero or negative subscript will also be zero. Hence eqns (5.63a) to (5.63e) give that

$$b_1 = 0, \quad c_1 = 0, \quad d_1 = E_1, \quad e_1 = F_1/d_1 = F_1/E_1,$$

$$f_1 = H_1/d_1 = H_1/E_1,$$

and

$$b_2 = 0, \quad c_2 = \frac{D_2}{1 + \alpha f_1} = \frac{D_2 E_1}{(E_1 + \alpha H_1)},$$

$$d_2 = E_2 + \alpha c_2 f_1 - c_2 e_1 = \text{etc.}$$

The implementation of the iteration

In practice, the iteration defined by eqn (5.57), namely

$$(\mathbf{A}+\mathbf{N})\mathbf{u}^{(n+1)} = (\mathbf{A}+\mathbf{N})\mathbf{u}^{(n)} + (\mathbf{q}-\mathbf{A}\mathbf{u}^{(n)}),$$

where $(\mathbf{A}+\mathbf{N}) = \bar{\mathbf{L}}\bar{\mathbf{U}}$, is dealt with as follows.

Let $\mathbf{d}^{(n)} = \mathbf{u}^{(n+1)} - \mathbf{u}^{(n)}$ and $\mathbf{R}^{(n)} = \mathbf{q} - \mathbf{A}\mathbf{u}^{(n)}$.

Then by (5.57) a complete cycle of the iteration consists of the solution of

$$\bar{\mathbf{L}}\bar{\mathbf{U}}\mathbf{d}^{(n)} = \mathbf{R}^{(n)}, \tag{5.64}$$

followed by

$$\mathbf{u}^{(n+1)} = \mathbf{u}^{(n)} + \mathbf{d}^{(n)},$$

which is the iterative refinement described on p. 258. Equation (5.64) would, of course, be solved by the forward and backward substitutions

$$\bar{\mathbf{L}}\mathbf{y}^{(n)} = \mathbf{R}^{(n)}$$

and

$$\bar{\mathbf{U}}\mathbf{d}^{(n)} = \mathbf{y}^{(n)}.$$

An additional acceleration parameter ω can also be introduced into the procedure by replacing (5.64) with

$$\bar{\mathbf{L}}\bar{\mathbf{U}}\mathbf{d}^{(n)} = \omega\mathbf{R}^{(n)},$$

as in reference 6. Further details concerning the calculation of α and the solution of the equations are given in Stone's paper, reference 26. In particular (Fig. 5.9), he recommends a double-sweep procedure in which for odd iterations the equations and unknowns are ordered row by row from left to right, moving upwards, (the natural ordering), but for even iterations the equations and unknowns are ordered row by row from left to right, moving downwards, (the reading ordering). This procedure introduces function values at the points $p-N+1$ and $p+N-1$ as before, and also at the points $p-N-1$ and $p+N+1$. His results indicate that the method is economical arithmetically in relation to older methods and that its rate of convergence is much less sensitive to the choice of iteration parameters than are the SOR and ADI methods.

A recent direct method

A *method for 'variables separable' equations*

The following method which depends upon the differential equation being 'variables separable', although that is not immediately obvious, was first proposed by Hockney 1966, reference 12, who considered the problem

$$\frac{\partial^2 U}{\partial x^2} + \frac{\partial^2 U}{\partial y^2} = g(x, y) \quad (x, y) \in D,$$

$$U = 0, \qquad (x, y) \in C,$$

where C is the boundary of the rectangular domain $D = \{(x, y) : 0 < x < a, 0 < y < b\}$. Using Fig. 5.9 and a square mesh, the five-point difference equations approximating this problem may be written in partitioned form as

$$\begin{bmatrix} \mathbf{B} & \mathbf{I} & & & \\ \mathbf{I} & \mathbf{B} & \mathbf{I} & & \\ & \mathbf{I} & \mathbf{B} & \mathbf{I} & \\ & & \cdot & \cdot & \cdot \\ & & & \mathbf{I} & \mathbf{B} \end{bmatrix} \begin{bmatrix} \mathbf{u}_1 \\ \mathbf{u}_2 \\ \mathbf{u}_3 \\ \\ \mathbf{u}_M \end{bmatrix} = \begin{bmatrix} \mathbf{b}_1 \\ \mathbf{b}_2 \\ \mathbf{b}_3 \\ \\ \mathbf{b}_M \end{bmatrix} \qquad (5.65)$$

where $\mathbf{u}_r$ is the vector of mesh values along $y = rh$, $r = 1(1)M$, $\mathbf{b}_r$ is a known vector corresponding to $\mathbf{u}_r$ and the $N \times N$ matrix $\mathbf{B}$ is

$$\begin{bmatrix} -4 & 1 & & \\ 1 & -4 & 1 & \\ & \cdot & \cdot & \cdot \\ & & 1 & -4 \end{bmatrix},$$

where N is the number of mesh points along a row parallel to Ox. By (5.65)

$$\mathbf{B}\mathbf{u}_1 + \mathbf{u}_2 = \mathbf{b}_1$$

$$\mathbf{u}_1 + \mathbf{B}\mathbf{u}_2 + \mathbf{u}_3 = \mathbf{b}_2 \qquad (5.66)$$

$$\cdot \quad \cdot \quad \cdot$$

$$\mathbf{u}_{M-1} + \mathbf{B}\mathbf{u}_M = \mathbf{b}_M.$$

Let $\mathbf{q}_r$ be an eigenvector of $\mathbf{B}$ corresponding to the eigenvalue λ_r. Then

$$\mathbf{B}\mathbf{q}_r = \lambda_r \mathbf{q}_r, \quad r = 1(1)M,$$

and this set of equations can be written as

$$\mathbf{BQ} = \mathbf{Q}\text{diag}(\lambda_1, \lambda_2, \ldots, \lambda_M),$$

where $\mathbf{Q}$ is the modal matrix $[\mathbf{q}_1, \mathbf{q}_2 \ldots \mathbf{q}_M]$. But $\mathbf{B}$ is symmetric, therefore the eigenvectors $\mathbf{q}_r$, $r = 1(1)M$, can be normalized so that $\mathbf{Q}^T\mathbf{Q} = \mathbf{I}$. Hence $\mathbf{Q}^T\mathbf{BQ} = \text{diag}(\lambda_1, \lambda_2, \ldots, \lambda_M) = \mathbf{\Lambda}$, say. Let

$$\bar{\mathbf{u}}_r = \mathbf{Q}^T\mathbf{u}_r \text{ and } \bar{\mathbf{b}}_r = \mathbf{Q}^T\mathbf{b}_r, \tag{5.67}$$

from which it follows that

$$\mathbf{u}_r = \mathbf{Q}\bar{\mathbf{u}}_r \text{ and } \mathbf{b}_r = \mathbf{Q}\bar{\mathbf{b}}_r. \tag{5.68}$$

Substituting from (5.68) into (5.66) and premultiplying throughout with $\mathbf{Q}^T$ leads to the equations

$$\mathbf{\Lambda}\bar{\mathbf{u}}_1 + \bar{\mathbf{u}}_2 = \bar{\mathbf{b}}_1$$

$$\bar{\mathbf{u}}_1 + \mathbf{\Lambda}\bar{\mathbf{u}}_2 + \bar{\mathbf{u}}_3 = \bar{\mathbf{b}}_2$$

$$\cdots$$

$$\bar{\mathbf{u}}_{M-1} + \mathbf{\Lambda}\bar{\mathbf{u}}_M = \bar{\mathbf{b}}_M. \tag{5.69}$$

Denote the ith components of $\bar{\mathbf{u}}_r$ and $\bar{\mathbf{b}}_r$ by $\bar{u}_{i,r}$ and $\bar{b}_{i,r}$ respectively and select the ith row of each of the equations (5.69). This gives the tridiagonal system of equations

$$\lambda_i \bar{u}_{i,1} + \bar{u}_{i,2} = \bar{b}_{i,1}$$

$$\bar{u}_{i,1} + \lambda_i \bar{u}_{i,2} + \bar{u}_{i,3} = \bar{b}_{i,2}$$

$$\bar{u}_{i,2} + \lambda_i \bar{u}_{i,3} + \bar{u}_{i4} = \bar{b}_{i,3}$$

$$\cdots$$

$$\bar{u}_{i,M-1} + \lambda_i \bar{u}_{i,M} = \bar{b}_{i,M}, \tag{5.70}$$

for $\bar{u}_{i,r}$, $r = 1(1)M$. All the components of $\bar{\mathbf{u}}_r$, $r = 1(1)M$ can clearly be found by solving N such sets of equations for $\bar{u}_{i,r}$, $i = 1(1)N$. The procedure is therefore:

(i) Calculate the eigenvalues and eigenvectors of $\mathbf{B}$. (These are well known for the problem considered. See p. 154).
(ii) Compute $\bar{\mathbf{b}}_r = \mathbf{Q}^T\mathbf{b}_r$.
(iii) Solve eqns (5.70), which is easily done.
(iv) Calculate $\mathbf{u}_r = \mathbf{Q}_r\bar{\mathbf{u}}_r$.

This method has been extended to more general self-adjoint 'variables separable' elliptic equations, to problems with derivative boundary conditions, and with irregular boundaries, but research on the method is still relatively recent.

Exercises and solutions

1. The function ϕ satisfies the equation

$$\frac{\partial^2\phi}{\partial x^2}+\frac{\partial^2\phi}{\partial y^2}+2=0$$

at every point inside the square bounded by the straight lines $x=\pm1$, $y=\pm1$, and is zero on the boundary. Calculate a finite-difference solution using a square mesh of side $\frac{1}{2}$. (The non-dimensional form of the torsion problem for a solid elastic cylinder with a square cross-section.)

Assuming the discretization error is proportional to h^2 calculate an improved value of ϕ at the point $(0,0)$. (The analytical solution value is 0.589.)

Solution

Because of the symmetry there are only three unknowns; ϕ_1 at $(0,0)$, ϕ_2 at $(\frac{1}{2},0)$, ϕ_3 at $(\frac{1}{2},\frac{1}{2})$. The equations are $8\phi_2-8\phi_1+1=0$, $4\phi_3+2\phi_1-8\phi_2+1=0$ and $4\phi_2-8\phi_3+1=0$, giving $\phi_1=0.562$, $\phi_2=0.438$ and $\phi_3=0.344$ to 3D.

A coarse mesh of side $h=1$ gives the finite-difference equation, $-4\phi_1+2=0$, so $\phi_1=0.5$. Hence the 'deferred approach to the limit' method gives an improved value of $\phi_1=0.562+\frac{1}{3}(0.062)=0.583$, which is very close to the exact value of 0.589 in spite of the crude mesh $h=1$.

2. The function U satisfies the equation

$$\frac{\partial^2 U}{\partial x^2}+\frac{\partial^2 U}{\partial y^2}-32U=0$$

at every point inside the square $x=\pm1$, $y=\pm1$, and is subject to the boundary conditions

(i) $U=0$ on $y=1$, $-1\leqslant x\leqslant1$,
(ii) $U=1$ on $y=-1$, $-1\leqslant x\leqslant1$;
(iii) $\partial U/\partial x=-\frac{1}{2}U$ in $x=1$, $-1<y<1$,
(iv) $\partial U/\partial x=\frac{1}{2}U$ on $x=-1$, $-1<y<1$.

Take a square mesh of side $\frac{1}{4}$ and label the points with coordinates $(0, \frac{3}{4})$, $(\frac{1}{4}, \frac{3}{4})$, $(\frac{1}{2}, \frac{3}{4})$, $(\frac{3}{4}, \frac{3}{4})$, $(1, \frac{3}{4})$, $(0, \frac{1}{2})$, $(\frac{1}{4}, \frac{1}{2})$, ... as $1, 2, 3, 4, 5, 6, 7, \ldots$ etc. (similar to Fig. 5.2). Using the simplest central-difference formulae, show that the thirty-five finite-difference equations approximating this problem can be written in matrix form as

$$\mathbf{Au} = \mathbf{b},$$

where $\mathbf{u}$ is a column vector whose transpose is $(u_1, u_2, u_3, \ldots, u_{34}, u_{35})$, $\mathbf{b}$ a (35×1) column vector whose transpose is $(0, 0, \ldots, 0, -1, -1, -1, -1, -1)$, and $\mathbf{A}$ a matrix which can be written in partitioned form as

$$\begin{bmatrix} \mathbf{B} & \mathbf{I} & & & & & \\ \mathbf{I} & \mathbf{B} & \mathbf{I} & & & & \\ & \mathbf{I} & \mathbf{B} & \mathbf{I} & & & \\ & & \mathbf{I} & \mathbf{B} & \mathbf{I} & & \\ & & & \mathbf{I} & \mathbf{B} & \mathbf{I} & \\ & & & & \mathbf{I} & \mathbf{B} & \mathbf{I} \\ & & & & & \mathbf{I} & \mathbf{B} \end{bmatrix},$$

where

$$\mathbf{B} = \begin{bmatrix} -6 & 2 & & & \\ 1 & -6 & 1 & & \\ & 1 & -6 & 1 & \\ & & 1 & -6 & 1 \\ & & & 2 & -6\frac{1}{4} \end{bmatrix}$$

and

$$\mathbf{I} = \begin{bmatrix} 1 & & & & \\ & 1 & & & \\ & & 1 & & \\ & & & 1 & \\ & & & & 1 \end{bmatrix}.$$

(This is a special case, in non-dimensional form, of the equation $\partial^2 U / \partial x^2 + \partial^2 U / \partial y^2 - 2H(U - U_0)/KD = 0$, which determines the steady temperature at points on a thin flat plate radiating heat from its surface into a medium at temperature U_0. D represents its thickness, K its thermal conductivity and H its surface conductance.)

Solution

Proceed as in Worked Example 5.2, after noting that the problem is symmetric with respect to $x = 0$.

3. The slow steady motion of viscous fluid through a cylindrical tube, whose cross section is the area S bounded by the closed curve C, can be calculated from a function ψ that satisfies Laplace's equation at all points of S and equals $\frac{1}{2}r^2$ on the curve C, where (r, θ) are the polar co-ordinates of a point in the plane of S. Calculate a finite-difference solution for flow through a circular sector bounded by the lines $\theta = 0$, $\theta = 0.8$ radians, and the circle $r = 1$, at the mesh points defined by $r = \frac{1}{3}i$, $(i = 1, 2)$, $\theta = 0.2j$, $(j = 1, 2, 3)$.

Solution

The problem is symmetrical about $\theta = 0.4$ so there are four unknowns, ψ_1 at $(\frac{1}{3}, 0.2)$, ψ_2 at $(\frac{1}{3}, 0.4)$, ψ_3 at $(\frac{2}{3}, 0.2)$ and ψ_4 at $(\frac{2}{3}, 0.4)$. The boundary values, and the polar co-ordinate finite-difference form of Laplace's equation, give the equations

$$-52\psi_1 + 25\psi_2 + 1\tfrac{1}{2}\psi_3 + 1\tfrac{7}{18} = 0,$$
$$50\psi_1 - 52\psi_2 + 1\tfrac{1}{2}\psi_4 = 0,$$
$$\tfrac{3}{4}\psi_1 - 14\tfrac{1}{2}\psi_3 + 6\tfrac{1}{4}\psi_4 + 2\tfrac{1}{72} = 0,$$
$$\tfrac{3}{4}\psi_2 + 12\tfrac{1}{2}\psi_3 - 14\tfrac{1}{2}\psi_4 + \tfrac{5}{8} = 0.$$

Their solution is

$$\psi_1 = 0.0705, \quad \psi_2 = 0.0756, \quad \psi_3 = 0.2591, \quad \psi_4 = 0.2704.$$

4. The solution domain for the two-dimensional Laplace equation $\nabla^2 \phi = 0$ is the area bounded by the closed curve C on which the values of ϕ are known. Derive the formula for the residual at the mesh point O when the curve C intersects both of the perpendicular mesh lines of length h that pass through O.

A semicircular lamina of radius $2h$, and uniform conductivity, has its diameter kept at a temperature of $0°$ and its circumference at a temperature of $100°$. Calculate a finite-difference solution to the steady-state temperatures at the nodal points of a square mesh of side h.

Solution

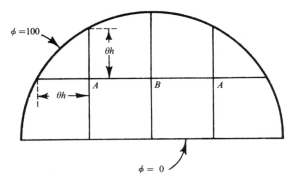

Fig. 5.13

$$\theta = \sqrt{3} - 1.$$

Hence

$$R_A = \frac{400}{(\sqrt{3}-1)\sqrt{3}} + \frac{2\phi_B}{\sqrt{3}} - \frac{4\phi_A}{\sqrt{3}-1}, \quad R_B = 2\phi_A + 100 - 4\phi_B$$

$$R_A = R_B = 0 \text{ gives } \phi_A = 70.5°, \quad \phi_B = 60.2°.$$

Although this coarse mesh is used only to provide a simple exercise it does, in fact, give the temperatures correct to within almost 1°. The analytical solution to this problem by the method of separation of the variables, and in terms of polar co-ordinates (r, θ), is

$$\phi = \frac{400}{\pi} \sum_{n=1}^{\infty} \frac{1}{(2n-1)} \left(\frac{r}{2h}\right)^{2n-1} \sin(2n-1)\theta,$$

and gives $\phi_A = 70.5°$, $\phi_B = 59°$.

5. Prove that the truncation error of the five-point finite-difference formula approximating Laplace's equation at the point (x_i, y_i), for a square mesh of side h, can be written as

$$\tfrac{1}{12}h^2\left\{\frac{\partial^4}{\partial x^4} U(\xi, y_j) + \frac{\partial^4}{\partial y^4} U(x_i, \eta)\right\},$$

where $x_i - h < \xi < x_i + h$, $y_j - h < \eta < y_j + h$, and it is assumed that the first-, second-, third- and fourth-order partial derivatives of U with respect to x and y are continuous throughout these intervals respectively.

Solution

$$T_{i,j} = \frac{1}{h^2}(U_{i+1,j} + U_{i,j+1} + U_{i-1,j} + U_{i,j-1} + 4U_{i,j}) - \left(\frac{\partial^2 U_{i,j}}{\partial x^2} + \frac{\partial^2 U_{i,j}}{\partial y^2}\right).$$

By Taylor's expansion,

$$U_{i+1,j} = U_{i,j} + h\frac{\partial U_{i,j}}{\partial x} + \frac{1}{2}h^2\frac{\partial^2 U_{i,j}}{\partial x^2} + \frac{1}{6}h^3\frac{\partial^3 U_{i,j}}{\partial x^3} + \frac{1}{24}h^4\frac{\partial^4}{\partial x^4}U(\xi_1, y_j),$$

where $x_i < \xi_1 < x_i + h$, etc.

6. The elliptic equation

$$\frac{\partial^2 U}{\partial x^2} + \frac{\partial^2 U}{\partial y^2} + d\frac{\partial U}{\partial x} + e\frac{\partial U}{\partial y} + fU = 0$$

is satisfied by the function U at every point of an area S. Prove that a non-constant value of U cannot assume a positive maximum or a negative minimum inside S when f is negative.

Solution

Assume U has a positive maximum at the point P in S. Then at P, $\partial U/\partial x = \partial U/\partial y = 0$, $\partial^2 U/\partial x^2 \leq 0$ and $\partial^2 U/\partial y^2 \leq 0$. Hence the left side of the equation is negative, in which case U cannot be a solution. Similarly for U a negative minimum. This result also holds for $f = 0$, but the proof is more difficult.

7. Derive eqns (5.5) for S_1, S_2 and S_3. Deduce that,

(a) $\quad \nabla^2 U_0 = \dfrac{4S_1 + S_2 - 20U_0}{6h^2} - \frac{1}{12}h^2\nabla^4 U_0$

$$\qquad\qquad\qquad - \frac{1}{180}h^4(\tfrac{3}{2}\nabla^6 U_0 + \mathcal{D}^4\nabla^2 U_0) + O(h^6);$$

(b) $\quad \nabla^4 U_0 = \dfrac{1}{h^4}(S_3 + 2S_2 - 8S_1 + 20U_0) + O(h^2).$

8. Use Exercise 7(a) to show that when f is a constant, Poisson's equation,

$$\nabla^2 U + f = 0,$$

can be approximated at the central point point '0', Fig. 5.5, by the nine-point finite-difference equation

$$\begin{bmatrix} 1 & 4 & 1 \\ 4 & -20 & 4 \\ 1 & 4 & 1 \end{bmatrix} u + 6h^2 f = 0,$$

and that the truncation error is of order h^6.

Use this result to solve Exercise 1, the torsion problem for a square section.

Solution

In terms of the notation of Exercise 1, the equations are

$$\phi_1 + 8\phi_2 - 20\phi_3 + 3 = 0,$$

$$4\phi_1 - 18\phi_2 + 8\phi_3 + 3 = 0,$$

$$-20\phi_1 + 16\phi_2 + 4\phi_3 + 3 = 0.$$

Hence $\phi_1 = 0.590$, $\phi_2 = 0.459$, $\phi_3 = 0.363$ to 3D.

As the analytical value of ϕ_1 is 0.589 it is seen that the nine-point formula gives a very accurate solution in this case.

9. The simply connected open-bounded domain D with closed boundary curve C is covered with a square mesh defined by the lines $x_i = ih$, $y_i = jh$, i, $j = 0$, ± 1, ± 2, ... The set of mesh points interior to D is denoted by D_h and the set on C by C_h. The function $w_{i,j}$ defined on $D_h \cup C_h$ is such that $Lw_{i,j} \geq 0$ for all $(i, j) \in D_h$, where $Lw_{i,j} = (w_{i+1,j} + w_{i-1,j} + w_{i,j+1} + w_{i,j-1} - 4w_{i,j})/h^2$.

Prove that $\max_{D_h} w_{i,j} \leq \max_{C_h} w_{i,j}$, i.e. that the maximum of $w_{i,j}$ is on C_h.

Solution

The proof is by contradiction. Using the numbering of Fig. 5.5 with P replacing 0, $P \in D_h$, $Lw_p = (w_1 + w_2 + w_3 + w_4 - 4w_p)/h^2 \geq 0$ by hypothesis. Therefore $w_P \leq \frac{1}{4}(w_1 + w_2 + w_3 + w_4)$. Assume now that $w_{i,j}$ has a maximum value M at P, i.e. $M = w_p \geq w_{i,j}$, $(i, j) \in$

D_h and $M > w_{i,j}$, $(i, j) \in C_h$. In this case the preceding inequality can hold only as an equality with $w_1 = w_2 = w_3 = w_4 = M$. By choosing P as the point 1 and repeating the argument, etc., it follows that $w_{i,j} = M$ at all points of D_h and C_h. This contradicts the assumption $w_{i,j} < M$ on C_h.

10. The function U satisfies Laplace's equation at the points of the square $0 < x < 4$, $0 < y < 4$ and the Dirichlet boundary conditions:

   (i) $U = 0$ along $x = 0$ and $y = 0$.
   (ii) $U = x^3$ along $y = 4$, $0 \le x \le 4$,
   (iii) U linear and continuous along $x = 4$, $0 \le y \le 4$.

Use a square mesh of side 1 and write down the five-point difference equations approximating this problem at the nine internal mesh points. Either write and run a programme to solve these equations by the SOR method or carry out two SOR iterations using a hand-calculator.

Solution

By p. 285,

$$\rho(\mathbf{B}) = \frac{1}{2}\left(\cos\frac{\pi}{4} + \cos\frac{\pi}{4}\right) = \frac{1}{\sqrt{2}}.$$

Hence $\omega = 2/\{1 + \sqrt{(1 - \rho^2)}\} = 1.1716$. Labelling the mesh points 1, 2, 3, ..., 9 from left to right and from $y = 1(1)3$, a consistent ordering is obtained by taking the equations in the same order. The corresponding SOR equations are

$$u_1^{(n+1)} = 0.2929(u_2^{(n)} + u_4^{(n)}) - 0.1716u_1^{(n)}$$

$$u_2^{(n+1)} = 0.2929(u_1^{(n+1)} + u_3^{(n)} + u_5^{(n)}) - 0.1716u_2^{(n)}$$

$$u_3^{(n+1)} = 0.2929(u_2^{(n+1)} + u_6^{(n)} + 16) - 0.1716u_3^{(n)}$$

$$\cdots \cdots \cdots \cdots \cdots$$

$$u_9^{(n+1)} = 0.2929(u_6^{(n+1)} + u_8^{(n+1)} + 75) - 0.1716u_9^{(n)}.$$

Taking all initial values as zero the solution to $4D$ requires 11 iterations and $(u_1, u_2, \ldots, u_9)$ is (2.6071, 5.9643, 10.5, 4.4643, 10.75, 20.0357, 4.5, 12.5357, 26.8929). The values of the second iteration are $(0, 1.3726, 7.4316, 0.0858, 4.2860, 18.5255, 0.9792, 11.0331, 26.1934)$.

11. (i) Deduce that the eigenvalue λ_1 of largest modulus of a real $m \times m$ iteration matrix $\mathbf{G}$ is real if $|\lambda_1| > |\lambda_i|$, $i = 2(1)m$.

(ii) Assuming that the conditions of (i) hold, show that $\mathbf{d}^{(n)} \simeq \lambda_1 \mathbf{d}^{(n-1)}$, for sufficiently large n, where the displacement vector $\mathbf{d}^{(n)} = \mathbf{x}^{(n+1)} - \mathbf{x}^{(n)}$.

(iii) Assuming that $1 > |\lambda_1| > |\lambda_i|$, $i = 2(1)m$, derive Lyusternik's acceleration process $\mathbf{x} \simeq \mathbf{x}^{(n)} + \mathbf{d}^{(n)}/(1 - \lambda_1)$ by writing the solution $\mathbf{x}$ as the limiting value of the infinite series

$$\mathbf{x} = \mathbf{x}^{(n)} + \mathbf{d}^{(n)} + \mathbf{d}^{(n+1)} + \dots$$

Solution

(i) The elements of $\mathbf{G}$ are real so complex roots occur in conjugate pairs. Hence λ_1 is real.

(ii) As per text.

(iii) By (ii), $\mathbf{d}_{(n+1)} \simeq \lambda_1 \mathbf{d}^{(n)}$, $\mathbf{d}^{(n+2)} \simeq \lambda_1^2 \mathbf{d}^{(n)}$, etc. Hence

$$\mathbf{x} \simeq \mathbf{x}^{(n)} + (1 + \lambda_1 + \lambda_1^2 + \dots)\mathbf{d}^{(n)}$$

$$= \mathbf{x}^{(n)} + \frac{\mathbf{d}^{(n)}}{(1 - \lambda_1)}, \text{ since } 0 < \lambda_1 < 1.$$

12. The following numbers to $5D$ are the third components of the fifth, sixth, and seventh iteration vectors respectively of a Jacobi iterative solution.

$$0.41504, \quad 0.45874, \quad 0.49500.$$

Use Aitken's method to calculate an improved value for this component. Verify that Lyusternik's method gives the same value to $4D$ when λ_1 is approximated by $d_3^{(6)}/d_3^{(5)}$.

Solution

$x_3 \simeq 0.495 - (0.03626)^2/(-0.00744) \simeq 0.67172$.

$d_3^{(5)} = 0.0437$, $d_3^{(6)} = 0.03626$, $\lambda_1 = 0.8297$. By Lyusternik's method $x_3 \simeq x_3^{(6)} + d_3^{(6)}/(1 - \lambda_1)$ gives $x_3 \simeq 0.6717$ to $4D$.

(The problem from which these figures came required 65 iterations for $4D$ accuracy. $x_3 = 0.6555$.)

13. The Crank–Nicolson equations approximating $\partial U/\partial t = \partial^2 U/\partial x^2$, where U is known at $x = 0$ and $x = Nh$, may be written as

$$u_i = \rho(u_{i-1} + u_{i+1}) + c_i, \quad i = 1(1)(N-1),$$

where u_i denotes the unknown $u_{i,j+1}$, c_i denotes a known number for all $i = 1(1)(N-1)$, $\rho = r/2(1+r)$ and $r = k/h^2$. Prove that the Gauss–Seidel iterative procedure for solving these equations converges for all positive values of r, it being assumed that the $(n+1)$th iterative values are calculated systematically from $i = 1(1)(N-1)$.

Solution

The iteration is defined by the equation

$$u_i^{(n+1)} = \rho(u_{i-1}^{(n+1)} + u_{i+1}^{(n)}) + c_i.$$

The simplest way to obtain the Gauss–Seidel iteration matrix is to substitute for $u_{i-1}^{(n+1)}$ in terms of $u_{i-1}^{(n)}$ from the preceding equation. Then

$$u_1^{(n+1)} = \rho u_2^{(n)} + (\rho u_0 + c_1) = \rho u_2^{(n)} + c_1',$$

where c_1' is known because u_0 is a boundary value.

$$\begin{aligned}
u_2^{(n+1)} &= \rho u_3^{(n)} + \rho u_1^{(n+1)} + c_2 \\
&= \rho u_3^{(n)} + \rho^2 u_2^{(n)} + (\rho c_1' + c_2) \\
&= \rho u_3^{(n)} + \rho^2 u_2^{(n)} + c_1', \text{ etc.}
\end{aligned}$$

In matrix form,

$$\mathbf{u}^{(n+1)} = \begin{bmatrix}
0 & \rho & 0 & & & \\
0 & \rho^2 & \rho & 0 & & \\
0 & \rho^3 & \rho^2 & \rho & 0 & \\
\vdots & & & & & \\
0 & \rho^{N-1} & \rho^{N-2} & & & \rho^2
\end{bmatrix} \mathbf{u}^{(n)} + \mathbf{c}$$

The second column gives the largest column sum of the elements of the iteration matrix. If λ_i represents an eigenvalue of the matrix then by Gerschgorin's first theorem,

$$\max_i |\lambda_i| \leqslant \rho + \rho^2 + \ldots + \rho^{N-1} < \frac{\rho}{1-\rho} = \frac{r}{2+r} < 1.$$

14. For the equations

$$x_1 + 2x_2 + 4x_3 = 1,$$

and
$$\tfrac{1}{8}x_1 + x_2 + x_3 = 3$$

$$-x_1 + 4x_2 + x_3 = 7,$$

prove that the Jacobi iteration is convergent but that the Gauss–Seidel iteration is divergent.

Solution

$$\mathbf{A} = \mathbf{D} - \mathbf{L} - \mathbf{U} = \begin{bmatrix} 1 & 2 & 4 \\ \tfrac{1}{8} & 1 & 1 \\ -1 & 4 & 1 \end{bmatrix}.$$

The eigenvalues μ of the Jacobi iteration matrix are the μ roots of

$$\det(\mu\mathbf{D} - \mathbf{L} - \mathbf{U}) = 0 = \det\begin{bmatrix} \mu & 2 & 4 \\ \tfrac{1}{8} & \mu & 1 \\ -1 & 4 & \mu \end{bmatrix} = \mu(\mu^2 - \tfrac{1}{4}).$$

Hence $\mu = 0, \pm\tfrac{1}{2}$, so the iteration converges.

The eigenvalues λ of the Gauss–Seidel iteration matrix are the λ roots of

$$\det(\lambda\mathbf{D} - \lambda\mathbf{L} - \mathbf{U}) = 0 = \det\begin{bmatrix} \lambda & 2 & 4 \\ \tfrac{1}{8}\lambda & \lambda & 1 \\ -\lambda & 4\lambda & \lambda \end{bmatrix} = \lambda(\lambda^2 + \tfrac{7}{4}\lambda - 2).$$

Hence $\lambda = 0, 0.788, -2.538$, so the iteration diverges.

15. A function satisfies Laplace's equation at every point inside the square bounded by the straight lines $x = \pm 1$, $y = \pm 1$, has known boundary values, and is symmetrical with respect to Ox and Oy. Write down the simplest finite-difference equations giving an approximate solution at the nodal points of a square of side $\frac{1}{2}$. Prove that the Jacobi and Gauss–Seidel iterative processes for their solution both converge. Hence verify that the asymptotic rate of convergence of the Gauss–Seidel iteration is twice that of the Jacobi iteration.

Solution

Denote the pivotal values at the points $(0, 0)$, $(\frac{1}{2}, 0)$ and $(\frac{1}{2}, \frac{1}{2})$ by a, b, c, respectively. Then the finite-difference equations at these points, in the order given, are $a - b = \text{constant}$; $-\frac{1}{4}a + b - \frac{1}{2}c = \text{constant}$, and $-\frac{1}{2}b + c = \text{constant}$. The matrix of coefficients is

$$\mathbf{A} = \begin{bmatrix} 1 & -1 & 0 \\ -\frac{1}{4} & 1 & -\frac{1}{2} \\ 0 & -\frac{1}{2} & 1 \end{bmatrix} = \mathbf{I} - (\mathbf{L} + \mathbf{U}) = \mathbf{I} - \begin{bmatrix} 0 & 1 & 0 \\ \frac{1}{4} & 0 & \frac{1}{2} \\ 0 & \frac{1}{2} & 0 \end{bmatrix}.$$

Hence the eigenvalues of the Jacobi iteration matrix $(\mathbf{L} + \mathbf{U})$ are the μ roots of

$$\det \begin{bmatrix} -\mu & 1 & 0 \\ \frac{1}{4} & -\mu & \frac{1}{2} \\ 0 & \frac{1}{2} & -\mu \end{bmatrix} = 0 = \mu(\mu^2 - \frac{1}{2}).$$

Therefore the spectral radius $\rho(\mathbf{B})$ is $1/\sqrt{2} < 1$, so the iteration converges.

The eigenvalues λ of the Gauss–Seidel iteration matrix are the λ roots of $\det(\lambda \mathbf{D} - \lambda \mathbf{L} - \mathbf{U}) = 0$, i.e.

$$\det \begin{bmatrix} \lambda & -\lambda & 0 \\ -\frac{1}{4}\lambda & \lambda & -\frac{1}{2} \\ 0 & -\frac{1}{2}\lambda & \lambda \end{bmatrix} = \lambda^2(\lambda - \frac{1}{2}) = 0,$$

proving $\rho(\mathbf{G}) = \frac{1}{2}$.

As the asymptotic rate of convergence is the modulus of the natural logarithm of the spectral radius, the result follows.

16. The coefficient matrix of the linear equations $\mathbf{A}\mathbf{x} = \mathbf{b}$ is defined by

$$\mathbf{A} = \begin{bmatrix} 1 & c & 1 \\ c & 1 & c \\ -c^2 & c & 1 \end{bmatrix}, \quad c \text{ real and non-zero.}$$

Find the range of values for c for which the Gauss–Seidel iteration converges. Will the Jacobi iteration converge for $c = 2$?

Solution

The eigenvalues λ of the Gauss–Seidel iteration matrix are the λ roots of $\det(\lambda\mathbf{D}-\lambda\mathbf{L}-\mathbf{U})=0$, where $\mathbf{A}=\mathbf{D}-\mathbf{L}-\mathbf{U}$, i.e. of

$$\det\begin{bmatrix} \lambda & c & 1 \\ \lambda c & \lambda & c \\ -\lambda^2 c & \lambda c & \lambda \end{bmatrix}=0=\lambda(\lambda^2-c^4).$$

Hence $|\lambda|=0$ or c^2. For convergence $|\lambda|<1$, implying $0<c^2<1$, since $c\neq 0$. Hence $-1<c<1$.

The eigenvalue μ of the Jacobi iteration matrix are the μ roots of $\det(\mu\mathbf{D}-\mathbf{L}-\mathbf{U})=0$, i.e. of

$$\det\begin{bmatrix} \mu & 2 & 1 \\ 2 & \mu & 2 \\ -4 & 2 & \mu \end{bmatrix}=0=\mu^3-4\mu-12=f(\mu).$$

As $f(2)=-12$ and $f(3)=3$ there is a real root >2, implying divergence.

17. An $m\times m$ non-singular matrix $\mathbf{A}$ with elements a_{ij} and non-zero diagonal elements a_{ii} is written as $\mathbf{D}-\mathbf{L}-\mathbf{U}$, where $\mathbf{D}$ is a diagonal matrix and $\mathbf{L}$ and $\mathbf{U}$ are strictly lower and upper triangular matrices, respectively. A variation of the Gauss–Seidel iterative method for solving the linear system $\mathbf{A}\mathbf{x}=\mathbf{b}$ is defined by the equation

$$a_{ii}x_i^{(n+1)}=\omega b_i-\sum_{j=1}^{i-1}a_{ij}x_j^{(n+1)}+(1-\omega)\sum_{j=1}^{i}a_{ij}x_j^{(n)}$$

$$-\omega\sum_{j=i+1}^{m}a_{ij}x_j^{(n)},\quad i=1(1)m,\ n=0,1,2,\ldots,$$

where ω is a real non-zero positive parameter.

Show that the iteration can be expressed as

$$\mathbf{x}^{(n+1)}=\mathbf{M}(\omega)\mathbf{x}^{(n)}+\mathbf{c},$$

where the iteration matrix $\mathbf{M}$ is given by

$$\mathbf{M}(\omega)=\omega(\mathbf{D}-\mathbf{L})^{-1}\mathbf{U}-(\omega-1)\mathbf{I}.$$

Deduce that the eigenvalues λ of $\mathbf{M}(\omega)$ are the λ-roots of the equation

$$\det\{(\lambda + \omega - 1)(\mathbf{D} - \mathbf{L}) - \omega\mathbf{U}\} = 0.$$

Given that matrix $\mathbf{A}$ is block tridiagonal, derive the relationship $\mu^2\omega = \lambda + \omega - 1$ between λ, ω, and μ, where μ is an eigenvalue of the Jacobi iteration matrix corresponding to $\mathbf{A}$.

If $\mathbf{A}$ is real and symmetric, deduce that λ is real. Given in this case that the Jacobi iteration converges, find the range of values for ω for which the Gauss–Seidel variant converges.

Solution

In matrix form the iteration is defined by

$$\mathbf{D}\mathbf{x}^{(n+1)} = \omega\mathbf{b} - (-\mathbf{L})\mathbf{x}^{(n+1)} + (1-\omega)(-\mathbf{L}+\mathbf{D})\mathbf{x}^{(n)} - \omega(-\mathbf{U})\mathbf{x}^{(n)},$$

giving

$$\mathbf{x}^{(n+1)} = \{(1-\omega)\mathbf{I} + \omega(\mathbf{D}-\mathbf{L})^{-1}\mathbf{U}\}\mathbf{x}^{(n)} + \omega(\mathbf{D}-\mathbf{L})^{-1}\mathbf{b} = \mathbf{M}(\omega)\mathbf{x}^{(n)} + c.$$

The λ are the roots of $\det(\lambda\mathbf{I} - \mathbf{M}) = 0$, where

$$\lambda\mathbf{I} - \mathbf{M} = (\lambda + \omega - 1)\mathbf{I} - \omega(\mathbf{D}-\mathbf{L})^{-1}\mathbf{U}$$

$$= (\lambda + \omega - 1)(\mathbf{D}-\mathbf{L})^{-1}(\mathbf{D}-\mathbf{L}) - \omega(\mathbf{D}-\mathbf{L})^{-1}\mathbf{U}$$

$$= (\mathbf{D}-\mathbf{L})^{-1}\{(\lambda + \omega - 1)(\mathbf{D}-\mathbf{L}) - \omega\mathbf{U}\}.$$

As $\det(\mathbf{D}-\mathbf{L})^{-1} = 1/a_{11}a_{22}\ldots a_{mm} \neq 0$, therefore

$$\det\{(\lambda + \omega - 1)(\mathbf{D}-\mathbf{L}) - \omega\mathbf{U}\} = 0.$$

When $\mathbf{A}$ is block tridiagonal then so is

$$(\lambda + \omega - 1)(\mathbf{D}-\mathbf{L}) - \omega\mathbf{U}.$$

Hence

$$\det\{(\lambda + \omega - 1)\mathbf{D} - \alpha(\lambda + \omega - 1)\mathbf{L} - \alpha^{-1}\omega\mathbf{U}\} = 0.$$

Let $\alpha(\lambda + \omega - 1) = \alpha^{-1}\omega$, giving $\alpha = \omega^{\frac{1}{2}}(\lambda + \omega - 1)^{-\frac{1}{2}}$. Then

$$\{\omega(\lambda + \omega - 1)\}^{m/2}\det\mathbf{D}\det\{(\lambda + \omega - 1)^{\frac{1}{2}}\omega^{-\frac{1}{2}}\mathbf{I} - \mathbf{D}^{-1}(\mathbf{L}+\mathbf{U})\} = 0.$$

Since $\mathbf{D}^{-1}(\mathbf{L}+\mathbf{U})$ is the Jacobi iteration matrix this equation states that μ is related to λ by $\mu = (\lambda + \omega - 1)^{\frac{1}{2}}\omega^{-\frac{1}{2}}$.

When $\mathbf{A}$ is real and symmetric, then so is $\mu\mathbf{D} - (\mathbf{L}+\mathbf{U})$, the matrix giving μ. Hence μ is real. As $\lambda = 1 + \omega(\mu^2 - 1)$, λ is real.

The Jacobi iteration converges when $0 \le \mu^2 < 1$. Hence $0 \le (\lambda + \omega - 1)/\omega < 1$, $\omega > 0$, giving $0 \le \lambda + \omega - 1 < \omega$. The right-hand inequality gives $\lambda < 1$. The left-hand inequality gives $\lambda \ge 1 - \omega$. For convergence $\lambda > -1$, which will be satisfied by $1 - \omega \ge -1$, implying $\omega < 2$. Therefore $0 < \omega < 2$.

18. A function satisfies Poisson's equation at the points of the rectangle $0 < x < ph$, $0 < y < qh$ and has known values on its boundary. Show that the matrix of the five-point difference equations approximating this problem at the mesh points defined by $x_i = ih$, $i = 1(1)(p-1)$, and $y_j = jh$, $j = 1(1)(q-1)$, can be written in block partitioned form as

$$\mathbf{A} = \begin{bmatrix} \mathbf{B} & \mathbf{I} & & & \\ \mathbf{I} & \mathbf{B} & \mathbf{I} & & \\ & \mathbf{I} & \mathbf{B} & \mathbf{I} & \\ & & \cdot & \cdot & \cdot \\ & & & \mathbf{I} & \mathbf{B} \end{bmatrix} \quad \text{of order } (p-1)(q-1), \text{ where}$$

$$\mathbf{B} = \begin{bmatrix} -4 & 1 & & & \\ 1 & -4 & 1 & & \\ & 1 & -4 & 1 & \\ & & \cdot & \cdot & \cdot \\ & & & 1 & -4 \end{bmatrix} \quad \text{is of order } (p-1) \text{ and } \mathbf{I}$$

is the unit matrix of order $(p-1)$. (This assumes the equations are ordered row by row from left to right or right to left.)

Use the theorem on p. 148, or otherwise, to show that the eigenvalues $\lambda_{i,j}$ of $\mathbf{A}$ are given by

$$\lambda_{i,j} = -4 + 2\left(\cos\frac{i\pi}{p} + \cos\frac{j\pi}{q}\right), \quad i = 1(1)(p-1), \quad j = 1(1)(q-1).$$

Deduce that the spectral radius of the corresponding Jacobi iteration matrix is

$$\frac{1}{2}\left(\cos\frac{\pi}{p} + \cos\frac{\pi}{q}\right).$$

Solution

As $\mathbf{B}$ is real and symmetric it has $(p-1)$ linearly independent eigenvectors $\mathbf{v}_i$ which may be taken as the eigenvectors of $\mathbf{I}$

because $\mathbf{Iv}_i = 1\mathbf{v}_i$. Hence the theorem may be used. The eigenvalues μ_i of B are $\mu_i = -4 + 2\cos(i\pi)/(p)$, $i = 1(1)(p-1)$. By the theorem, the eigenvalues $\lambda_{i,j}$ of $\mathbf{A}$ are

$$\lambda_{i,j} = \mu_i + 2\cos\frac{j\pi}{q}, \quad j = 1(1)(q-1).$$

Therefore

$$\lambda_{i,j} = +4 + 2\left(\cos\frac{i\pi}{p} + \cos\frac{j\pi}{q}\right).$$

The Jacobi iteration matrix is $\mathbf{D}^{-1}(\mathbf{L} + \mathbf{U})$ where $\mathbf{D} = \mathrm{diag}(-4, -4, \ldots, -4)$ and $\mathbf{A} = \mathbf{D} - \mathbf{L} - \mathbf{U}$, i.e., $\mathbf{L} + \mathbf{U} = \mathbf{D} - \mathbf{A}$. The eigenvalues of $\mathbf{L} + \mathbf{U} = -4 - \lambda_{i,j}$. The eigenvalues of

$$\mathbf{D}^{-1}(\mathbf{L} + \mathbf{U}) = (-\tfrac{1}{4})(-4 - \lambda_{i,j}) = \frac{1}{2}\left(\cos\frac{i\pi}{p} + \cos\frac{j\pi}{q}\right).$$

The largest value of this is

$$\frac{1}{2}\left(\cos\frac{\pi}{p} + \cos\frac{\pi}{q}\right).$$

19. Let μ_i represent an eigenvalue of the Jacobi iteration matrix **B** associated with the matrix **A** and let $\mathbf{v}_i$ represent the corresponding eigenvector of **B**. Prove that the Jacobi iteration matrices associated with all possible re-orderings of the equations $\mathbf{Ax} = \mathbf{b}$ have the same set of eigenvalues μ_i and corresponding eigenvectors $\mathbf{v}_i'$ where the components of $\mathbf{v}_i'$ are those of $\mathbf{v}_i$ re-ordered the same way as the equations.

Solution

Make the coefficient of every diagonal term unity. Then the equations can be written as $(\mathbf{I} - \mathbf{L} - \mathbf{U})\mathbf{x} = \mathbf{c}$ and the associated Jacobi iteration matrix is $(\mathbf{L} + \mathbf{U})$. A re-ordering of the equations and unknowns can be written as $\{\mathbf{P}(\mathbf{I} - \mathbf{L} - \mathbf{U})\mathbf{P}^T\}\mathbf{Px} = \mathbf{Pc}$, giving that $\mathbf{Px} = \{\mathbf{P}(\mathbf{L} + \mathbf{U})\mathbf{P}^T\}\mathbf{Px} + \mathbf{Pc}$ since $\mathbf{PP}^T = \mathbf{I}$. The associated Jacobi iteration matrix is $\mathbf{P}(\mathbf{L} + \mathbf{U})\mathbf{P}^T$, which is a similarity transformation of $(\mathbf{L} + \mathbf{U})$. Hence the result concerning eigenvalues. Let $\mathbf{v}'$ represent the eigenvector of $\mathbf{P}(\mathbf{L} + \mathbf{U})\mathbf{P}^T$ corresponding to the eigenvalue μ. Then $\mathbf{P}(\mathbf{L} + \mathbf{U})\mathbf{P}^T\mathbf{v}' = \mu\mathbf{v}'$. Premultiply by $\mathbf{P}^T$ and use $\mathbf{P}^T\mathbf{P} = \mathbf{I}$ to obtain $(\mathbf{L} + \mathbf{U})(\mathbf{P}^T\mathbf{v}') = \mu(\mathbf{P}^T\mathbf{v}')$. This equation states that the eigenvector of $(\mathbf{L}+\mathbf{U})$ corresponding to μ is $\mathbf{P}^T\mathbf{v}'$, i.e. that $\mathbf{P}^T\mathbf{v}' = \mathbf{v}$. Premultiply by $\mathbf{P}$ to give that $\mathbf{v}' = \mathbf{P}\mathbf{v}$. Therefore the components of $\mathbf{v}'$ are those of $\mathbf{v}$ re-ordered the same way as the equations.

20. Given that

$$
\begin{bmatrix}
a_{11} & a_{12} & \cdots & a_{1n} \\
a_{21} & a_{22} & \cdots & a_{2n} \\
\vdots & & \ddots & \vdots \\
a_{n1} & a_{n2} & & a_{nn}
\end{bmatrix}
=
\begin{bmatrix}
l_{11} & & & \\
l_{21} & l_{22} & & \\
\vdots & & \ddots & \\
l_{n1} & l_{n2} & & l_{nn}
\end{bmatrix}
\times
\begin{bmatrix}
1 & u_{12} & \cdots & \cdots & u_{1n} \\
0 & 1 & u_{23} & \cdots & u_{2n} \\
& & \ddots & & \\
& & & & 1
\end{bmatrix}
$$

where $l_{ij} = 0$, $i < j$ and $u_{ij} = 0$, $i > j$, show that

$$l_{ij} = a_{ij} - \sum_{k=1}^{j-1} l_{ik}u_{kj}$$

and

$$u_{ij} = \frac{a_{ij} - \sum_{k=1}^{i-1} l_{ik}u_{kj}}{l_{ii}}.$$

Develop formulae to calculate the components of the vectors $\mathbf{y}$ and $\mathbf{x}$ when the solution of the equations $\mathbf{L}\mathbf{U}\mathbf{x} = \mathbf{b}$ is obtained by solving $\mathbf{L}\mathbf{y} = \mathbf{b}$ for $\mathbf{y}$ by forward substitutions then $\mathbf{U}\mathbf{x} = \mathbf{y}$ for $\mathbf{x}$ by backward substitutions.

Solution

$$
\begin{aligned}
a_{ij} &= \sum_{k=1}^{n} l_{ik}u_{kj} \\
&= \sum_{k=1}^{j-1} l_{ik}u_{kj} + l_{ij}u_{ji} + \sum_{k=j+1}^{n} l_{ik}u_{kj}.
\end{aligned}
$$

As $u_{jj} = 1$ and $u_{kj} = 0$ for $k > j$ it follows that

$$l_{ij} = a_{ij} - \sum_{k=1}^{j-1} l_{ik} u_{kj}.$$

Again,

$$a_{ij} = \sum_{k=1}^{n} l_{ik} u_{kj}$$
$$= \sum_{k=1}^{i-1} l_{ik} u_{kj} + l_{ii} u_{ij} + \sum_{k=i+1}^{n} l_{ik} u_{kj}.$$

As $l_{ik} = 0$ for $i < k$ it follows that

$$u_{ij} = \frac{a_{ij} - \sum\limits_{k=1}^{i-1} l_{ik} u_{kj}}{l_{ii}}.$$

The ith component of $\mathbf{Ly} = \mathbf{b}$ gives that

$$b_i = \sum_{k=1}^{n} l_{ik} y_k$$
$$= \sum_{k=1}^{i-1} l_{ik} y_k + l_{ii} y_i + \sum_{k=i+1}^{n} l_{ik} u_k.$$

As $l_{ik} = 0$ for $i < k$ it is seen that

$$y_i = \frac{b_1 - \sum\limits_{k=1}^{i-1} l_{ik} y_k}{l_{ii}}, \quad i = 1(1)n.$$

Therefore

$$y_1 = \frac{b_1}{l_{11}}, \quad y_2 = \frac{b_2 - l_{21} y_1}{l_{22}}, \text{ etc.}$$

The ith component of $\mathbf{Ux} = \mathbf{y}$ gives that

$$y_i = \sum_{k=1}^{i-1} u_{ik} x_k + u_{ii} x_i + \sum_{k=i+1}^{n} u_{ik} x_k.$$

As $u_{ii} = 1$ and $u_{ik} = 0$ for $i > k$, we obtain

$$x_i = y_i - \sum_{k=i+1}^{n} u_{ik} x_k, \quad i = (n-1)(1)1.$$

Therefore

$$x_n = y_n, \quad x_{n-1} = y_n - u_{n-1,n}x_n, \quad \text{etc.}$$

21.

$$\mathbf{A} = \begin{bmatrix} a_{11} & a_{12} & 0 & a_{14} & 0 \\ a_{21} & a_{22} & a_{23} & 0 & a_{25} \\ 0 & a_{32} & a_{33} & 0 & 0 \\ a_{41} & 0 & 0 & a_{44} & a_{45} \\ 0 & a_{52} & 0 & a_{54} & a_{55} \end{bmatrix}$$

(i) Show that matrix $\mathbf{A}$ is 2-cyclic.

(ii) Write down two ordering vectors for $\mathbf{A}$ in terms of the numbers 0 and 1. Is $\mathbf{A}$ consistently ordered with respect to either of them? If not, use one of them to re-order $\mathbf{A}$ into a consistently ordered matrix $\mathbf{B}$.

(iii) Verify that $\mathbf{A}$ is consistently ordered with respect to the ordering vector $\boldsymbol{\gamma}^{(3)} = (0, 1, 2, 1, 2)$. Use $\boldsymbol{\gamma}^{(3)}$ to re-order $\mathbf{A}$ into a consistently ordered block tridiagonal matrix $\mathbf{C}$. Write down, in matrix notation, the equation giving the re-ordering of $\mathbf{A}\mathbf{x} = \mathbf{b}$ corresponding to $\mathbf{C}$.

Solution

(i) As in Worked Example 5.4, $S = \{1, 3, 5\}$, $T = \{2, 4\}$.

(ii) $\boldsymbol{\gamma}^{(1)} = (0, 1, 0, 1, 0)$, $\boldsymbol{\gamma}^{(2)} = (1, 0, 1, 0, 1)$. Consider $\boldsymbol{\gamma}^{(1)}$ and the element a_{32}. As $\gamma_3^{(1)} - \gamma_2^{(1)} = -1$, the matrix $\mathbf{A}$ is not consistently ordered with respect to $\boldsymbol{\gamma}^{(1)}$. Similarly for $\boldsymbol{\gamma}^{(2)}$ and a_{12}. Consider $\boldsymbol{\gamma}^{(2)}$. The interchange of the first and fourth components gives the vector $(0, 0, 1, 1, 1)$ which is an ordering vector for the consistently ordered matrix

$$\mathbf{B} = \mathbf{P}_{14}\mathbf{A}\mathbf{P}_{14} = \begin{bmatrix} a_{44} & 0 & 0 & a_{41} & a_{45} \\ 0 & a_{22} & a_{23} & a_{21} & a_{25} \\ \hdashline 0 & a_{32} & a_{33} & 0 & 0 \\ a_{14} & a_{12} & 0 & a_{11} & 0 \\ a_{54} & a_{52} & 0 & 0 & a_{55} \end{bmatrix}$$

(iii) Consider a_{12} and a_{32}, $\gamma_2^{(3)} - \gamma_1^{(3)} = 1 = \gamma_3^{(3)} - \gamma_2^{(3)}$. Similarly for the remaining non-zero elements. The interchange of the third and fourth components of $\gamma^{(3)}$ gives $(0, 1, 1, 2, 2)$ which is an ordering vector for

$$\mathbf{C} = \mathbf{P}_{34}\mathbf{A}\mathbf{P}_{34} = \begin{bmatrix} a_{11} & a_{12} & a_{14} & 0 & 0 \\ a_{21} & a_{22} & 0 & a_{23} & a_{25} \\ a_{41} & 0 & a_{44} & 0 & a_{45} \\ 0 & a_{32} & 0 & a_{33} & 0 \\ 0 & a_{52} & a_{54} & 0 & a_{55} \end{bmatrix}$$

The re-ordered equations are $(\mathbf{P}_{34}\mathbf{A}\mathbf{P}_{34})\mathbf{P}_{34}\mathbf{x} = \mathbf{P}_{34}\mathbf{b}$.

22. With the usual notation show that the eigenvalues of the Jacobi and SOR iteration matrices are respectively the roots of the equations
  (i) $\det(\mu\mathbf{D}-\mathbf{L}-\mathbf{U})=0$, and
  (ii) $\det(k\mathbf{D}-\lambda\omega\mathbf{L}-\omega\mathbf{U})=0$, where $k = \lambda + \omega - 1$.
Show that the matrix

$$\mathbf{A} = \begin{bmatrix} 4 & 0 & 0 & -1 \\ -1 & 4 & -1 & 0 \\ 0 & -1 & 4 & 0 \\ -1 & 0 & 0 & 4 \end{bmatrix}$$

is consistently ordered with respect to the ordering vector $(0, 1, 2, 1)$. Prove, for this particular matrix, that the eigenvalues λ of the associated SOR iteration matrix are related to the eigenvalues μ of the corresponding Jacobi iteration matrix by $(\lambda + \omega - 1)^2 = \lambda\omega^2\mu^2$. Comment on this result.

Solution

  (i) The eigenvalues μ of $\mathbf{D}^{-1}(\mathbf{L}+\mathbf{U})$ are the roots of
$$\det\{\mu\mathbf{I}-\mathbf{D}^{-1}(\mathbf{L}+\mathbf{U})\} = \det\mathbf{D}^{-1}\{\mu\mathbf{D}-\mathbf{L}-\mathbf{U}\}$$
$$= \det\mathbf{D}^{-1}\det\{\mu\mathbf{D}-\mathbf{L}-\mathbf{U}\}=0.$$

By hypothesis the elements of $\mathbf{D}$ are non-zero. Hence the result. Similarly for (ii).

Replace the non-zero elements of $\mathbf{A}$ by a_{ij} and apply the definition of consistent ordering as in Exercise 21. The eigenvalues μ are given by

$$\det \begin{bmatrix} 4\mu & 0 & 0 & -1 \\ -1 & 4\mu & -1 & 0 \\ 0 & -1 & 4\mu & 0 \\ -1 & 0 & 0 & 4\mu \end{bmatrix} = 0 = (16\mu^2 - 1)^2,$$

i.e. $\mu = \frac{1}{4}, \frac{1}{4}, -\frac{1}{4}, -\frac{1}{4}$.

The eigenvalues λ are given by

$$\det \begin{bmatrix} 4k & 0 & 0 & -\omega \\ -\lambda\omega & 4\kappa & -\omega & 0 \\ 0 & -\lambda\omega & 4k & 0 \\ -\lambda\omega & 0 & 0 & 4k \end{bmatrix} = 0 = (16k^2 - \lambda\omega^2)^2.$$

Therefore $k^2 = \frac{1}{16}\lambda\omega^2$, where $k = \lambda + \omega - 1$. Hence the result. The matrix $\mathbf{A}$ is not block tridiagonal. Because, however, it is consistently ordered the relationship $(\lambda + \omega - 1)^2 = \lambda\omega^2\mu^2$ still holds.

References for supplementary reading

1. Albasiny, E. L. (1960). On the numerical solution of a cylindrical heat-conduction problem. *Q. J. Mech. & Applied Math.* **13**, 374–84.
2. Carslaw, H. S. and Jaeger, J. C. (1959). *Conduction of heat in solids* (2nd edn), Clarendon Press, Oxford.
3. Courant, R., Friedrichs, K. and Lewy, H. (1928). Über die partiellen differenzengleichungen de mathematischen Physik. *Mathematische Annalen* **100**, 32–74.
4. Crank, J. and Nicolson, P. (1947). A practical method for numerical evaluation of solutions of partial differential equations of the heat conduction type. *Proc. Camb. Phil. Soc.* **43**, 50–67.
5. Crank, J. (1975). *Mathematics of diffusion* (2nd edn), Clarendon Press, Oxford.
6. Dupont, T., Kendall, R. P. and Rachford, H. H. (1968). An approximate factorization procedure for solving self-adjoint elliptic difference equations. *SIAM J. Num. Anal.* **5**, 559–73.
7. Forsythe, G. E. and Wasow, W. R. (1960). *Finite-difference methods for partial differential equations*, John Wiley, New York.
8. Fox, L. (ed.) (1961). *Numerical solution of ordinary and partial differential equations*, Pergamon Press, Oxford.
9. Fox, L. (1964). *An introduction to numerical linear algebra*, Clarendon Press, Oxford.
10. Gourlay, A. R. and Morris, J. Ll. (1980). The extrapolation of first order methods for parabolic partial differential equations, II. *SIAM J. Num. Anal.* **17**, 641–55.
11. Gustafsson, B. (1975). The convergence rate for difference approximations to mixed initial boundary value problems. *Math. Comp.* **29**, 396–406.
12. Hockney, R. W. (1965). A fast direct solution of Poisson's equation using Fourier analysis. *J. Assoc. Comp. Mach.* **12**, 95–113.
13. Khaliq, A. Q. M. (1983). Numerical methods for ordinary differential equations with applications to partial differential equations. Brunel University, England, Ph.D. thesis.
14. Lawson, J. D. and Morris, J. Ll. (1978). The extrapolation of first order methods for parabolic partial differential equations, I. *SIAM J. Num. Anal.* **15**, 1212–25.
15. Lax, P. D. (1954). Weak solutions of non-linear hyperbolic equations and their numerical computations. *Comm. Pure Appl. Math.* **7**, 157–93.

16. Lees, M. (1966). A linear three level difference scheme for quasi-linear parabolic equations. *Math. Comp.* **20**, 516–22.
17. Milne-Thomson, L. M. (1949). *Theoretical hydrodynamics* (2nd edn), Macmillan, London.
18. Mitchell, A. R. and Griffiths, D. F. (1980). *The finite difference method in partial differential equations*, John Wiley, New York.
19. Morton, K. W. (1980). Stability of finite-difference approximations to a diffusion–convection equation. *Int. J. Num. Meth. Eng.* **12**, 899–916.
20. N.P.L. (1961). *Notes on Applied Science*, **16**, Modern computing methods, London, HMSO.
21. O'Brien, C. G., Hyman, M. A. and Kaplan, S. (1951). A study of the numerical solution of partial differential equations. *J. Math. Phys.* **29**, 223–51.
22. Papamichael, N. (1970). Property (A) and consistent orderings in the iterative solution of linear equations. Brunel University, England, M.Tech. dissertation.
23. Papamichael, N. and Smith, G. D. (1974). The determination of consistent orderings for the SOR iterative method. *J. Inst. Math. Applic.* **15**, 239–248.
24. Peacemann, D. W. and Rachford, H. H. (1955). The numerical solution of parabolic and elliptic differential equations. *J. Soc. Indust. Applied Maths.* **3**, 28–41.
25. Richtmyer, R. D. and Morton, K. W. (1967). *Difference methods for initial-value problems*, Interscience Publishers, New York.
26. Stone, H. L. (1968). Iterative solution of implicit approximations of multidimensional partial differential equations. *SIAM J. Num. Anal.* **5**, 530–58.
27. The Open University (1974). *Mathematics Course* M321, Units 11–14, The Open University Press, Milton Keynes, England.
28. Twizell, E. H. and Khaliq, A. Q. M. (1982). L_0-stable methods for parabolic partial differential equations. *Technical Report TR/02/82*, Brunel University, Department of Mathematics, England.
29. Varga, R. S. (1962). *Matrix iterative analysis*, Prentice-Hall, New Jersey.
30. Wilkinson, J. H. (1965). *The algebraic eigenvalue problem*, Clarendon Press, Oxford.
31. Wilkinson, J. H. (1963). *Rounding errors in algebraic processes*, N.P.L. *Notes on Applied Science*, **32**, London, HMSO.
32. Wilkinson, J. H. and Reinsch, C. (eds) (1971). *Handbook for automatic computation*, Vol. 2, Springer-Verlag, Berlin.
33. Young, D. M. (1971). *Iterative solution of large linear systems*, Academic Press, London and New York.
34. Young, D. M. (1954). Iterative methods for solving partial differen-

tial equations of elliptic type. *Trans. Am. Math. Soc.* **76**, 92–111, 218, 242–72, 355, 394.

35. Twizell, E. H. and Khaliq, A. Q. M. (1983). Backward difference replacements of the space derivative in first order hyperbolic equations. *Technical Report TR*/01/83, Brunel University, Department of Mathematics, England.

Index